AF739032

Interdisciplinary Evolution Research

Volume 6

Series Editor

Nathalie Gontier, AppEEL, Centre for Philosophy of Science, University of Lisbon, Lisbon, Portugal

Editorial Board Members

Michael Bradie, Department of Philosophy, Bowling Green State University, Bowling Green, OH, USA

Maria Botero, Department of Psychology and Philosophy, Sam Houston State University, Huntsville, TX, USA

Ian Davidson, Faculty of Humanities, Arts, Social Sciences, University of New England, Armidale, NSW, Australia

Francisco Dionisio, Centre for Ecology, Evolution and Environmental Change, University of Lisbon, Lisbon, Portugal

Charbel El-Hani, History, Philosophy, and Biology Teaching, Federal University of Bahia, Salvador, Bahia, Brazil

Arantza Etxeberria, IAS-Research Centre for Life, Mind and Society, University of the Basque Country, San Sebastián, Spain

Roslyn M. Frank, European Society for Astronomy in Culture, University of Iowa, Iowa City, IA, USA

Francesco Ferretti, Department of Philosophy and Communication, Roma Tre University, Rome, Roma, Italy

Augusta Gaspar, Centre for Psychological Research and Intervention, Catholic University of Portugal, Lisbon, Portugal

Benedikt Hallgrimsson, Department of Cell Biology and Anatomy, University of Calgary, Calgary, AB, Canada

Misato Hayashi, Primate Research Institute, Kyoto University, Inuyama, Aichi, Japan

Lydia Hopper, Lester E. Fisher Center for the Study and Conservation of Apes, Lincoln Park Zoo, Chicago, IL, USA

John Hunter, Department of Comparative and Digital Humanities, Bucknell University, Lewisburg, PA, USA

Mary Lee Jensvold, Fauna Foundation, Carignan, QC, Canada

Carl Knappett, Department of History of Art, University of Toronto, Toronto, ON, Canada

David Leavens, Department of Psychology, University of Sussex, Brighton, UK

David Morrison, Sequence Alignment and Phylog Networks, Systematic Biology, Uppsala University, Uppsala, Sweden

Roland Mühlenbernd, Leibniz-Zentrum Allgemeine Sprachwissenschaft (ZAS), Berlin, Germany

April Nowell, Department of Anthropology, University of Victoria, Victoria, BC, Canada

Mark Pagel, School of Biological Sciences, University of Reading, Reading, UK

Elena Pagni, Department of Psychology, Federal University of Juiz de Fora (UFJF), Juiz de Fora, Brazil

Anna Prentiss, Department of Anthropology, University of Montana, Missoula, MT, USA

Timothy Racine, Department of Psychology, Simon Fraser University, Burnaby, BC, Canada

Eugenia Ramirez-Goicoechea, Department of Social and Cultural Anthropology, National University of Distance Education, Madrid, Spain

António J. Santos, William James Centre for Research, ISPA - University Institute, Lisbon, Portugal

Peter Saunders, Department of Mathematics, Kings College, London, UK

James A. Shapiro, Department of Biochemistry and Molecular Biology, University of Chicago, Chicago, IL, USA

Chris Sinha, Department of Cognitive Science, Hunan University, Changsha, China

Sune Vork Steffensen, Department of Language and Communication, University of Southern Denmark, Odense, Denmark

David Suárez Pascal, Department of Evolutionary Biology, UNAM, National Autonomous University of Mexico, Mexico City, Distrito Federal, Mexico

Franceso Suman, Pikaia -The Evolution Portal, University of Padua, Padova, Italy

Sandra Swart, History Department, Stellenbosch University, Stellenbosch, Western Cape, South Africa

Monica Tamariz, PPLS, Dougal Stewart Building, The University of Edinburgh, Edinburgh, UK

Ian Tattersall, Division Anthropology, Central Park, American Museum of Natural History, New York, NY, USA

Natalie Uomini, Language Origin Society, Max Planck Institute for the Science of Human History, Jena, Germany

Natasha Vita-More, Information Science Department, University of Advancing Technology, Tempe, AZ, USA

Slawomir Wacewicz, Department of Linguistics, Nicolaus Copernicus University Torun, Torun, Poland

Douglas Zook, Boston School for the Environment, University of Massachusetts Boston, Boston, MA, USA

More information about this series at http://www.springer.com/series/13109

Elena Pagni • Richard Theisen Simanke
Editors

Biosemiotics and Evolution

The Natural Foundations of Meaning and Symbolism

Springer

Editors
Elena Pagni
Department of Philosophy
Federal University of Juiz de Fora (UFJF)
Juiz de Fora, Brazil

Richard Theisen Simanke
Department of Psychology
Universidade Federal de Juiz de Fora (UFJF)
Juiz de Fora, Brazil

ISSN 2199-3068 ISSN 2199-3076 (electronic)
Interdisciplinary Evolution Research
ISBN 978-3-030-85264-1 ISBN 978-3-030-85265-8 (eBook)
https://doi.org/10.1007/978-3-030-85265-8

© Springer Nature Switzerland AG 2021
This work is subject to copyright. All rights are reserved by the Publisher, whether the whole or part of the material is concerned, specifically the rights of translation, reprinting, reuse of illustrations, recitation, broadcasting, reproduction on microfilms or in any other physical way, and transmission or information storage and retrieval, electronic adaptation, computer software, or by similar or dissimilar methodology now known or hereafter developed.
The use of general descriptive names, registered names, trademarks, service marks, etc. in this publication does not imply, even in the absence of a specific statement, that such names are exempt from the relevant protective laws and regulations and therefore free for general use.
The publisher, the authors, and the editors are safe to assume that the advice and information in this book are believed to be true and accurate at the date of publication. Neither the publisher nor the authors or the editors give a warranty, expressed or implied, with respect to the material contained herein or for any errors or omissions that may have been made. The publisher remains neutral with regard to jurisdictional claims in published maps and institutional affiliations.

This Springer imprint is published by the registered company Springer Nature Switzerland AG.
The registered company address is: Gewerbestrasse 11, 6330 Cham, Switzerland

Preface

This book resulted from the intersection of two different philosophical and epistemological concerns that somehow converged into the focus constituted by the evolutionist and biosemiotic approaches to meaning and symbolism. One of us (Pagni) came from a philosophical education centered on the phenomenological tradition. The interface between phenomenology and the epistemology of the life sciences led her to the philosophical and biological problem of individuality (Pagni 2015), which was the object of a postdoctoral research stage at the Federal University of Juiz de Fora (Brazil), where our collaboration effectively began. From this project, the problem of meaning in nature, already envisaged in the work of philosophers such as Maurice Merleau-Ponty, Jan Patočka, and Renaud Barbaras, among others, became part of Pagni's research interests, bringing with it the questions raised by the biosemiotic paradigm in the life sciences (Pagni and Simanke 2016; Pagni 2016, 2020). The other of us (Simanke) had been working predominantly in the history and philosophy of psychology and psychoanalysis and became interested in the phenomenology of nature to reassess the meaning of scientific naturalism and account for Freudian psychoanalysis' unique naturalistic epistemology (Simanke 2011a, 2011b).

The common ground of such different projects was challenging the still widely accepted dichotomy that divides scientific practice between the natural and the human sciences. Psychoanalysis and phenomenology have traditionally been aligned with the latter, although the authors and doctrines that interested us in both disciplines tended to reconnect them to the former. However, it was not the case of simply subordinating our respective fields of interest to the naturalist models prevailing in the history of science during the last decades, still plagued by the assumptions tirelessly criticized by phenomenologists and humanists: objectivism, mechanism, physicalism, and reductionism, among others. Instead, we intended to join the alternative program of reintroducing mind, meaning, history, and the qualitative dimension of experience into the universe of natural determinations and seize this opportunity to revaluate the meaning of scientific naturalism and, in a broader perspective, the concept of nature as such.

Indeed, we never thought that the issues that interested us—the phenomenology of life or the psychoanalytic model of the mind—would suit better in the context of the natural sciences than the human or social ones. Instead, we believed that the very nature of the problems and the parameters for approaching and solving them called into question the *necessity* of that duality. Our common working hypothesis was that this dichotomy, far from self-evident, reveals insurmountable methodological, epistemological, and metaphysical impasses when analyzed in depth. Moreover, it emerged in a circumscribed historical and philosophical context so that its widespread and longstanding acceptance, whose effects one can still feel today, resulted from contingent external circumstances with no intrinsic relation to its content or original meaning. We cannot detail these circumstances, which are better approached from the viewpoint of historical sociology of science. Suffice it to say that dividing the field of scientific activity between human and natural sciences no longer accurately represents what is currently accomplished in contemporary science. The most philosophically *and* scientifically instigating research programs to emerge in recent times—such as evolutionary psychology, human ethology and ecology, environmental sociology, behavioral genetics, and cognitive neuroscience, to name but a few—have in common overcoming or even completely disregarding this borderline.

These assumptions form the broader background to the essays brought together in this collection, which take the intersection between biosemiotics and evolutionary biology as a privileged field for erasing the dividing line that for so long separated the understanding of human life and the study of nature. This book thus hopefully contributes to reflecting on these privileged research fields that discuss biosemiotics' reception in disciplines such as biology, psychology, and sociology. In one way or another, all these disciplines approach living beings and the development of their cognitive faculties—the same ones through which life itself emerged and persisted.

Knowing—an activity related to learning and actions—is only possible through semiosis (Kull 2014). Acknowledging this essential feature points to the necessity of exploring how biosemiotics can provide a comprehensive framework to integrate and support biological and bioinformatic research (see Chaps. 2 and 5 below).

Biological semiotics has its roots in Charles Peirce's triadic conception of sign, as Reyes Cardenas remarks in his introductory chapter (Chap. 1). The doctrines that constitute the particular variant of pragmatism developed by Peirce are part of a naturalist research program on human cognition to recover the conceivable practical and experimental purposes of rational life (Peirce 2005b, p. 448) as the modus operandi of natural, social, and ethical sciences. As Giovanni Maddalena observes, the fundamental problem from which Peirce's reflection begins is the passage from *substance*—such as it appears, with no connotation and only as an actual phenomenon ("it")—to *being*, understood as the predication of substance within a proposition (Peirce 2005a, p. 9).

Peirce designates the categories defining the whole phenomenological field[1] (MSS. 305-306/CP. 1.284-287; 5.41-56) as "firstness,"[2] "secondness,"[3] and "thirdness."[4] They fundamentally refer to "to the scientifically examinable (. . .) relations of possibility, existence and law" (Favareau 2007, p. 30). Firstness refers to the "given state" of a fact, to *its subject being positively such as it is regardless of anything else*. In turn, secondness is the "experience," i.e., what it is and what it is like for another being. Finally, thirdness includes and expresses a firstness and a secondness; in other words, what makes them a sign of something else within a range of possibilities. This sign determines the horizon of significance against which its reception becomes something habitually acquired by an interpreters' community (Peirce 2005a).

Hence, every thought beyond an immediate perception is a sign (MS.318, p. 23). Intellectual concepts, says Peirce, "those upon the structure of which arguments concerning objective fact may hinge" (MS.318, p. 10) are concepts that "essentially carry some implication concerning the general behaviour of some conscious being or of some inanimate object, and so convey more, not merely than any feeling, but more, too, than any existential fact, namely, the 'would-acts' of habitual behaviour; and no agglomeration of actual happenings can ever completely fill up the meaning of a would-be" (MS.318, pp. 11–12). Thus, intellectual concepts convey a context and allow for an action characterized by relevance and significance through the affective mediation of the interpretants:

> Now the problem of what the "meaning" of an intellectual concept is can only be solved by the study of the interpretants, of proper significate effects, of signs. (. . .) The first proper significate effect of a sign is a feeling produced by it. There is almost always a feeling which we come to interpret as evidence that we comprehend the proper effect of the sign, although

[1] Peirce says: "(. . .) what we have to do, as students of phenomenology, is simply to open our mental eyes and look well at the phenomenon and say what are the characteristics that are never wanting in it, whether that phenomenon be something that outward experience forces upon our attention, or whether it be the wildest of dreams, or whether it be the most abstract and general of the conclusions of science" (CP. 5.41). Or elsewhere: "The faculties which we must endeavour to gather for this work are three. The first and foremost is that rare faculty, the faculty of seeing what stares one in the face, just as it presents itself, unreplaced by any interpretation, unsophisticated by any allowance for this or for that supposed modifying circumstance. (. . .) The second faculty we must strive to arm ourselves with is a resolute discrimination which fastens itself like a bulldog upon the particular feature that we are studying, follows it wherever it may lurk, and detects it beneath all its disguises. The third faculty we shall need is the generalising power of the mathematician who produces the abstract formula that comprehends the very essence of the feature under examination purified from all admixtures of extraneous and irrelevant accompaniments" (CP.5.42).

[2] "Firstness is the mode of being which consists in its subject being positively such as it is regardless of aught else. That can only be a possibility. For as long as things do not act upon one another, there is no sense or meaning in saying that they have any being unless it be that they are such in themselves that they may perhaps come into relation with others" (CP.1.25).

[3] "(. . .) a mode of being of one thing which consists in how a second object is. I call that Secondness" (CP.1.24).

[4] "(. . .) the mode of being which consists in the fact that future facts of Secondness will take on a determinate general character, I call a Thirdness" (CP.1.26).

> the formulation of truth in this is frequently very slight. This "emotional interpretant", as I call it, may amount to much more than that feeling of recognition; and in some cases, it is the only proper significate effect that the sign produces. Thus, the performance of a piece of concerted music is a sign. It conveys and is intended to convey, the composer's musical ideas; but these usually consists merely in a series of feelings. If a sign produces any further significate effect, it will do so through the mediation of the emotional interpretant, and such further effect will always involve an effort. (MS.318, pp. 32–33)

Therefore, the object of a sign is in itself necessarily unexpressed. What Peirce means by semiosis is the "action of a sign" (MS. 318, p. 28), namely, an "influence, which is, or involves, a cooperation of three subjects, such as a sign, its object, and its interpretant, this tri-relative influence not being anyway resolvable to actions between pairs" (MS.318, pp. 49–50).

The binary scheme of the sign relation proposed by Fernand de Saussure (2017) defined the sign as the unity of the signified (*signifié/sens*) and the signifier (*signifiant/phonations*). Contrarily, for Peirce, a signification results from a semiotic triadic system formed by a sign, an object, and the interpretant. Biosemiotic research transfers this conceptual framework to biology to provide new theoretical perspectives for understanding the dynamic underlying ontogenetic and phylogenetic evolutionary processes. Based on Peirce, one could say that the biosemiotics' purpose is restoring the *pragmatic meaning* of life, relating this meaning to the living beings' fundamental capacity for coding and decoding information, feeling, and acting without necessarily understanding these processes intellectually or anthropomorphically.

"*Plus la conscience s'intellectualise, plus la matière se spatialise*," said Bergson in *L'évolution créatrice* (Bergson 2013, p. 190). In other words, the tendency of the spirit to determine itself as intelligence leads to a spatialized conception of space and matter, hence to an abstract and geometric representation of nature (ibid., p. 191). This attitude opens an unbridgeable gap between the inorganic and the organic, the inanimate and the living (ibid., p. 187). According to Bergson, the only way to re-establish contact and continuity between these domains and their respective models of knowledge is to renounce "the artificial unity that the understanding imposes on nature from the outside" (ibid., p. 200). Ultimately, this proposal means dismantling the very idea of knowledge that, since Plato, regards reason as a powerful instrument for unification, i.e., for the conceptual organization of multiplicity—an assumption that one can find in both metaphysical and empirical realism.

In opposition to this realist and representational model of knowledge, biosemiotics aims at restoring a certain continuity between different epistemic levels at which one can consider a given state of affairs—mental experience and biological organization; life and the law-like processes proper to inanimate matter; among others (Favareau 2007, p. 2). Biosemiotics also attempts to re-establish the meaningful quality of life by postulating the underlying signs and mechanisms as the foundation from which biological existence evolution emerges (See Chap. 4 below). Accordingly, life as such knows itself through signs. Expression thresholds of meaning, symbolic communication, and interpretation manifest paradigmatically in human semiosis (see Chap. 3) but are not exclusive to it, being also present in the human bio-cultural evolution (Chap. 7) and other species' social and sexual behavior (Chap. 9).

In the biosemiotic theoretical model, signs are understood as natural phenomena in absolute contrast to intelligent design pseudoscientific theories. They underlie the functioning, reproduction, expression, and receptivity of any organic or linguistic code. Biosemiotics universal principle is that life is semiosis, and semiosis, in turn, is life's fundamental process (Barbieri 2007). Indeed, in its broadest sense, biosemiotics conceives "all the realm of living as a process of signs interpretation" (Gálik 2013, p. 859). Ultimately, one can say the fundamental unity of life is the sign, not the molecule (Hoffmeyer 1996; Barbieri 2003). Accordingly, biosemiotics emphasizes the central role of the interpretant "in the process of semiosis, as it can lead to a change in the disposition of the organism for different behaviour" (Huber and Schmid-Tannwald 2007, p. 461).

Returning to the Peircean vocabulary, biosemiotics places at the basis of life's evolution the semiotic relation between an agent emitting a sign and a patient interpreting it within a *context* of adjacent collateral observations. "Context" is here understood as the set of conditions that makes it possible for living beings to actively explore their *Umwelt* and "be compared with the interactive possibilities it offers" (Giorgi 2017, p. 140).

The need to reformulate and reconceptualize evolutionary theories is a direct consequence of biosemiotics' first central thesis, according to which "life is semiosis." Two other crucial considerations follow from this need: first, that biology is incomplete as a science in the absence of explicit semiotic grounding (2nd central thesis of biosemiotics) and, secondly, that semiosis is a central concept for biology that requires a more exact definition (3rd central thesis of biosemiotics) (Kull et al. 2009; Gálik 2013).

In consequence, the conditions for adaptive and teleonomic evolution are not only genetic but also include the tendency to achieve meaningful mutual exchanges within a given environment. The objective of biosemiotics is to reintroduce semiosis into the study of meaning-making occurring in living systems, emphasizing that a semiogenesis lies at the life's core and not necessarily as a purposeful "intentional fact" (Giorgi 2017). Instead of regarding the DNA structure as a program to be followed and achieved, genetic information represents "the memory of the achieved state and the selective path to reach it" (Giorgi 2017, p. 78). As Giorgi remarks, "defining the intention as the result of a process of sensory-motor adaptation to the environment is not yet sufficient to explain how the agentive subject can become aware as he is able to act in a contextually more suitable (that, in turn, is a fundamental stage for the development of self-awareness)" (Giorgi 2017, p. 253).[5] According to Marcello Barbieri, organic codes and memories are biological realities

[5]Chapter 10 below presents a detailed philosophical analysis of the concept of intentionality through a comparison between mental and physical notions of intentionality, pointing out the most significant characteristics of the paradigmatical shift from a phenomenology of the inert to a phenomenology of the living.

and not semantic metaphors, as advocated by physicalism (Barbieri 2003, 2007, 2008a, 2008b).

Evolution is fundamentally regarded as an ecological phenomenon emerging from the engagement of teleonomic agents with their affordances.[6] In turn, these affordances are defined as opportunities or impediments in pursuing a goal and have to do with the system's agency. They comprise two significant features: (1) the ability to experience those conditions as affordances; and (2) an adaptive repertoire. In sum, "that is to say that on any occasion, there must be a range of possible outcomes or activities that the system or its parts could implement" (Walsh 2015, pp. 163–164).

Biosemiotics is primarily concerned with the evolutionary processes of bioinformation. Explaining the transition from inorganic to increasingly complex forms of organic life requires the introduction, in biology, of the idea of information as *in*-formation, i.e., active information (Fernández 2016)—information as something in the process of being assembled, and not as the cause of what has already been shaped (Giorgi 2017, p. 78). Even the intercellular signalling processes can be conceived as a repertoire of possible actions (different DNA expressions) among those phenotypically significant to the cellular environment.[7]

Biosemiotics' anti-reductionism rejects mechanic causation, for which the causal laws describing the functional characteristics at a certain level of complexity (such as behavioral traits) must be reduced to the components and relations of the structurally lower layers. As Giorgi (2017) emphasized, the risk here is to accept the statistical validity of this correlation uncritically and deterministically and understand any emergent behavior as caused by lower-level factors. Moreover, as an epistemic model, reductionism attributes ontological precedence to mechanisms and enhances the ontological value of the predefined properties expressed by the system's interacting parts to the detriment of their relations and actual interaction.

As Giorgi (2017) argues, from a biosemiotic perspective, the situation is wholly reverted: the properties expressed by interacting networks are secondary to establishing their relationship. It is only from the inside that one can establish some interactive properties as selectively favorable. According to biosemiotics, the relationship is significant when considered as emergent among several possible or available alternatives. Conversely, its mechanical explanation is caused by predefined properties and excludes all alternatives paths (Giorgi 2017, p. 143). The problem here is what ontological value one can attribute to mechanisms. Is a particular mechanism necessary for the emergence of behavior, or is this emergence a contingent result of exploring such possible relations? (Giorgi 2017, p. 144).

[6]One must remind that teleonomy is not the same as teleology. Teleonomy refers to those phenomena that one can explain without reference to the subject's intentionality, despite being directed towards a goal. For example, all the bodily structures and functions assure the organism's health and survival.

[7]Chapter 11 below provides an excellent analysis of carcinogenic and cell death mechanisms (apoptosis and necrosis) from the point of view of cancer medical research based on applied biosemiotics.

In conclusion, we believe that biosemiotics sheds new light on the classical philosophical dilemma of the relationship between experience and truth, whose implications to empirical scientific research should be more than evident. It also provides an original perspective to approximate human sciences and the life sciences. These two contributions are not entirely independent if one considers that natural sciences seek the truth about the world (regardless of the different possible underlying concepts of truth), and human sciences explore and describe the human experience in this world. This dilemma dates back from ancient Greek philosophy and is still with us, now and then passing by and catching us so evidently and regrettably unprepared.

Juiz de Fora, Brazil
June 26, 2021

Elena Pagni
Richard Theisen Simanke

Bibliography

Barbieri M (2003) The organic codes. An introduction to semantic biology. Cambridge University Press, New York
Barbieri M (2007) Introduction to biosemiotics. The new biological synthesis. Springer Netherlands
Barbieri M (2008a) The codes of life. The rules of macroevolution. Springer
Barbieri M (2008b) Biosemiotics: a new understanding of life. Naturwissenschaften 95:577–599. https://doi.org/10.1007/s00114-008-0368-x
Barbieri M (2009) A short history of biosemiotics. Biosemiotics 2(2):221–284
Bergson H (2013) L'évolution créatrice. Presses Universitaires de France, Paris. (Original work published in 1907)
de Saussure F (2017) Corso di Linguistica generale. Editori Laterza, Roma-Bari [Original work published in 1922]
Favareau D (2007) The evolutionary history of biosemiotics. In: Barbieri M (ed) Introduction to Biosemiotics. The new biological synthesis. Springer Netherlands, pp 1–68
Favareau D (2009) Essential readings in biosemiotics. Springer Netherlands, Dordrecht
Fernández E (2016) Synergy of energy and semiosis: cooperation climbs the tree of life. Biosemiotics 9(3):383–397
Gálik D (2013) Biosemiotics: a new science of biology? Filozofia 68(10):859–867
Giorgi F (2017) Un mondo di relazioni. Tangram Edizioni Scientifiche, Trento
Harney M (2017) Peirce and phenomenological naturalism. In: Gare A, Hudson W (eds) For a new naturalism. Telos Publishing, Candor, NY, pp 124–143
Hoffmeyer J (1996) Signs of meaning in the universe. University of Indiana Press, Bloomington, IN

Huber J and Schmid-Tannwald I (2007) A biosemiotic approach to epigenetics: constructivist aspects of oocyte-to-embryo transition. In: Barbieri M (ed) Introduction to biosemiotics. The new biological synthesis. Springer Netherlands, pp 457–471
Kull K et al (2009) Theses on biosemiotics: prologomena to a theoretical biology. Biological Theory 4(2):167–173
Kull K (2014) Zoosemiotics is the study of animal forms of knowing. Semiotica 198:47–60. https://doi.org/10.1515/sem-2013-0101
Longo G, Montévil M (2014) Perspectives on organisms. Biological time, symmetries and singularities. Springer, Berlin
Pagni E (2015) I fondamenti dell'indeterminatezza ontologica: l'organismo vivo tra identità e negazione. In: Lanfredini R (ed) Figure della materia. Mimesis, Milano, pp 133–145
Pagni E (2016) Why explicit semiotic grounding is essential to biology as a science? The point of biosemiotics. Humana.Mente J Philos Stud 31:57–72. ISSN:1972-1293
Pagni E, Simanke RT (2016) Natureza, vida e animalidade no pensamento de Merleau-Ponty: uma reavaliação de seu alcance e implicações. In Caminha IO, Nóbrega TP (eds) Compêndio Merleau-Ponty. Editora LiberArs, São Paulo
Peirce ChS Pragmatism: autograph manuscript. Charles S. Peirce papers, MS Am 1632, (318). Houghton Library, Harvard College Library. https://id.lib.harvard.edu/ead/c/hou02614c00333/catalog. Accessed 20 June 2021
Peirce ChS (2005a) Pragmatismo [MS.318]. In: Maddalena G (ed) Scritti Scelti. Unione Tipografico-Editrice Torinese, Torino, 591–625
Peirce ChS (2005b) La fenomenologia [MSS.305-306]. In: Maddalena G (ed) Scritti Scelti. Unione Tipografico-Editrice Torinese, Torino, 445–465
Simanke RT (2011a) Freudian psychoanalysis as a model for overcoming the duality of natural and human sciences. In: Krause D, Videira AA (eds) Brazilian studies in history and philosophy of science: an account of recent works. Springer Verlag, Dordrecht, pp 211–223
Simanke RT (2011b) Freudian psychoanalysis as depth psychology: rereading Freud's theory of the mental apparatus in light of Merleau-Ponty's concept of depth. Journal of the British Society of Phenomenology 42(3): 255–289
Walsh DM (2015) Organisms, agency, and evolution. Cambridge University Press, Cambridge

Acknowledgments

We want to thank the Program of Graduate Studies in Psychology of the Federal University of Juiz de Fora (UFJF) for all the support received during Elena Pagni's postdoctoral fellowship in that institution (2014–2015), as well as the CAPES (Brazilian funding institute for academic education and research) for the scholarship granted to her project. We also thank the Philosophy Department and the Program of Graduate Studies in Philosophy of the Federal University of Juiz de Fora for their kind reception and productive collaboration during Pagni's time there as Foreign Visiting Professor (2018–2021). Special thanks are due to the UFJF International Office's staff for all their friendship and administrative support.

We also address special thanks to Natalie Gontier and Paniel Reyes Cárdenas for the exceptional efforts they have put in during the COVID-19 pandemic to make this book possible, for their supportive collaboration and openness to dialogue.

Richard Theisen Simanke thanks the CNPq (Brazil's National Council for Technological and Scientific Development) for supporting this project in the form of a Scholarship for Productivity in Research.

Contents

Editors and Contributors

About the Editors

Elena Pagni Graduated in Philosophy (University of Pisa—Italy) with a thesis on Aristotle and his view of perception, and was awarded a PhD in Philosophy from the University of Florence. She has been a member of the Research Laboratory "Epistemologica" (University of Firenze, Italy), directed by Prof. Roberta Lanfredini. She got a one-year postdoc (Research in Paris fellowship) in philosophy of biology at the Ens of Paris (Centre Cavaillès), under the supervision of Dr. Giuseppe Longo, did a postdoc scholarship at the PPG in Psychology at the Federal University of Juiz de Fora (scholarship of the PNPD/Capes Program) under the direction of Prof. Richard Theisen Simanke, and a postdoc at the Faculdade de Filosofia, Ciências e Letras de Ribeirão Preto, Departamento de Psicologia, Brazil, under the direction of Prof. Reinaldo Furlan. From 2018 to 2021, she has been a visiting professor, with exclusive dedication, at the Department of Philosophy of the Institute of Human Sciences at the Federal University of Juiz De Fora.

Research fields: Ancient Philosophy, Phenomenology, Biosemiotics, evolutionary processes of biological in-formation, biological problem of individuality, problem of meaning in nature, biosemiotic paradigm in the life sciences.

Richard Theisen Simanke Graduated in Philosophy and Methodology of Science from the Federal University of São Carlos (Brazil) and was awarded a PhD in Philosophy from the State University of São Paulo (Brazil). He was Professor of History and Philosophy of Psychoanalysis in the Department of Philosophy and Methodology of Science at the Federal University of São Carlos, Brazil until 2012 and is currently Professor of History and Philosophy of Psychology in the Department of Psychology at the Federal University of Juiz de Fora, Brazil. His fields of expertise are the history and philosophy of psychoanalysis, history and philosophy of psychology, history and philosophy of science, and science and phenomenology. His secondary research interests are philosophy of biology, history and philosophy of psychiatry, and the history of sexuality.

Contributors

Arthur Araujo Department of Philosophy, Federal University of Espírito Santo, Vitória, Brazil

Paniel Reyes Cárdenas UPAEP University, Puebla, Mexico
The University of Sheffield, Sheffield, UK

Giulia Degl'Innocenti Department of Humanities, University of Studies of Florence, Florence, Italy

Marta Facoetti Applied Evolutionary Epistemology Lab, Centro de Filosofia das Ciências, Departamento de História e Filosofia das Ciências, Faculdade de Ciências, Universidade de Lisboa, Lisboa, Portugal

Rogério Estevam Farias Laboratory of Immunopathology and Experimental Pathology, Federal University of Juiz de Fora, Juiz de Fora, Brazil

Jonathan Luís H. Ferreira Department of Biological Sciences, Federal University of Juiz de Fora, Juiz de Fora, Brazil

Franco Giorgi University of Pisa, Pisa, Italy

Nathalie Gontier Applied Evolutionary Epistemology Lab, Centro de Filosofia das Ciências, Departamento de História e Filosofia das Ciências, Faculdade de Ciências, Universidade de Lisboa, Lisboa, Portugal

Philippe Huneman Institut d'Histoire et de Philosophie des Sciences et des Techniques, Paris, France
Université Paris I Panthéon Sorbonne, Paris, France

Roberta Lanfredini Department of Humanities (DILEF), University of Studies of Florence, Florence, Italy

Juan Mendoza-Collazos Cognitive Semiotics, Center for Language and Literature, Lund University, Lund, Sweden
Universidad Nacional de Colombia, Bogotá, Colombia

Alin Olteanu University of Tartu, Tartu, Estonia
Kaunas University of Technology, Kaunas, Lithuania

Elena Pagni Department of Philosophy, Human Sciences Institute (ICH), Federal University of Juiz de Fora (UFJF), Juiz de Fora, Brazil

Vinicius Romanini School of Communication and Art, University of São Paulo, São Paulo, Brazil

Richard Theisen Simanke Department of Psychology, Human Sciences Institute (ICH), Federal University of Juiz de Fora (UFJF), Juiz de Fora, Brazil

Göran Sonesson Cognitive Semiotics, Center for Language and Literature, Lund University, Lund, Sweden

Jordan Zlatev Research Education for Cognitive Semiotics, Center for Language and Literature, Lund, Sweden

Introduction: Biosemiotics and Evolution

Paniel Reyes Cárdenas

Abstracts One of the most exciting endeavours and opportunities presented to a researcher lies in the possibility of being at the hinges of historical change, especially if the historical turn guides to a more positive outcome and we have reflective control over our participation in such quest. The present book is such precious opportunity, and the editors, Richard Simanke Theisen and Elena Pagni, have bravely positioned themselves at the crossroads of development and consolidation of a new discipline, or, shall we say, a new paradigm of biology: Biosemiotics. In exploring the natural foundations of meaning and symbolism, Biosemiotics must effect a turn to be an even more open inter-disciplinary and trans-disciplinary discussion; this discussion encounters the most prominent explanation of living systems on earth: evolution. The resources for our dialogue come from semiotics, the rigorous science of signs, but are not limited to it. In this volume, the foundations of symbolism, as the title of the present book alludes, can span beyond into a phenomenological, technological, philosophical, psychological discussion that will enrich our knowledge of those foundations and establish an engaging exchange with all synthesis of evolution. In this introduction we will position the reader to a short reminder of what the subject of Biosemiotics is about, and through that start preparing ourselves to the depth of the discussions to come. After this, I will offer a very short comment on what discussions are going to be present throughout the book and how these discussions are articulated with the project of the present volume, highlighting its unity and the outstanding work of the editors.

Keywords Biosemiotics · Evolution · Semiotics · Symbolism · Naturalism · Evolution theory

P. Reyes Cárdenas (✉)
UPAEP University, Puebla, Mexico

The University of Sheffield, Sheffield, UK
e-mail: panielosberto.reyes@upaep.mx

© Springer Nature Switzerland AG 2021

E. Pagni, R. Theisen Simanke (eds.), *Biosemiotics and Evolution*, Interdisciplinary Evolution Research 6, https://doi.org/10.1007/978-3-030-85265-8_1

Kalevi Kull has expressed that Biosemiotics is "the science of signs in living systems" (Kull 1999, 386). Biosemiotics is a surrogated discipline of the rigorous science of signs *qua* signs, i.e., of semiotics. Semiotics acquired the status of a rigorous logical science when Charles Sanders Peirce described the triadic logic of relations. Semiotics renders understanding of the functioning of signs, thus, "the sign is a medium of information that professes to represent its object in some aspect so as to produce an effect, which is its interpretant" (Romanini and Fernández 2014, 2). Once we understand the appropriate structure of sign-relations we are apt to inquire into living systems, the origins of life, the diversification, evolution and functioning of these systems, as well as their interaction with their environment.

Donald Favareau (2009) has done a wonderful job recapitulating the prehistory, the history, and the trends of Biosemiotics, in his wonderful review he truly delves deeply within the history of philosophy and biology. He brings the help of John Deely's and Alastair McIntyre amongst others to convey an understanding as to how the sign was at the very beginning of inquiry into living systems. He starts with Plato and Aristotle, and how, however concomitantly, the medieval thought on sign *qua* sign came to a pinnacle in the work of John Poinsot (1589–1644) and his *Tractatus de signis* (1632). He also makes a very compelling case of how the Cartesian divide of body/mind escalated to become the prejudice of separating meaning and interpretation from signs, constituting a paradigm of separation between materiality and cognition. Descartes, however, was not the only one to endorse such divide, it is ubiquitous in much rationalist and empiricist philosophy, and particularly poignantly taken for granted in the development of physics from Galileo and Newton onwards.

Barbiery (2008) recalls that contemporary biology is indeed affected by the above-mentioned Cartesian division that not only prejudices philosophy, but scientific practice: a first dichotomy of particular concern is the duality genotype/phenotype conceived as a sort of software/hardware divide, that leaves meaning out of the picture, along with a satisfactory explanation of how living systems function in sustaining themselves. Secondly, there is the always lurking problem of a physicalist metaphysics that reduces all realities to materialism of objects and their quantifiable properties, leaving aside everything that is not of such kind (including the laws of nature themselves!). Physicalism is a kind of obsessive reductionism, and for biology is disastrous, since downplays the best explanations of living systems reducing them to their material properties that are present in the sciences below them, for instance, to the chemistry and physics of part of their material molecular conformation. Finally, the belief that every biological novelty has been brought to existence by natural selection also constitutes one of the reductionisms of contemporary biology, presenting an incomplete and insufficient story of evolutionary development that leaves the code to understand it comprehensibly outside the narrative of biological explanation.

The achievement of having a field of study consolidated was no easy accomplishment, it is the result of an active strive for dialogue that owes much to Thomas A. Sebeok. Sebeok plays a very important part in recalling the stages of Biosemiotics in its conformation as a new paradigm for biology; let us have a quick review of the importance of his work:

Thomas A Sebeok started his career as a linguist, but soon enough his explorations of languages took him to the task of joining sign science with life science. Sebeok was initially aware of how a rigorous science of signs can and ought to be applied to the animal world first and then extending the world of meaning to all living systems. Sebeok's spirit was a free one, and even though he lived through the times of a divide in the academic world between the western world and the eastern communist world, this was no impediment for him to establish significant connections and develop a community of eastern and western researchers during the Cold War.

While Sebeok needed a rigorous science of signs, he soon enough realised that this could not be found in the structuralist tradition of Ferdinand de Saussure, but in the triadic conception of the sign proper to the semiotic created by Charles Sanders Peirce. Sebeok's synthesis of Charles Sanders Peirce's semiotics accounted not only of his semiotic ideas, but the architectonic system of his thought, including his triadic system of categories. The attention he lent to Peirce's seminal brilliance was very fertile, and this had a liberating effect in Sebeok's openness to new ideas. This openness also led Sebeok to the thought of Jakob von Uexküll and his idea of the *Umwelt*. Sebeok's synthesis of Jakob von Uexküll's *Umwelt* with Peirce's thought became the axis of a lasting contribution to biosemiotics: the discovery of the primacy of signs and meaning in living systems. But Sebeok not only contributed to establish the discipline in a theoretical way, he actively campaigned to break the disciplinary borders of science and to create a project of mass cross-fertilisation that Biosemiotics was called to be. Sebeok's legacy and the continuation of the biosemiotic project were free of the initial defence mechanisms of the circle of established academics and open a fruitful niche.

In the same wise we have to acknowledge the parallel work and efforts that came to biosemiotics from the biological research, which is the case of Jesper Hoffmeyer and the Danish School. Sebeok brought the semiotical thought to biology, and Hoffmeyer and his colleagues came to semiotical thought by the effect of finding an answer to the most important conundrums of biology and scientific inquiry. The convergence and consolidation of the two projects into a common programme shows that fruitfulness that Peirce wished for a community of inquiry. It is as though diverse ecosystems of researchers, prompted by an unrestricted desire to understand and overcome the limited regionalisms and borders of a discipline, came to flourish. Part of this conformation was definitely given by the International Gathering in Biosemiotics that rendered opportunities for that genuine exchange. Thus, scholars, researchers, and minds from different backgrounds found in Biosemiotics the opportunity to contribute to the brand new science. Just to mention outstanding examples we ought to remember how the Copenhagen–Tartu connection of scholars was consolidated, how the hermeneutical way of understanding Biosemiotics came to the picture with the work of Cvrcková and Markoš with their 'Readers of the book of life' (2002). Physicists joined the liberating effect of biosemiotics and we can briefly mention Edwina Taborsky (1998), Peter Voetmann Christensen (2000) and started taking seriously Peirce's categories and semiotic in the field of physical cosmology and its corresponding research. Howard Pattee (1965) is an important biologist that understood the need of semiotics to understand nature appropriately with the categories of semantic closure and epistemic cut. Biosemiotics is also an appropriate environment for concepts pertaining Maturana and Varela's 'autopoiesis' (1973); Ilya Prigogine's 'dissipative structure' (1969), and many more, that steadily called for an overcoming of reductionism, physicalism, and eliminativism not only in biology but in science at large. Of course, the discipline will have developments and challenges unforeseeable to us, but from the 2000s to the date (2021), we can confidently witness a consolidation of the field. Two important examples are the official conformation of the International Society for Biosemiotic Study and the foundation of the *Journal of Biosemiotics* in 2005.

Since 1981, Marcello Barbieri came across the semiotic importance of the organic code for the study of Biosemiotics, according to Barbiery (2008, 1),

> Biosemiotics is the idea that life is based on semiosis, i.e., on signs and codes. This idea has been strongly suggested by the discovery of the genetic code, but so far it has made little impact in the scientific world and is largely regarded as a philosophy rather than a science. (2008, 1)

Barbieri has a very clear conscience of the need to improve the paradigm of evolutionary theory, and in this way explains that evolution cannot be fully understood only considering the role of natural selection, but the idea of natural convention is needed to explain the continuities and discontinuities in the processes of macroevolution. For Barbieri, Biosemiotics has to focus its efforts in the paradigm of the organic code, perhaps entering its final stage of conformation. Elena Pagni (2016) has stressed that a common approach to Biosemiotics suggests that semiosis is embedded in evolution and because of that requires meaningful processes. Though Barbieri is hugely enthusiastic about the importance of the code, Biosemiotics ought not to be reduce to it, as the evidence of the presence of meaning is everywhere to be found in biological research, and not only in genetics.

After our quick log of recapitulation of the evolution of Biosemiotics one naturally would ask about the future of Biosemiotics. A natural response is that the evolution of a new scientific paradigm is under way and that some clear trends guarantee us, in virtue of the evidence of its integrated origins, that Biosemiotics will continue to be a necessarily inter- and trans-disciplinary study. However, there seem to be no need of establishing limits on the breadth and depth of this study. Of course, biology and semiotics will always define the principles of the programme, but this book is a token of evidence that the explorations into new discussions are unlimited and will become the seeds of more breakthroughs for the discipline and for the understanding of meaning by and large. However, it is this book the first one that endeavours exploring the relationship between Biosemiotics and Evolution: for the first time we not only have a presentation of different approaches to living systems as living signs and codes, but as evolving signs and evolving codes. Indeed, understanding living systems from Biosemiotics opens a very important door of understanding those living systems as dynamical systems, and the most important dynamism at play is evolution. The reader, of course, will ask what account of evolution might be brought to the dialogue: in short, we can answer that an open dialogue between Biosemiotics and Evolution requires the most comprehensive and experimentally open account of evolution, and here the modern synthesis falls short. In my opinion, we need a powerful and open account, one such as Extended Evolutionary Synthesis, which is a set of theoretical concepts argued to be more comprehensive and successful in integrating evolutionary developmental biology, the role of configurations, genomic structures, dimensionality of fitness landscapes, multilevel selection, epigenetic inheritance, developmental plasticity, phenotypic novelty, punctuated equilibrium, and a host of phenomena successfully developed: Biosemiotics and Extended Evolutionary Synthesis (as defended for example, by Pigliucci and Müller 2010 and Jablonka and Lamb 2005, 2014), allows a predictive and experimental status that can also be a semiosis of living systems, as some of the chapters of this book allow us to see.

This collection goes beyond much of what has been produced before in the field of Biosemiotics, as we will appreciate below, many of the contributions of this volume put Biosemiotics in new and novel spaces of dialogue, particularly accounting for evolutionary transformations, creating a series of roads for future contributions. As D. Favareau (2009) and Barbiery (2008) optimistically point out, Biosemiotics is a science in the making, and at this early stage we cannot predict what future outcomes and scientific breakthroughs will be made in the way. However, we are certain that this book represents a definite way forward that the discipline is taking: we have here a consistent and open dialogue with other disciplines and paradigms that opens more and more the possibilities of the discipline. Our authors have not only generated a dialogue with the consolidated sciences, but even with the frontier research of contemporary scientific inquiry as well as philosophical research.

The book is divided in three parts: the first one, 'Life, meaning, and information' includes contributions that establish the terminology of our dialogue that spans across Biosemiotics to Evolution.

The second chapter of our book is a great exploration on the philosophical and scientific foundation of meaning entitled 'Exploring the Philosophical Background and Scientific Foundations of Naturalist Approaches to Meaning and Symbolism' tailored by our editors Elena Pagni and Richard Simanke Theisen. The chapter delineates a historical, philosophical, and scientific framework that is needed to understand the fascinating connections between biological sciences, phenomenology, and Biosemiotics, and thus, these are a handy as well as important clarification of the framework of the whole volume. Perhaps there are little works as good as this to emphasise how the biosemiotic paradigm is apt to generate an epistemological pluralism that helps overcome unjustified impasses in the history of thought and pushes science and philosophy forward.

Franco Giorgi contributes with a hugely interesting work: "Life Sciences and the Natural History of Signs" that is subtitled with the question 'Can the Origin of Life Processes Coincide with the Emergence of Semiosis?' In this work Giorgi explores the concept of life as expressed in evolution and a dynamical interaction to the environment (*Umwelt*), and shows us that the role played by signs in the emergence of forms and functions is the best way of channelling and understanding the unity and establishment of evolutionary trends.

Alin Olteanu's contributed chapter has the title: 'A Proposal for a Biosemiotic Approach to Digitalization: Literacy as Modelling Competence', and constitutes a bold exploration from Biosemiotics to the representational possibilities of the digital systems. The proposal is a potential merging of communicational networks that can revolutionise the way we approach digital systems: he focuses in multimodality as opposed to Glottocenticism and simultaneous participativity as opposed to the feedback processes of current digital systems. The result of such a proposed transformation is a more natural interaction between users and digital systems that will significantly improve the possibilities of digital communication as a more effective process of human semiosis.

The research on the foundations of meaning for the living systems keeps deepening in the work of Arthur Araujo: 'Threshold, Meaning and Life'. In his chapter he argues that meaning is an activity that distinguishes life systems from lifeless ones. Furthermore, that meaning and life are overlapping processes that hold continuity, this continuity has no strict boundaries, it is a process without a discreet tipping point. The author uses Peirce's anti-Cartesian view of cognition to explain that the threshold is not a specific moment in time, but a continuity that makes meaning and life convertible.

As we entered the depth of this volume, it is turn to Vinicius Romanini to offer us the work: 'How Information Gets Its Meaning'. The aim of the chapter is to clarify, using Peirce's semiotics and Peirce's pragmatist philosophy, the concept of information in the appropriate dynamism through communication. Romanini demonstrates that Peirce's clarification on the universal semiosis is well needed in biological explanations. Information then, should be embodied in icons, expressed by indices and communicated by symbols.

The second part of the book takes us to the dialogue of Semiosis and Evolution proper. Such dialogue shows the different models from the modern synthesis to the extended one, and from models of evolution to the applied evolutionary epistemology that deals with then. The contributions of this part open the Biosemiotic interpretation of evolution by approaching the fine grained theories.

In chapter seven, Philippe Huneman enriches the span of the discussion with a work entitled: 'Inclusive Fitness Teleology and Darwinian Explanatory Pluralism: A Theoretical Sketch and Application to Current Controversies'. In this text Huneman considered the uses of a view of evolutionary theory based on Formal Darwinism, but he re-activates the Aristotelian quadripartite teleological and etiological causality so as to show their potential fruitfulness in expanding explanatory practices in evolutionary biology.

We move onto: 'The Origins and Evolution of Design: A Stage-based Model', the authors Juan Mendoza-Collazos, Jordan Zlatev, and Göran Sonesson offer us a model that suggest the harmonious integration of continuities and relative discontinuities in the evolution of design, in such project they take on account human bio-cultural evolution and Biosemiotics of living systems, and use Donald's theory of the evolution of the mind to account for the resources of design.

Nathalie Gontier and Marta Facoetti's chapter 'Biosemiotics and Applied Evolutionary Epistemology: A Comparison' present us a wonderful systematic comparison between Biosemiotics and the results of the programme of Applied evolutionary epistemology. In their exploration, they find the important complementarity present in research programmes as well as the convergence in their interest, they also venture to establish the connection altogether explicit by touching the issues of adaptation and finality as well as the evolutionary nature of knowledge, cognition, and meaning.

The following chapter is presented to us by Jonathan Luís H. Ferreira: "Extended Synthesis and Jablonka and Lamb's Four-Dimensional View of Evolution: Some Remarks". Ferreira explores the four-dimensional view of evolution given by the paradigm of Extended Evolutionary Synthesis as articulated by Jablonka and Lamb (Jablonka and Lamb 2005, 2014). The author believes that the extended view helps liberate the constrains of what he calls 'reproductionism' and then helps us, along with the enriched conceptions of the extended view. In short, a phenomenological attitude with biosemiotics helps us to articulate a richer view of sexual behaviour that is open as an expression of meanings.

From this point we enter the third part of our volume: Physics, medicine, and bioenergetics. In this final section we can appreciate the impressive potential of application of our dialogue between Biosemiotics and Evolution by concrete applications in different experimental approaches.

'Physical Intentionality. The Phenomenological Roots of Biosemiotics' is a contribution by Roberta Lanfredini, in this chapter she shows us the important phenomenological roots that are modelled in Biosemiotics. She uses the notion of 'physical intentionality' to express the need of properties to be equivalent to meaning, then producing a freeing effect that liberates from physicalism and helps us to come across a phenomenology of the living.

Chapter twelve is a contribution by Rogério Estevam Farias called 'Cancer and Cell Death: A Biosemiotic Perspective' which is a splendid proof of the superiority of Biosemiotics as a paradigm of explanations. In this case the understanding of Cancer and carcinogenic cells is on point: the author explains us that the two systems by means of which the cell is indicated to die away and thus sustain the stability of a multicellular system: the processes are apoptosis and necrosis. The first one is given within the cell by its own singling resources that are transferred by code signalling. Farias explains us that an increase of disorder in the signalling processes of apoptosis called biosemiotic entropy inhibits the coding interpretation of the cell systems, producing a tumoural-like cell that has lost communication with the system. Though this is a short contribution, one cannot stop stressing the importance of this contribution, that offers the best explanation available of Cancer research based in its genomic evidence.

In our final contribution, Giulia Degl'innocenti contributes with the chapter 'Biosemiotics and Bioenergetics: Two Perspectives Compared'. The contributor explores an important common direction that emerges from both research programmes. On the one hand, according to Biosemiotics, the biological body is a place of signification as is generated, structured, and evolves. On the other hand, bioenergetic analysis shows that the body is inscribed in relational and affective experiences and fully characterised by meaning. Two different sources of inquiry into the living body lead, then, to a common conclusion with regard to the living body.

The above richness of fields in the explorations on the foundations of meaning and symbolism of living systems as evolving living signs and codes testifies to the importance of this volume, as well as reflects the characteristic spirit of collaboration proper of the community of inquirers that Biosemiotics is. Notwithstanding this book also shows the breadth and depth of the collaboration by its international and multidisciplinary character, that adds new layers of future bonds of thought and fertile discussion. We are certain that this is a hugely valuable step on the direction for a fruitful and rigorous dialogue between Biosemiotics and Evolution.

Acknowledgement I am immensely grateful to Elena Pagni for inviting me to be part of this amazing project and privileged enough to read the contributions to this magnificent collection. Thanks to Nathalie Gontier and Sabine Schwarz for their openness to value and support these wonderful discussions and managing the collection of which this book is a contribution.

Bibliography

Barbiery M (2008) Biosemiotics: a new understanding of life. Sci Nat 95(7):577–599

Christensen V (2000) Macro and micro-levels in physics. In: Andersen PB, Emmeche C, Finnemann NO, Christiansen PV (eds) Downward causation: minds, bodies and matter. Aarhus University Press, Århus, pp 51–62

Favareau D (2009) Essential readings in biosemiotics: anthology and commentary. Springer, New York

Jablonka E, Lamb M (2005, 2014) Evolution in four dimensions. MIT Press, Boston

Kull K (1999) Biosemiotics from the twentieth century: a view from biology. Semiotica 127 (1/4):385–414

Pagni E (2016) Why explicit semiotic grounding is essential to biology as a science? The point of view of biosemiotics. Humana Mente 16:52–72

Pattee H (1965) Experimental approaches to the origin of life problem. In: Nord FF (ed) Advances in enzymology, vol 27. Wiley, New York, pp 381–415

Peirce CS (1931–1935) Collected papers of Charles Sanders Peirce, vol 1–6. Harvard University Press, Cambridge

Peirce CS (1958) Collected papers of Charles Sanders Peirce, vol 7–8. Harvard University Press, Cambridge

Pigliucci M, Müller G (2010) Evolution: the extended synthesis. MIT Press, Boston

Poinsot J (1632/1985) Tractatus de Signis (With English translation by John Deely). University of California Press, Berkeley

Romanini V, Fernández E (2014) Peirce and biosemiotics: a guess at the riddle of life. Springer, New York

Taborsky E (ed) (1998) 1999. Semiosis, evolution, energy: towards a reconceptualization of the sign. Shaker Verlag, Aachen

Part I
Life, Meaning, and Information

Exploring the Philosophical Background and Scientific Foundations of Naturalist Approaches to Meaning and Symbolism

Richard Theisen Simanke and Elena Pagni

Abstract This introductory chapter delineates a general historical, philosophical, and scientific framework for the individual studies collected in this book. It emphasizes how the biosemiotic paradigm and related approaches in biology—especially current evolutionary thought—contribute to an epistemological pluralism that allows for a more productive dialogue between natural sciences, on the one hand, and social sciences and humanities, on the other hand, ultimately pointing out to the artificial character of this long-standing traditional duality. For this purpose, we first outline some essential philosophical and historical elements presupposed in the paradigm shifts that marked the life sciences in the nineteenth and twentieth centuries. Next, an overview of the fundamental connections between biological sciences, phenomenology, and biosemiotics is presented.

Keywords Biosemiotics · Evolutionary biology · History of life sciences · Philosophy of biology · Phenomenology

1 Introduction

The scientific literature collected in this book is characterized by theoretical pluralism and the design to keep open the dialogue between natural and social sciences, as well as between the humanities and the scientific field as a whole. The general underlying assumption is that this dialogue is possible and profitable for both fields, especially by questioning their respective object domains, methods, objectives, and theoretical foundations. This theoretical pluralism is arguably one of the best ways to

---

R. T. Simanke
Departament of Psychology, Human Sciences Institute (ICH), Federal University of Juiz de Fora (UFJF), Juiz de Fora, Brazil

E. Pagni (✉)
Federal University of Juiz de Fora (UFJF), Human Sciences Institute (ICH), Department of Philosophy, Juiz de Fora, Brazil

© Springer Nature Switzerland AG 2021
E. Pagni, R. Theisen Simanke (eds.), *Biosemiotics and Evolution*, Interdisciplinary Evolution Research 6, https://doi.org/10.1007/978-3-030-85265-8_2

prevent the imposition of one field's epistemological and methodological commitments onto the other. It is also intended to avoid the opposing risks of a dogmatic reductive naturalism, on the one hand, and the post-modern dissolution of science into a universe of equivalent socially constructed discourses, on the other. In the first case, a unitary and inflexible notion of scientific truth would result. In the latter, the ultimate vanishing of any idea of truth whatsoever would take place. More specifically, the theoretical stance adopted here aims to integrate the dimensions of meaning and symbolism in all its forms—once widely regarded as typical, if not exclusive, features of the semiotics and social sciences' object domain—into the biological sciences and related research fields (Cobley et al. 2011; El-Hani et al. 2008; Emmeche et al. 2002; Gare 2019; Hoffmeyer 1997). This focus understandably brings to the foreground the biosemiotic paradigm,[1] which assumes the inherence of semiosis in life and its evolution (Sharov et al. 2016) as its main *tour de force*.

From the nineteenth century onward, the scientific views of life have undergone a radical transformation with the emergence of evolutionary theories that turned change and the forms of becoming into indispensable categories to study living beings. A posteriori reasoning allows us to understand the motivation forces that led up to this conceptual discontinuity in the history of life sciences and the cultural factors that, over the last twenty-three centuries (from the fourth century BCE), have prepared the ground for the development and acceptance of the new determinations and meanings underlying evolutionary patterns. Be that as it may, evolutionary theories—in particular, Darwinian and Neo-Darwinian theoretical models and their revisions and complements—have become one of the cornerstones of the scientific investigation of life. Evolutionary thought (especially Darwinism) and biosemiotics have not always been on good terms historically speaking. However, the relevance of both paradigms in the life sciences' contemporary scenario calls for a discussion of their relations and attempts to integrate their respective viewpoints. These circumstances explain why evolutionary issues also form either the main or the subsidiary topics of many of the following chapters, dealing with the problem of how the meaning-making and symbolic capacities of the living beings have evolved.

Although most contributions in this volume come from the scientific field, philosophical problems are inevitably involved in approaching problems such as meaning and symbolism, especially concerning nonhuman life forms and the general picture of the evolution of the species on Earth. Even when these philosophical problems are not explicitly addressed—sometimes they are—they often underlie the arguments the authors present and the scientific views they maintain. Hence, the need to sketch the general philosophical background and provide a historical framework to situate the particular issues examined in the following chapters in the field's discussion and controversies.

For this purpose, in the second section of this chapter, we outline some essential philosophical and historical elements presupposed in the paradigm shifts that marked

[1]For general accounts of biosemiotics and its history, see Barbieri (2007) and Gare (2019).

the life sciences in the nineteenth and twentieth centuries. The third section addresses an overview of the fundamental phenomenological connection between biological sciences, phenomenology, and biosemiotics.

2 Philosophical and Historical Background

Aristotle was arguably the first philosopher to pay life and biological becoming systematic attention and give them a scientific and theoretical treatment. His attitude was made possible by discovering a rational *logos* suitable to accompany and provide answers to the philosophical interrogation of life, contrarily to Plato, who was skeptical about the possibility to turn the approach to physical and perceptual phenomena into a true *episteme*.

Phenomenology usually attributes the origins of a crisis of thought to metaphysic's decline, especially in the sense of ontology. In this crisis, the being was finally led to confront its finitude in the face of totality. In this context, phenomenology interprets the sciences' naturalistic attitude as an extreme application of the classic metaphysics' *ratio*. As a result, the pervasiveness (*ubiquitas*) of the forma mentis proper to mechanistic rationalism extends to all knowledge fields as a kind of radical ontologization—a process that began with the Socratic "*ti esti*" (what is it?) in the history of philosophy. Indeed, the thing discovered by answering the "*ti esti*" covers for the function of the question itself.

We believe that this interpretation is only partially accurate. The crisis can be traced back to what happens *before* this ontologization, that is, before Socrates' "*ti esti*" question. It relates to bringing back into human life the splitting between the finite and the infinite, the continuous and the discrete, the visible and the invisible, the rational and the irrational (let us remind, for example, the divided line described in the Book VI of Plato's *Republic*). In Plato, these divisions manifest in the idea of a soul that is born already internally split. Plato's metaphysics thus situates within humankind the original split between the divine and the mortal, the transcendent and the immanent—a rupture that, from the mythical narratives to the classical Greek tragedy, had been represented as the struggles and torments of mortal beings facing invincible gods.

Plato made things hugely switch because, before being outside and identified as a principle or idea, the divine exists inside human beings as a possibility experienced through a noetic intellectual vision capable of overcoming itself and its limits by successive degrees of understanding. Philosophy was born from placing human beings in the face of their limits. From then on, humans must deal with their finitude while realizing, at the same time, that the conditions to transcend this finitude are within themselves.

In the context of Aristotle's essentialist conception of human nature, the "*ti esti*" question, when addressed to human *physis*, leads to a discourse (*logos*) about life, such as in *De Anima*. Here, "essentialist" stands for substantial, that is, the idea that an essence reveals a substance. Aristotle's biology must be understood as a rational discourse (*logos*) about life through which one can express life's essential properties.

It is well known that biology was only born in the early nineteenth century as an independent science, as the study of a dimension of being with specific objects and methods. However, Aristotle opened the way to the regional ontologies through which life expresses itself and can be rationally and empirically understood and explained. Aristotelian natural philosophy articulates the first spatializing view of life in Western thought, the first instance of life knowing itself through a *logos*'s rational norm. One can describe his approach as consisting of two simple logical operations that can be expressed as two straightforward questions: (A) What is this? (B) How can one speak about it? These two questions roughly correspond to the ontological and epistemological aspects of any scientific inquiry, respectively. In other words, one asks what kind of *logos* is possible for each specific category of objects. It is not only a question of what objects are proper to each science—or, to put it in less realistic terms, what objects the emergence of each science originated—but also a question about their respective epistemic needs. The truth values "true" and "false," for example, apply differently to science and technique. In the former case, they are justified by the knowledge or ignorance of indemonstrable firsts principles; in the latter, they are based on conformity to the procedure. Moreover, the categories of the *possible* and the *debatable* as applied in ethics and politics are proper to the technical or practical discourse. This is where one can choose since no one deliberates on what cannot be changed—on what is necessarily the way it is—as is the science's case. Therefore, through the *logos* inherent in its nature, the rational *psychê* divides, separates, and cuts off different domains of being. Within this framework, there is no need to postulate a subject that posits itself as the ultimate foundation from which interrogation begins. The ways and inclinations of the logos determine the appearance or manifestation of a particular field of experience—a *logos* conceived as an ahistorical and impersonal category. For Aristotle, the human being is not a subject: it is a substance among other substances. What distinguishes it from other beings, and other living beings, in particular, is that it possesses the *logos*. As for the rest, they may share many functions, such as reproduction, perception, and movement.

With this framework, Aristotle carries out an essential cultural operation. The truth about the essence of life manifests through the path of *logos* applied explicitly to biology. The way is open for regional ontologies that, as mentioned above, allow life to be expressed, understood, and explained: "to live is said in many senses; we affirm that a being lives if only one of these characteristics belongs to it, namely the intellect, the sensation, motion and stillness in the place, and also the change in the sense of nutrition, decrease, growth" (*De Anima*, 413a 23). Aristotle's *De Anima* is a pioneering form of rational *logos* on life. It likely presents the first instance of spatialization of being through language, that is, discursive knowledge. The continuous flowing of becoming is analyzed into discrete units of understanding, i.e., into a sequence of thematized practices or actions (thinking, perceiving, reproducing) through language mediation.

Through the logos, life knows itself, leading us to the first and ultimate essence (entelechy) of human beings, represented by the *psychê* or soul. For Aristotle, the essence of each thing determines the origins and limits of knowledge (*intelligere* in

Latin). In his *Metaphysics* (V, 7), Aristotle established the meanings of limit (πέρας) and claimed that one could also call the substance and essence of each thing a limit. In fact, they are the limit of knowledge, and, as such, they also define the limit of the thing, i.e., its proper way of being.

Aristotelian physics (*physike episteme*) is the contemplative science (*theoria*) of mobile natural substances, the ones subjected to becoming. In turn, Aristotelian psychology is the study of the principle of life and becoming of sensible natural substances and, as such, is part of the physical investigation. It concerns those functions of the *psychê* that relate to matter and the properties of the body. Aristotle compares the *psychê*'s condition concerning those functions with a straight-line tangent to a sphere (*De Anima*, I, 1).

Therefore, Aristotle is the first philosopher in history to manifest this primary naturalistic attitude. Nevertheless, what does the term "naturalistic" mean here? To answer this question, of course, one must first understand what Aristotle means by "nature." In *Metaphysics* (V, 4), among multiple other meanings of the φύσις, one can read that it is "*the principle of the first movement which is in each of the natural beings and which exists in each of them, precisely because it is natural.*" *It is a principle of* "*unity or an organic continuity.*" Hence, such principle—nature—is the substance of natural beings. In other words, nature is what gives substantiality to the being. The world presents itself to the human *logos* as characterized by movement and forces whose legality is intrinsic to the quality of the different beings' substantiality. As the Czech philosopher Jan Patočka states, the world was first discovered as *physis* (nature), and so the first science in the world was physics (Patočka 1992, p. 15).

Aristotle's *De Anima* provides the epistemological context to grasp the *psychê* concept as a generating principle of continuity and discontinuity. Here, continuity means a substantial unity that confers indivisibility (unity of being) to the individual and the species. Discontinuity, in turn, refers to this being differential relations to all other life forms that, in this respect, do not pertain to or manifest the same essence apprehended by an indivisible act of the intellect. Ultimately, the Aristotelian categories specifying the concept of *the psychê*—the principle of life and its vegetative, perceptual, and cognitive functions—are inherent in Greek rationality and manifest the continuous interdependence between the organic/physical world, on the one hand, and thought and language, on the other.

Massimo Cacciari regards *psychê* and *logos* as fatal terms that inform all human existence and civilization. Hence, without proper knowledge of their origin and etymology, nothing can be known of this civilization (Cacciari 2009, 2014, 2019). In turn, Massimo Barale remarks that the ancient concept of logos is tributary to an idea of rationality as underlying and sustaining things, hence, to the idea of their own substantiality. The *physis* must then be located into the Cosmos' aegis (order, beauty, harmony) wherefrom it obtains the power of generation and expression. *Logos* is reason, of course, but, more precisely, it is the reason as converging with the possibility of coherent discourse and as reflecting the order in which and by which it exists. According to Barale, the eventual collapse of such an ontological and cosmological assumption has only made others emerge. The seventeenth-century revolution in scientific thought brought out a rationality model that could no longer manifest a thing's hidden essence. Along the phenomenon manifesting itself, Barale argues, the conditions of its appearance are also produced. One can derive from here the idea of a procedural reason accompanying the phenomenon in its given presence and manifestation (Barale et al. 2001, p. 20).

Modern science undoubtedly introduced a remarkable change of perspective in the study of nature. The concept of inertia (at least as it was conceived until the nineteenth century) then replaced the Aristotelian idea of movement as interdependent of the bodies' qualitative substantial properties and form. Besides Descartes' ontological dualism, the Newtonian concept of *vis insita* (innate force) strongly contributed to this change.

The concepts of gravity and inertia introduced by Galilei and mathematically formalized by Newton also strengthen this view of inertia as a fundamental property of the physical world. In Galilei's *Dialogue over the two greatest systems of the world* (1632), the main argument Salviati employs against Simplicio in the section corresponding to the dialogue's first day is that motion does not depend on the qualitative properties of bodies, especially their generation/corruptibility or non-generation/incorruptibility:

> *Every body constituted in a state of rest but naturally capable of motion will move when set at liberty only if it has a natural tendency toward some particular place; for if it were indifferent to all places it would remain at rest, having no more cause to move one way than another. Having such a tendency, it naturally follows that in its motion it will be continually accelerating. Beginning with the slowest motion, it will never acquire any degree of speed (velocità) without first having passed through all the gradations of lesser speed—or should I say of greater slowness?* (Galilei 1632/1967, p. 20)

Therefore, inertia is the default state of physical bodies, and physical reality becomes the domain of mathematics and geometry. Aristotle based his physics on the concept of substance and the modalities of motion derived from it. For example, the movement of natural (earthly) substances is different from the heavenly ones. The theory of the natural places fits perfectly this qualitative approach to natural bodies. However, from modern science's birth onward, a homogenization of nature around the concept of gravity occurs—a nature from which not even humans can escape, being themselves also subjected to external, coercive, and universal forces. Let us consider, for example, Definition III and the first law of motion in Newton's *Mathematical Principles of Natural Philosophy* (1687), as well as the concept of *vis insita* as expressed in all physical reality. The latter's defining feature is its disposition to resist and maintain its state of inactivity (rest) or uniform rectilinear motion unless it is forced to change by the action of external forces.

This scientific paradigm shift is at the origins of the bifurcation, never again to be reverted between natural and human science. In fact, this opposition, such as it is widely employed even today, took shape much later. It basically dates back to late nineteenth-century German neo-Kantianism, even though one can regard these neo-Kantian views as the culmination of a division established by modern science from its very outset. Many of the most distinguished neo-Kantian philosophers attempted to propose alternatives to the generalization of the Newtonian model for all sciences. Besides physics, by the late nineteenth century, this model had been adopted, with the necessary adjustments, by chemistry and some branches in the biological sciences, for example, as these sciences emerged and consolidated as independent disciplines. However, in the neo-Kantian philosophers' views, the epistemological commitments in the natural sciences were inadequate to explain the phenomena related to human action and its products, thus requiring new categories to describe the scientific field. This recategorization took different forms in the various schools (Baden, Marburg) and philosophers (Heinrich Rickert, Wilhelm Windelband, Wilhelm Dilthey) who discussed the contrasts between the two forms of scientific inquiry. There are terminological variations in the designation of the class opposing the natural sciences: *Geisteswissenschaften* (sciences of the spirit or mind) or *Kulturwissenschaften* (sciences of culture), for instance. However, the main disagreements concern perhaps the character of this distinction: ontological (a difference in the nature of objects), epistemological (a difference in the nature and aims of knowledge), or methodological (a difference in how knowledge is obtained).[2] While circumscribed to a specific intellectual context and a particular philosophical agenda, this distinction became canonical and remained so even after the dissolution of neo-Kantianism in the years after World War I. This duality's permanence has been especially conspicuous in the scientific and philosophical currents resisting the positivist-like programs for a unitary science—and, more than any other, in the human sciences, for apparent reasons.

Nevertheless, the development of scientific research in recent decades has made this opposition between natural and social sciences (or humanities) considerably obsolete and attempts to integrate the naturalistic and humanistic "cultures" have become frequent since the 1970s at least (Simanke 2011, 2014). Biosemiotics is an essential part of this movement. The emergence of different forms of evolutionism in the life sciences and evolutionary cosmologies that represented nature itself as a

[2]In the nineteenth-century German-speaking world, the distinction between the natural sciences (*Naturwissenschaften*) and "the sciences of the spirit" (*Geisteswissenschaften*) was usual. However, there was considerable ambiguity as to what "*Geist*" (spirit or mind) exactly meant in this expression. In the tradition of German idealism, specifically, "*Geist*" could often refer to the cultural world as a whole. Windelband, for example, disliked the term *Geisteswissenschaft* because he considered that the objects of history and the cultural sciences did not always involve the human mind. According to him, the distinction between the historical and the natural sciences could not be simply reduced to a material difference in their respective objects. Instead, he proposed his famous epistemic and methodological distinction between nomothetic (natural) and idiographic (historical) sciences. Rickert, who had studied under Windelband, resumed and developed the latter's views and engaged in a lifelong controversy with Dilthey on the subject. Dilthey argued for the legitimacy of the *Geisteswissenschaft* notion and considered that a renewed psychology could represent for the sciences of the spirit the same foundational role that mathematics had played in the natural sciences. Rickert, in turn, accepted the term "*Geist*" as long as understood as synonymous with "culture," but often preferred the less ambiguous term *Kulturwissenschaft* as opposed to the natural sciences (Rickert 1899/2014). In the English-speaking world, the expression "human sciences" appears since Locke at least, but "moral sciences" became more frequent after Hume (Flew 1986). *Geisteswissenschaften* was commonly used to translate "moral sciences" into German. For an overview of the neo-Kantian tradition, see Köhnke (1991) and Beiser (2014).

historical process (Collingwood 1945/2014) has also contributed significantly to nuance the opposition between natural and historical sciences underlying neo-Kantianism's categories. Hence, one can hardly exaggerate the significance of evolutionary thought to overcome this duality.[3]

Since modern science's inception, nature ceases to be the subject matter of natural philosophy (*Philosophia naturalis*) and its speculative philosophical *logos*. Instead, it becomes a symbolic system allowing the interpretation of physical bodies as formal relationships expressed through mathematical equations. Substances are replaced by processes (Barale et al. 2001, p. 23). These changes beginning in the seventeenth century culminate in the virtual dissolution of the category of substance in science in the late nineteenth and early twentieth centuries. With the emergence of quantum physics, relativist physics, and new mathematical tools such as non-Euclidean geometries, set theory, and mathematical intuitionism, the ancient ontology of substance was definitively replaced by relational ontologies privileging processes and functions to the detriment of stable entities enduring through time.[4]

The seventeenth century represents a rough slap in the face of classical metaphysics, which turns out deprived of its proper object, namely, physics. Metaphysics finds itself orphaned and dispossessed, so to say, excluded from physics and astronomy, even though it preserves its right to eschatological reflections concerning theological, ethical, and political matters. Kant's and the positivist critiques of metaphysics are among the most influential arguments against classical speculative metaphysics' scientific status, even though they approach this problem from entirely different perspectives and with different objectives. However, the separation of metaphysics and science was a long process. The emergence of modern science represented much more the abandonment of a particular (classical and medieval) metaphysics and worldview and the endorsement of others (the Cartesian and Newtonian ones, for example). One can doubt that the program for science without metaphysics is completely accomplished even today or that it can ever be accomplished. As Schrenk (2016, p. ix) points out:

> At the beginning of the twentieth century metaphysics as a whole fell into disrepute. Or, we should say, fell once again into disrepute. Early in the last century, the Logical Empiricists revived their eighteenth-century ancestors' metaphysics critique. They, as much as the ancestors, saw no merit in non-scientific theorizing about the fundamental features of reality. Metaphysics, when understood as the philosophical investigation into the most basic, ultimate, fundamental features and structures of a reality that goes beyond what can be known via (experimental) observations and sensory experiences was thought to be pointless. (. . .) It turned out, however, that satisfactory explications of this [empirical] kind are not so

[3]Prominent evolutionary biologists have approached integrating the natural sciences with the social sciences and humanities (Gould 2003/2011; Wilson 1999). Genetic data's employment in social research has also contributed to attenuate the opposition of biological and cultural anthropology (Goodman et al. 2003). Schaeffer (2007) argued that recent developments in natural sciences are progressively rendering obsolete the idea that the human world requires a distinct and exclusive form of scientific knowledge. Many other examples could be provided.

[4]These developments were the object of classical philosophical analysis, like Bachelard's (1934/1971) and Cassirer's (1910/1953), among others.

easily available and that it is probably not possible for a multitude of scientific terms without making assumptions that were pejoratively labelled 'nonsensical metaphysics' by the Empiricists.[5]

On the other hand, with the scientific revolution and its aftermath, matter loses its soul (in the Aristotelian sense) and, ultimately and somewhat paradoxically, de-materializes in the twentieth-century scientific developments that leave behind the last remnants of substantial thinking in science (Hanson 1962).[6] Processes, rather than substances, now characterize nature and becoming. These processes must be understood as relations of numerical invariance that assign things a common ground of intersubjective receptiveness. Being no longer exists in terms of form, purpose, matter, and motion (i.e., the Aristotelian causes), and nature ceases to be conceived of as a static collection of classes or types. There is a single rationality mode expressing the functional laws of facts, whereas, for Aristotle, the rational and cognitive *psychê* manifests through three primary and irreducible inclinations: technique/poietic, practical, and theoretical. Language and the physical world cease to be linked by a substantial unity through an indivisible act of the intellect. Procedural rationality is now the unit of measurement that encompasses and relates things to each other, even if this measure is arbitrary, and its adoption is purely conventional.

The twentieth-century updates of the nineteenth-century original positivistic program eventually reject the last remains of substantialism in the mechanistic view ("*mechanism transposed into mathematics*," Longo [2020b]). Rationality is now a tool for predicting chains of events through the concepts of relation and function. A purification of experience occurs as positivism struggles to eliminate the dichotomy between the physical and the subjective world. Both worlds are then reduced to neutral data representing how phenomena take shape through reciprocal relations. However, positivism thus perpetuates some fundamental mistakes of the mechanistic paradigm. First of all, it assumes that reality consists of mathematical relations, failing to subject the scientific consciousness operating under this assumption to radical criticism. Secondly—and more importantly, perhaps—it also fails to consider that certainty and predictability only manifest on the plane of the subject.

In the meantime, biology's emergence as an independent science in the early nineteenth century brings another variable to the equation. Some technological advances, such as the invention of the microscope, made it possible to criticize the uniformity attributed to physical and biological bodies by observing the infinite variety of living forms in the world of microorganisms. With these advances, protozoology and bacteriology could be finally born.

---

[5]Burtt (1925/2003) was one of the first twentieth-century philosophers to criticize the anachronical view that modern science was born by unconditionally rejecting any metaphysical world picture. The relationship between metaphysics and the empirical sciences has been a constant issue in the philosophy of science ever since. See also Dilworth (2007).

[6]See Simanke (2016) for an argument on how these developments concerning the modern concept of matter open a path for bringing some trends in phenomenology closer to the philosophy of natural sciences.

In the history of the life sciences, the most notable paradigm shift allowing a different narrative on life and organisms was offered by Charles Darwin with his 1859 publication of *The Origin of Species* and the introduction of the natural selection hypothesis. Species, incipient species, and varieties can then be considered arbitrary if not conventional concepts.[7] Life can no longer be grasped through its essence since it takes place in time autonomously and with no final cause as its guiding principle. Darwin's theory of species transmutation certainly owes much to a long history of previous evolutionary thought,[8] but it also articulates an original and innovative view if compared to those earlier approaches. Darwin's historical situation and context may not be as evident as it is often presented, especially if considered from a pluralistic theoretical viewpoint. However, even so, one can say that, with him, the ancient dream of establishing a "human," "animal," or "vegetal" nature, according to an idealistic and essentialist view of nature as a parameter to a *scala naturae*, was definitively over. Many features of Darwinian thought contribute to this consequence. As mentioned above, species and varieties become arbitrary concepts employed by convention. Life is conceived of as producing itself and reproducing without the interference of a substantialist principle.[9] The truth of life can no longer be grasped through the intellectual intuition of its essence (in the Aristotelian sense) because the essence's causal action cannot be directly or indirectly appreciated. Essence becomes an unfalsifiable principle that does not fit into a logos (or a *forma mentis*) directed to producing an episteme, that is, a scientific approach to life. In turn, the fossil record reveals a close-knit network of ties and kinships binding all living beings (including extinct taxa) to one another. Hence, there is no longer ground for an idealistic belief in purely formal, unconditional, and goal-directed development of kinds, classes, or species, that is, the belief in a telos toward which any individual or group tend, in a transcendent plan to which they all are subjected.

Darwin argued that currently existing species are historically and phylogenetically related to ancestral species. He illustrated this hypothesis through a diagram showing how the gradual divergence of characters (perpetuated by natural selection)

---

[7]The problems related to the species problem are complex and unlikely to find a simple final solution. At first, transformism seems to turn "species" into an arbitrary or conventional concept: a name given by human observers to a group of organisms for the purposes of scientific investigation. Nevertheless, realist species concepts have been formulated in evolutionary biology, most famously by Ernst Mayr ("the biological species concept") and Theodosius Dobzhansky, and based on the reproductive isolation criterium. One can also mention the phylogenetic or cladistic species concept as examples of a realist stance on this matter. Ultimately, the species concept's realist or nominalist character may be a metaphysical problem raised by evolutionary biology rather than a properly scientific one (Ghiselin 1997; Stamos 2003).

[8]The history of pre-Darwinian evolutionary theories includes works by Diderot, Buffon, Erasmus Darwin, and Lamarck, among many others (Mayr 1982/1985; Bowler 1983/2003).

[9]Darwin's theory of evolution does not necessarily imply a mechanistic view of life, even though it is compatible with one. However, the philosophical reception of Darwinian scientific ideas has often interpreted the biological philosophy implicit in them as a form of mechanism. Bergson's critique (1907/2013, 2016) has been influential in this respect.

would give origin to varieties stemming from a single species that constitutes their common ancestor in a few thousand generations. Within tens of thousands of generations, these varieties or incipient species—life forms progressively distinct from their ancestors—would be subjected to further transformation and diverge even more into the significant differences that separate one species from another (reproductive isolation, for instance). Whole new genres would then result from this multiplication process, originating the kinship network that binds all living beings together:

> I have attempted to show, that the arrangement of all organic beings throughout all time in groups under groups—that the nature of the relationships by which all living and extinct organisms are united by complex, radiating, and circuitous lines of affinities into a few grand classes,—the rules followed and that difficulties encountered by naturalists in their classifications,—the value set upon characters, if constant and prevalent, whether of high or of the most trifling importance, or, as with rudimentary organs, of no importance,—the wide opposition in value between analogical or adaptive characters, and characters of true, affinity; and other such rules;—all naturally follow if we admit the common parentage of allied forms, together with their modification through variation and natural selection, with the contingencies of extinction. (Darwin 1859/2009, p. 402)

Even though causes may remain unknown, Darwin (1859/2009) argues that one cannot assume spontaneous, contingent, and unavoidable variation as the sole mechanism of species transmutation. In *The Origin of Species'* fifth chapter, he addresses the laws of variation and discusses the organization's developmental relationships and correlations of an organism's parts and structures during embryogenesis and epigenesis up to the adult stage. Such developmental *constraints*[10] entail diminished degrees of freedom in change and variety (Deacon 2011). However, the most important of these constraints—the DNA molecule—remained evidently unknown to Darwin. As Longo (2020a) says, DNA is the most evident physico-chemical trace of evolution. For example, most living beings have at least one symmetry plane (spherical, radial, bilateral) that can also change during their development's different stages. In general, movement favors a bilateral symmetry: gravitation's vertical axis and the horizontal axis fix a symmetry plane. Other contingencies may make the large pincers of certain crabs prevail—a channeling effect at work on that level. Ceteris paribus, growth in plants may give rise to radial or bilateral symmetry in leaves, but not to a general symmetry in shrubs, roots, and stems. The wind or sunlight is sometimes enough to disturb growth symmetry. It is open to discussion if one can speak of proper movement in microorganisms, but one must take into account that locomotory organelles like cilia and flagella are also symmetrical.[11] In animals, symmetries are determined by movement; in plants, the conditions differ due to the photosynthesis' activity that induces the extension of surfaces and may offer an obstacle to maintaining symmetrical growth plans.

[10] For a more in-depth analysis of the distinction between the concepts of natural law and constraint, see Pattee and Rączaszek-Leonardi (2012) and Longo and Montévil (2014).

[11] Giuseppe Longo, personal communication, October 23, 2020c.

From a Darwinian perspective, the lifeworld and the biological becoming are directly or indirectly characterized by *the common kinship of the related form: "each organic being is either directly or indirectly related in the most important manner to other organic beings"* (Darwin 2009, p. 135). Other distinguishing features are *the modification through variation, heredity, and a posteriori natural selection in addition to the contingencies of extinction.* The population thinking here involved is founded on the virtually infinite and constant variability in the organisms' populations that drastically opposes the substantialist idea of a principle of absolute continuity across the species. Later, however, certain remnants of essentialism returned, disguised as different systems of determination, in molecular biology's central dogmas, at least in some interpretations of the binary equation one gene/one protein. Indeed, the gene-centered view's core principles are the unidirectional and exact transmission of information (Longo 2020a). According to the one-gene/one-enzyme hypothesis, the gene is a unit of function (recombination, mutation) (El-Hani et al. 2008, p. 15).[12] The twenty-first century saw the emergence of a definitive countertrend thinking in opposition to the gene-centered perspective that dominated the life sciences' history in the twentieth century and conceived DNA as driving the organism in the ecosystem (Longo 2020a). One can find the origin of this gene-centered paradigm in the hypothesis of a precise correspondence between the so-called germ-cell and the visible character (genotype vs. phenotype) proposed by Wilhelm Johannsen, even if this correspondence was only established at the ontological level and not at the logical one (El-Hani et al. 2008).

The problem of arbitrariness concerning the "species" category has to do with the empirical and methodological difficulties in establishing a clear-cut borderline between individual variance within species and small varieties or between subspecies and incipient species.[13] Nevertheless, it also results from breaking with the traditional metaphysical view that infers truth from the intellectual faculty to apprehend an ideal unity in the constant physical multiplicity flow.

The epistemological context from which the natural selection hypothesis emerged allows for reconstructing the ontological assumptions underlying the empirical

[12] "With Watson and Crick's discovery of the double helix, the skeptical attitude of most physicists and biologists disappeared. Life became accepted as just ordinary physics and chemistry" (Pattee and Rączaszek-Leonardi 2012, p. 6). For a recent problematization of the gene-centered view applied to biology, both in its metaphorical and non-metaphorical meaning, see Longo (2020a) and Longo and Mossio (2020). On the most recent perspectives in biology after the genome, see Keller (2000) and Richardson and Hallam (2015).

[13] Although criteria may vary, subspecies are usually regarded as the first stage of speciation where interbreeding occasionally occurs, but many sterile male offspring may result. Incipient species (or semi-species) is the next stage of speciation, where interbreeding is rare, and all male offspring is sterile. However, "incipient species" is also applied to all variation within a species resulting in reduced fertility or gene flow and, hence, to all intermediate stages leading to a new species' emergence. There is little terminological and conceptual consensus in this field. General criteria are also very unlikely to apply to all individual empirical cases. Hybridization, viral quasi-species, and ring species represent other challenges to the biological species concept based on the reproductive isolation criterium (Zachos 2016).

inquiry into the biological becoming. Life is no longer passed on from generation to generation as an immutable form, conceived according to the genus/kind principle and the postulate of species unity. These principles made it possible to maintain stable relations of logical predication, such as preconized in scholastic logic (see, for example, Thomas de Aquino's *De ente et essentia*).

Lamarck was one of the first naturalists to oppose fixism and elaborate a systematic theoretical model to sustain the possibility of species' transformation, thereby challenging the metaphysical discontinuity of classical thinking. However, Lamarck still maintained that the organisms' development and evolutionary change necessarily represented progress towards complexity, equilibrium, and formal perfection. He also had to rely on the hypothesis of spontaneous generation to account for the emergence and early evolution of the simplest life forms. Darwin himself could not avoid some Lamarckian views that gained greater importance in his theory's later revisions (Pievani 2012). These views included the inheritable effects of use and disuse, the transmission of acquired characteristics, and the role of environmental conditions in modulating the organisms' variability and function. These revisions gave the concept of variation a more substantial impact.

As Darwin reflects on *the reciprocal affinities between living beings*, including *their embryological relationships* and *geographical distribution*, it becomes clear that the currently existing species are not substantially distinct and separate classes, but manifest variable, albeit continuous, degrees of kinship evidencing their descent from common ancestors. Thus, his new epistemic categories explain how the continuous changes in individuals' successive generations give origin to different groups or populations within the ancestral species that eventually diversify into a new one. These categories comprise spontaneous inheritable variation, struggle for life, the differential rate of survival and reproductive success ("survival of the fittest"), and, more prominently, natural selection. They were later complemented with additional ones, such as genetic drift and the role of migration, and other causes of populational geographic isolation, among many others.[14]

Thus, the theoretical challenges in biology depend on understanding the relationship between change mechanisms and adaptations, both in morphophysiological and in behavioral terms. Since biology addresses a historical object, it must be clear that one can only relate mechanisms and adaptations and understand the biological value of behaviors and functions a posteriori. There are many difficulties in the way, some of them already acknowledged by Darwin from the beginning. They include the incompleteness of the fossil record, the geological data's fragmentary condition, and the problem of relating the complexity of current organic structures to the origins of variability, as well as the contingency that distinguishes the lifeworld and its transformations through time.

[14]From a Darwinian perspective, natural selection is necessary but not sufficient for speciation. Other factors must intervene, such as geographic isolation in allopatric speciation. Although it is not clear if the concept of allopatric speciation is already formulated in Darwin's works, he widely discussed the role played by geographic isolation in evolution, most famously concerning the Galapagos finches (Berlocher 1998).

The role of phylogenetic systematics or cladistics in reconstructing the origins and evolution of analogous and homologous characters in living beings is a crucial field of interest in biosemiotics. As "*the synthesis of sign science and life science*" (Favareau 2008, p. iii), biosemiotics inquiries into the meaning relationships present in biological systems at different levels of organization (cell, tissue, organism, species). More specifically, it regards semiotic processes as one of the main distinguishing features of life and a source of its evolutionary mechanisms. The study of signs, their interpretations, and effects is thus extended to their pre-linguistic ground, encompassing Jakob von Uexküll's *Umwelt* (the surrounding or self-world) as the field within which the experience and observation of life phenomena can take place. The *Umwelt* is the world where a given biological system effectively acts and performs its functions. In other words, it is the segment of the environmental factor to which each living being actually relates according to what is determined and made possible by its internal organization. In this manner, the organism shapes its surroundings and makes it intrinsic to a certain extent. Peirce defined a sign as that which, in a certain way or manner, represents something to someone (Peirce 2015). Thus, for example, biosemiotics regards mimicry (both Batesian and Müllerian) as a privileged and concrete field for the scientific inquiry into the manifestation of such semiotic relationships (Maran 2017). This epistemological attitude establishes a border or an interface between semiotics and the phenomenological tradition's philosophical context (Merleau-Ponty 2000).

Having all this in mind, one can understand why the French philosopher Maurice Merleau-Ponty believes that "we can follow up the history of Darwinism with the attitude of science in front of the problem of Being" (Merleau-Ponty 2003, p. 244), despite being a harsh critic of the Modern Synthesis and its implicit mechanistic and reductive trends. Accordingly, Merleau-Ponty explored the evolutionary tendencies in modern biology in his attempts to address fundamental ontological problems. His critique stressed the incapacity of a substantial segment of the life sciences—as well as philosophy—to maintain its adherence to the world on experiential and conceptual levels.

Merleau-Ponty and Patočka share the idea that the organism's vital activity belongs to a third-order history. For this reason, one cannot ascribe this activity either to the sphere of subjective consciousness or to the field of objective facts. The biological agency goes beyond the traditional distinction between the pure activity of an already constituted consciousness (providing the law for the world's intelligibility) and the pure passivity of a mass of inanimate things.

According to phenomenology, causal thought, and the rationalist attitude that feeds it are insufficient to explain the world's experience in the multifaced forms it can assume. This experience never reveals the thing in itself nor the mechanisms producing it. Instead, it continually pushes the subject to question itself about what constitutes its intentional, emotional, and symbolic-cultural bond with that experience. It is impossible, even for the sake of simplification, to reduce such experience to representation or to anything solely based on the natural properties of nervous phenomena that can change because of physical and chemical stimulation.

Unlike machines, in both human and nonhuman animals, the transition from signal to sense and meaning occurs. Machines are transmitter receivers and do not transform a meaning in their operations. The machine's function—Merleau-Ponty's tells us—has meaning, but this meaning is transcendent. It only exists in the user's or the designer's mind. The machine is a device that carries out a pre-established, finite number of possible operations. In contrast, organic beings live actual lives, i.e., they structure their world and their bodies. Meaning is at work within them as a result of a signification process. In fact, the living beings' macroscopic structure is also distinguished from artifacts by the autonomous and spontaneous character of their morphogenetic processes (Monod 1988)—this is the problem of form in biology. Morphogenesis also proceeds through self-organization processes that do not receive all their information from the DNA (Galleni 2010, p. 69).

Therefore, even pathologies and their symptoms become tangible manifestations of the continuous involvement of the individual with the world, with the surrounding environment (Merleau-Ponty 1942/1963). Merleau-Ponty's aim in *The Structure of Behavior* is precisely to understand this involvement, this intertwining: the relationship between consciousness and nature in all its forms through which living organisms express and process their contact with things. Physics and mathematics replace reality with a matrix of objective relations expressed in formalized language. Conversely, biology is classically inclined towards realism, both in mechanistic and in teleological views (those referring to an inner principle of self-organization).

In its mechanistic and finalistic variants, this realism leads to attributing the status of "a thing among things" (*partes extra partes*) to the biological body. It places the body within a geometric-analytical Cartesian paradigm. In the case of mechanism, as in molecular biology's gene-centered view, the body is reduced to fundamental and non-decomposable units whose relations are explained partly by chance and partly as instances of external natural laws (natural selection, for example). On the contrary, in the case of finalism, as in organism-centered and organizational theories, the living being's morphology is subjected to the geometric order of nature and its precise structures. In this case, one cannot automatically explain the regularity of forms and proportions by the natural selection's mechanics: "*The selection can only choose between ordered structures but not create order*" (Galleni 2010, pp. 69–70). Even here, self-organization is regarded as resulting from chemical–physical forces or some universal cosmic nature, as the tangential and radial forces at the origins of life's complexity postulated by the French paleontologist Pierre Teilhard de Chardin.

According to teleological views, the mechanic hypothesis is unable "to account for the complexity of psychic life" (du Noüy 1975, p. 51). However, at the opposite end, du Noüy's telefinalism, for example—the idea that evolution is impelled by a transcendent and ultimately spiritual cause guiding it through an ascendent path—formulates a limited, discontinuous, and irreversible law of evolution: "It was necessary to reach intermediate stages capable of further evolving to achieve the creation of a being that could host a thinking and abstract brain" (ibid., p. 141). Even if evolution's arrival point is given—and coincides with the perfect integration of structure and function—the means to achieve are not. Nature proceeds by groping, says Lecomte du Noüy, making thousands of unsuccessful attempts and maintaining its efforts toward the final goal.

When applied in mechanistic biology, the geometric paradigm represents life as a complex system of discrete units (nerves, neurons, genes, code symbols) whose relations and integration exhaust the organic phenomenon and its communication, symbolic expression, and purpose as Umberto Eco explains:

> [. . .] genetics deals with the coded transmission of hereditary characters, neurophysiology explains the sensory phenomena as the passage of signals from the peripheral nerve endings to the cortical area; and these disciplines make use of the tools provided by the mathematical theory of information, which was created to explain signal transmission phenomena at the level of machines (. . .); gradually, sciences such as cybernetics, which dealt with the control and management systems of automated systems or electronic computers, merged with biological and neurological research. (Eco 2016, p. 51)

It is implicit in this model that reducing complexity to discrete units is the only way to promote rational knowledge and intelligibility in this field. Thereby, one can understand why the organism–machine metaphor has risen and remained such a pervasive paradigm for more than half a century, applied to all fields where the living being's activity takes shape. This paradigm's emergence and widespread acceptance relates to the fact that it refers to the intelligibility proper to the analytic spirit that seeks truth in nature's laws operating inside the matter and independently from the observer.

Merleau-Ponty rejects both finalism and mechanist causal thinking as two forms of artificialism—the latter affirms a natural artificiality, and the former endorses artificial naturality. Merleau-Ponty was undoubtedly familiar with Husserl's deep interest in biology, through which he believed that the rediscovery of the *Lebenswelt*—the world before idealization—became possible (Husserl 2002). Indeed, a universal science of being could not refrain from studying the *Lebenswelt*, in Husserl's views. This view makes biology play a fundamental role in the development of his idea of philosophy. By approaching life, biological sciences can escape the call for a logical and abstract world-making, contrarily to mathematics and physics, and can finally overcome the physicalist prejudice. Husserl clearly states that by exploring the lifeworld—that, by definition, evades logical construction—biology avoids referring to it as an abstract structure.

However, abandoning such idealization has a counterpart that may represent a setback originating from the opposite side, i.e., from realism. Merleau-Ponty and Patočka claim that realism traces biology's cumbersome limits that may jeopardize its inherent original project: to reveal life in its making. The study of behavior and perception showed Merleau-Ponty the need to distance himself from Descartes. Phenomenology must take up once more the problem of consciousness' hybrid nature, that is, the nature of corporeality, perception, and behavior. These issues manifest a strong priority in Merleau-Ponty and Patočka: the soul inextricably refers to the body's affections and affects; one can no longer pigeonhole the subject of behavior and perception in the mutually excluding categories of *res cogitans* and *res extensa*.

In Merleau-Ponty's *The Visible and the Invisible* (1964), the fundamental ontological problem is expressed through the image of the flesh (*chair*). The passage through biology in the years 1956–1960 revealed to him the flesh as an element of the universal Being (an element in the pre-Socratic sense of the word, he says).[15] It is the cradle or the common ground of divisibility and indivisibility differentiation and undifferentiation from which the essentially reversible movement from the mute to the speaking (and spoken of) world, from the invisible to the visible, takes place and shape.

In principle, Merleau-Ponty precludes any possibility of referring to the flesh in terms of matter, substance, spirit, and any concept presupposing a pre-evolutionist, essentialist, and non-interrogative ontology concerning the structure and relations at work in the universal Being or underlying it. One can read the Collège de France courses on the concept of nature as part of a project for ontology's re-foundation, starting from biology and, of course, with no reductionist intent. After Husserl, it is possible to resort to biology as a source of theoretical tools and insights to renew ontology concerning the interrogation of being and the language and categories that can return the being to its proper place in philosophical and scientific thinking.

Therefore, the philosophy of the flesh is concretely committed to the idea that the encounter between the history of humankind and the history of nature is not only possible but inevitable and continuously renewed. Nature secretly feeds human action from within the biological organism. Merleau-Ponty emphasizes that higher animals' *Umwelt* is open instead of closed. In these lifeforms, phenomena such as mimicry or mating rituals require the ability to process signals, if not symbols. Here, the body in its completeness is a way of expression, the philosopher argues. Referring to interspecific relationships, he rejects submitting them to two principal explanatory models available: the one based on the utility principle and the other employing teleological or finalistic explanations. He proposed a third alternative in which one attributes morphogenesis with expressive intent to life. One needs—he claims—"to grasp the mystery of life in the way in which animals show themselves to each other" (Merleau-Ponty 2003, p. 188). Mimicry, for example, forces us to admit an internal relationship of similarity between the animal's morphology and the milieu. It also suggests that one can only define behavior through a perceptual relationship: the animal's existence implies being perceived.

Moreover, one cannot disregard the fact that many species manifest homologous or analogous traits revealing their historical phylogenetic relations, evolutionary distance, or proximity. Plants and animals, including the *Homo sapiens*, are never adaptive mechanisms and nothing else. Instead, they are open to their congeners and other species. They are articulated and structured so that biological existence ultimately means coexistence. Animals are configured to perceive and to exhibit themselves, to see and be seen. The symbolic agency is always at work and inherent

---

[15] "This that-is-openness to things, with participation on their part, or which carries them in its circuit, is properly *the flesh*" (Merleau-Ponty 2003, p. 223).

in living organisms. As Merleau-Ponty says, the animal is a logos of the sensible, and its body is not understandable apart from the function of being seen.

By its esthesiological and libidinal dimensions, the body manifests the natural roots of the being-for-others—that being is, ultimately, being-for-others, and not exclusively in the human world. This idea lays the foundation for an ontological rethinking of species. One must understand every organism primarily as a natural social or collective being. The species are already inscribed in the context of inter-corporeality and its manifestations (Merleau-Ponty 2002).

Acknowledging these claims is crucial to face the difficulties that phenomenology emphasizes in its attempts to build up an interface between biological and human sciences. The authentic world with which we are primarily and perceptively in contact is at the origins of the logical intelligence with which humans organize and order reality into classes of objects, relations, measures, laws, models, and institutions. According to Merleau-Ponty, this non-reflexive world is constituted by the constant and immediate presence of the flesh, that is, the presence of a preliminary *there is*—a vision that gives itself without thought. The fundamental problem of classical ontology is that the abstract model *represents* being instead of manifesting its actual *presence*, its own giving. Hence, vision is to be understood not as a way of thinking, or a presence in itself (Merleau-Ponty 2004). Cézanne's paintings "show the very birth of the landscape" (Merleau-Ponty 2002, p. 27) as a "space organically linked to us" (ibid., p. 28) in which any relationship of "domination between the sovereign spirit and the pieces of wax of Descartes' famous analysis ceases" (ibid., p. 39).

What interpretation can one assign to the expression of a worldview separated from thought? The lessons to be drawn from Merleau-Ponty's famous essay on Cézanne condense in this enigmatic and prophetic image: "The other spirits only offer themselves as incarnate and adhering to a face and gestures." (Merleau-Ponty 2004, p. 34). These words can be understood as announcing the rise of modern biosemiotics that took place only a few decades after Merleau-Ponty's death. In fact, biosemiotics teaches us that culture is inseparable from the sign that originated it and that, through signs, nature becomes visible, participable, and transmissible as signifying and signifiable.

Phenomenological attitude towards nature represents philosophical anticipation of some theoretical stances developed in biosemiotics, now as a scientific research program. However, this transition also entails reviewing many phenomenological tenets to establish a fruitful partnership between these two approaches to life and meaning. The challenges that biosemiotics poses to traditional phenomenology ultimately amount to changing the very notion of intentionality. Indeed, a relational epistemology of the living being, such as the one offered by biosemiotics, leads to de-substantializing the organism and de-essentializing any question about the nature of life and rethinking it in terms of an absent structure. Thereby, it may be possible to detach biology from its almost exclusive relation to the concept of information processing and its application in explaining the mind–brain relationship on the same grounds as the software–hardware relationship.

As Umberto Eco claimed, every discourse on structures presents two types of oscillations. The first one is between its ontological and epistemological aspects. Is the eidos a "given" or a "place"? Does one find it in the thing or apply it to make the thing intelligible? The second kind of oscillation is between "concrete aspect and abstract aspect, between object and model of the object, between individual and universal" (Eco 2016, p. 351). This oscillation mainly concerns the opposition between two trends: understanding structure as a substance or as a "network of relationships, a complex of relationships, an order that can remain constant even in the variation of terms between which occurs" (ibid., p. 351).

Among biosemiotics' principal claims, the following three deserve special consideration:

1. *The emphasis on "historicity" and "meaning" as fundamental features of biological systems.* A system's capacity for adaptive evolution is not exclusively genetic but also conveyed by a tendency to perform meaningful mutual exchanges with the environment—a process through which meaning itself is engendered. From this perspective, biosemiotics' project for reintroducing semiosis and communication through signs in natural events meets up with Merleau-Ponty's attempt to account for biological life as a result of meaning's "incarnation" or embodiment within the evolutionary process.
2. *The emphasis on the limitations of reductive and mechanistic paradigms in explaining living systems and their behavior* (Hoffmeyer 1996; Barbieri 2007; Henning and Scarfe 2013). These limitations justify the need to bring meaning back into the study of life processes under all its aspects. They also clarify why to revise widely accepted views on the conditions and possibilities to access the world—the lifeworld in particular—by reintroducing semiosis both in common sense and in the scientific worldview. As biosemiotics claims, one must consider that organisms are embedded in their environment, concerning which their actions can be concordant or discordant with the dynamics of achieving the end. For Merleau-Ponty, "since the perceived world is grasped only in terms of direction, we cannot dissociate being from orientated being" (Merleau-Ponty 2005, p. 227).
3. *The attempt to articulate an alternative "evolutionary history" of the relations between nature and nurture by including a matter-symbol complementarity* (Pattee 2015; Pattee and Rączaszek-Leonardi 2012). The sign concept is not restricted to a theory of mind but also concerns a theory of nature, as Harney (2007) posits. Thus, biosemiotics highlights the interpretant's central role "in the process of semiosis as it can lead to a change in the disposition of the organism for different behavior" (Barbieri 2007, p. 461). Accordingly, as Kull emphasizes, "biosemiotics means the study of living systems that interprets them as sign systems, or communicative structures" (Barbieri 2007, p. 175).

The relationship between biosemiotics as described by these central claims and life sciences as a whole (especially evolutionary biology) deserves closer consideration.

3 The Fundamental Phenomenological Connection Between Biosciences and Biosemiotics

Merleau-Ponty maintains that contact with ourselves is accomplished "always through culture, or at least through a language that we have received from outside and guides us in knowing ourselves" (Merleau-Ponty 1948/2002, p. 61). Patočka argues that Merleau-Ponty's work is permeated from end to end by attempts to build a bridge between *physis* and life and, consequently, between corporeal life and *logos*. One must understand *logos* in a broad sense, as a matrix and condition for expression. It also makes sense transmission possible—this sense disclosed in the body's life opening to the horizon of being's appearance that can be experienced, shared, and signified. Therefore, phenomenology engages in searching for a new language game expressing the relationships between two epistemic requirements that have often been placed at opposite ends in the line of knowledge: the world of perception and the world of expression.

Merleau-Ponty presents the organism's biological development as a construction—a work in progress never completed—participating in a horizon of shared symbols. One cannot understand the path taken by living beings only tracing it back to their instinctive and reproductive purposes, even though these purposes cannot be wholly neglected. Philosophy must start precisely from the reconciliation of these two horizons—natural and symbolic—to keep existing as a problematization of the human insertion in nature. In other words, the historical sedimentation underlying cultural and scientific revolutions are different moments or aspects of this problematic human/nature relationship and its intrinsic tension between rupture and reconciliation (Merleau-Ponty 2000).

Merleau-Ponty thus introduces a new, ambiguous, and unexpected ontological horizon. The objects and physical quantities in space and time are no longer raw facts and become "what is revealed to us through the senses and the practice of life" (Merleau-Ponty 1948/2002). In turn, perception is no longer regarded as direct knowledge of the world but as a common background against which every act (theoretical, axiological, practical, or technical) occurs. Perception is presupposed by each one of these acts: "I rediscover 'in me' as the permanent horizon of all my cogitations and as a dimension concerning which I am continually situating myself" (Merleau-Ponty 1945/2005, p. xii).

In particular, these words express the conditions for understanding, from a historical and evolutionary perspective, how this embodiment and openness are made possible and feasible. From the simplest to the most complex one, the history of each organism is an instance of this fundamental determination of life as such—its historicity. To survive and reproduce, life demands of every organism to relate, first of all, to its own being and, through it, to its challenging environment. It must respond to this environment actively, appropriately, and commensurately with its contingencies and circumstances, as required by the very concept of *historicity* (as opposed to *history*).[16]

What makes it possible to exchange matter and energy between organisms and their external environment to which they owe, for example, their perceptual capacities (perception of external temperature, the direction of the wind, proximity of a predator, among others)? How is the animal induced to take a stance or make a choice concerning that informative content? How can organisms and the surrounding environment *talk to each other* and *interact*? If physics and chemistry can explain interaction *mechanisms*, the key to understanding this interchange's meaningful and intentional aspects is clearly to be found elsewhere. For this, one must, first of all, have a clear idea of what is meant by information and communication (Pattee and Rączaszek-Leonardi 2012; Deacon 2011; El-Hani et al. 2008).

In his 1955 manuscript, Patočka (1988) tackles the question of objective biology and its claim to consider Cartesian dualism obsolete. This claim is based on the widespread application of the information metaphor in biology, which seems to offer a reliable and homogeneous ground for proper perception and a real understanding of life phenomena. The philosopher argues that assuming that reality lies in the *perceived* world, in the things given to the sense, is at the origins of all objectification. At the beginning of this essay, he states that objective biology presupposes a particular and peculiar concept of man—an organic conceptualization formulated in terms of information theory. In this view, man's characteristics and behavior are conceived and explained as an informational structure, so that human beings are regarded as not essentially different from a car, a computer, or a radio device. This informational image or metaphor allows for a computational view of organisms in contemporary biology. However, the information metaphor is only one among many. Science has resorted to others like this in its attempts to overcome the gap between thought and experience. The reference to a formalized language or code would then justify the alleged equivalence (or perfect resonance) between the ontological plane of being and the descriptive logical-epistemological plane.

As Lakoff and Johnson (1980) propose, most of our conceptual system is structured in a metaphorical form, i.e., most ideas commonly employed in everyday life are *conceptual metaphors* through which one field of experience can be referred to in terms of another (for example, when one express numerical quantity in terms of direction, as in "the prices are rising"). These metaphors are often so embedded in our ways of thinking and linguistic habits that they can hardly be perceived as such. Metaphors are also an essential part of theoretical and scientific discourse and, in this respect, have long been the object of philosophical inquiry (Black 1962). From this

[16] In a philosophical and phenomenological sense, *historicity* refers to the idea that concepts, practices, values, and, ultimately, reality itself have a historical origin and have developed through time (in its most extreme formulations, it maintains that being as such *is* time). Historicity is thus opposed to the belief that these entities are essential and exist universally or necessarily. It thus introduces an intrinsic element of contingency in ontological thinking and rejects the idea that the *history of being*, once occurred, could not have been otherwise. See Ricoeur (1983/1991) and Nuki (2020).

perspective, it becomes apparent that science, insofar as it operates through concepts, is not so distant from philosophy. Common to both disciplines is the claim to enrich and broaden our perspectives on the world, employing these images and metaphors that allow for more encompassing explanations of the processes constituting the reality—that is, the being. One may consider, for example, how the concept of *field* in physics is close to the notion of a "*sector of being*" as employed by Merleau-Ponty.

Patočka (1988) argues that the information metaphor, which is manifest in expressions such as "genetic information," "genetic code," "genetic message," "chemical," and "cell signaling"—in brief, the concept of genes as informational units (El-Hani et al. 2008)—can be profoundly deceptive. Unlike the human mind that designed it, a machine is permeated by an extrinsic purposefulness. Its goals are predetermined from the outside and only exist in the designer's or the user's minds. In turn, this transcendent teleology explains the machine's project, its configuration, the specific arrangement of its parts. Applying human, mechanical language to life phenomena prevents the consideration of its purpose and its very phenomenological existence as something different from inanimate matter. It is not enough to say that the meaning of biological information derives from the fact that both the signal carrying information and the response decoding it results from natural selection (El-Hani et al. 2008). This reply does not explain how genetic information can produce the biological (semantic and pragmatic) value of behavior, function, or character. The most significant problem in the deterministic interpretation of how information is genetically transmitted is considering organisms as mere vehicles or repositories of information units (Dawkins' [1976/2016] "replication machines"). According to El-Hani et al. (2008, p. 80), "the task of going beyond information metaphors in biology and building a theory of information in this science demands a semantic and pragmatic account."

Deacon dedicates an entire chapter of his book *Incomplete nature* to analyzing the information concept. He argues that, at the origin of the currently dominant technical view of information, lies the same misunderstandings that, in the nineteenth century, led to a substantive and non-relational conception of energy. According to him, "the presumed ethereal substance that conveyed heat and motive force from one context to another was an abstraction from a process—the process of performing work—not anything material" (Deacon 2011, pp. 351–352). One can find the same attitude when one thinks of information as some artifact or commodity quantifiable for transmission and storage. In Deacon's views:

> Information about an impending storm might cause one to close windows and shutters, information about a stock market crash might cause millions of people to simultaneously withdraw money from their bank accounts, information about some potential danger to the nation might induce idealistic men and women to face certain death in battle, and (if we are lucky) information demonstrating how the continued use of fossil fuels will impact the lives of future generations might affect our patterns of energy use worldwide. These not-quite-actualized, non-intrinsic relationships can thus play the central role in determining the initiation and form of physical work. In contrast, the material sign medium that mediates these effects (a darkening sky, a printed announcement, a stirring speech, or a scientific argument, respectively) cannot. The question is: *How could these non-intrinsic relationships*

> *become so entangled with the particular physical characteristics of the sign medium that the presence of these signs could initiate such non-spontaneous physical changes?* Answering this question poses a double problem. (...) *The currently dominant technical conception of information was important for the development of communication and computing technologies in response to the need to precisely measure information for purposes of transmission and storage. Unfortunately, it has also contributed to a tendency to focus almost exclusively on only the tangible physical attributes of information processes, even where this is not the only relevant attribute. The result is that the technical use of the term information is now roughly synonymous with difference, order, pattern, or the opposite of physical entropy* (...). To step beyond this impasse and make sense of the representational function that distinguishes information from other merely physical relationships, we will need to find a precise way to characterize its defining non-intrinsic feature—*its referential capacity*—and show how the content thus communicated can be causally efficacious, despite its physical absence. (Deacon 2011, pp. 352–353 our emphases)

The author proceeds:

> The relationship of present to absent forms of a sign medium embodies the openness of that medium to extrinsic intervention, whether or not any interaction has occurred. Importantly, *this also means that the possibility of change due to work, not its actual effect, is the feature upon which reference depends*. It is what allows absence itself, absence of change, or being in a highly probable state, to be informative. Consider a typo in a manuscript. It can be thought of as a reduction of referential information because it reflects a lapse in the constraint imposed by the language that is necessary to convey the intended message. Yet it is also information about the proficiency of the typist, information that might be useful to a prospective employer. (...) *In all these cases, the referential capacity of the informational vehicle is dependent on physical work that has, or could have, altered the state of some medium open to extrinsic modification.* (...) *The capacity to reflect the effect of work is the basis of reference*. (Deacon 2011, pp. 365–366, our emphases)

The key to understanding the difference between Shannon's classical information paradigm and classical thermodynamic theory lies in the shift from the reporting medium to the report,[17] "between what is and what could have been its state at any given moment," so that "the capacity to reflect the effect of work is the basis of reference" (Deacon 2011, p. 365). According to Deacon, it is necessary to combine these two approaches in a single theory to highlight information's referential function, which is ignored in the classical paradigm in favor of its substantive component. Nonetheless, the origin of referential processes (semiotic interactions) is closely related to the flow of historical singularities, as Hoffmeyer (2001) has argued throughout his work.

In turn, Patočka (1988) maintains that Jakob von Uexküll's biology underlines the practical aspects of existence: the world presents itself to the living being as a

---

[17] In Deacon's words, Shannon's view is that "the improbability of receiving a given sign or signal with respect to the background expectation of its receipt compared to other options defines the measure of potential information" (Deacon 2011, p. 368). From the thermodynamic perspective, "the improbability of being in some far-from-equilibrium state is a measure of its potential to do work, and also a measure of work that was necessarily performed to shift into this state. Inversely, being in a most probable state provides no information about any extrinsic influence, and indeed suggests that to the extent that this medium is sensitive to external perturbation, none was present that could have left a trace" (ibid., p. 368).

signal for action. However, one must consider that human action is marked by much greater variability than animal behavior. It mostly escapes the stereotypical character of instinctual action due to the constant interplay between the internal and the external world, the integration of meanings from the natural world with the subject's internal life project, which is never given, but always *created*. The Czech philosopher argues that materialism and idealism are different versions of the same objectivism. Materialism sets out to demonstrate that truth starts from the object, that is, some independent perceived reality. Idealism, in turn, emphasizes the lived experience in its relation to the object and attempts to establish the conditions for true experience starting from a particular interpretation of the Ego as substance. He says:

> (...) the phenomenology of the human immediate lived-through is indispensable, as well as the "subjective starting point", provided that one does not dogmatize it or grounds it on a substantialist metaphysics that believes to know what the "ego" is that subsists to the experience. (Patočka 1988, p. 156)

As discussed above, the ways of doing science rarely escape this tendency to fall into the nets of either materialism or idealism. For example, in genetics, both perspectives are clearly and simultaneously present, giving origin to difficult paradoxes. In this respect, Keller (2005) makes a crucial distinction between the "century of the gene" (the twentieth century) and "the century of genetics or genetic systems" (the twenty-first century):

> (...) if the 20th century was the century of the gene, the 21st would in all likelihood be the century of genetics, or rather, of genetic systems. The difference is important, so let me explain: Genetics, I take to be the study of the processing of DNA in the construction of phenotype; genes, I take to refer to the entities historically assumed to be the particulate of inheritance. The former, I take to refer to the biochemical interactions underlying the construction of actual organisms, the latter to a hypothetical conceptual scheme. (Keller 2005, p. 3)

According to Keller, the twentieth century is distinguished by the *unit's absolutization* (the gene as the discrete heredity unit) and by claiming its ontological and epistemic pre-eminence over the totality (the phenotype). This view results from a schematic and hypothetical, historically sedimented conceptualization. According to this conceptualization, genes are assumed as inheritance's basic building blocks determining the biological systems' structure and behavior. This operation allows for a vision and evaluation of life phenomena *at their very limit*, made possible by mathematical formalization. Conceiving living beings as machines and reducing their functions to effects of discrete units (genes as heredity atoms) facilitate their formal and theoretical understanding. However, this view emerges and subsists from a specific way of conceiving life that is still dependent on the metaphysical idea of substantiality and mechanisms deriving from the part/whole relationship.

On the contrary, the twenty-first century is more concerned with the relational mechanisms and gene expression involved in phenotype construction. There is something of a positivist legacy here, especially in how systems biology deals with objectivity. Even if systems are regarded as relational structures—and not as complex wholes made of self-subsisting parts—they remain objectively accessible and subject to control and prediction in their observable development regarding a given set of coordinates.[18] Leroy Hood states this view exemplarily, as Keller (2005, p. 5) comments:

> Leroy Hood, founder of System Biology Institute (SBI) in Seattle explains: "Unlike traditional biology that has examined single genes or proteins in isolation, systems biology simultaneously studies the complex interaction of many levels of biological information—genomic DNA, mRNA, proteins, functional proteins, informational pathways and informational networks—to understand *how they work together*". And for Hood, as for the vast majority of engineering scientists, understanding *how the parts work together is followed by "understanding how to control the system, and how to design the system"*.

In turn, the phenomenological paradigm that Patočka defends keeps distance from both materialism and idealism and does not purport to describe the natural world by creating formalized patterns and verifying their adequacy through sensible intuition:

> (...) the schema does not play here the role of a pure description, but that of a provocation that must incite a better, more detailed, more concrete intuition. Phenomenology's descriptive activity proceeds in a polemical way, always or almost always opposing its experiences to the expected experiences or to experiences constructed according to schemata. In phenomenology, too, as in all finite sciences, one sees the opposition between schematization and verification coming into play, a moving forward that never reaches its goal. (Patočka 1988, p. 166)

Patočka (1995a) suggests a phenomenological connection addressing bodily life as a mediator in a field of possibilities and actions. This field is not representational—it is a field of action(s), and, in this sense, it is essentially non-thematic. In other words, things become what one acts on, the situation where one's action takes place. Action is neither an immediate nor passive process. Instead, it is undertaken by accepting or rejecting the possibility of the subject being challenged by the world. From this perspective, representation (the object's meaning) and object (the means through which meaning is accomplished) presuppose a third dimension of being (bodily life) for which and through which meaning ascends to signification. Bodily life is thus the mediation through which one attributes meaning to the milieu. This embodiment leads to a form of meaning that, insofar as it is not thematized, implies or presupposes the infinite and indeterminate modes of the same thing's presence (Patočka (1995b). There is a deep resonance between the phenomenological strategy to relate representations to objects—i.e., investing them with semantic and pragmatic value—and the biosemiotic perspective on the organism's nature and behavior.

To show how pervasive informational language is in everyday life, Patočka (1988) remarks that the same rhetoric Husserl employs to describe the world as the situation horizon of one's acts and needs also underlies the metaphors expressing the subjects' relation to the world and their reactions to a changing environment through information processing. Nevertheless, Patočka adds, Husserl is right in

[18]Keller (2005) also discusses the differences between positivism and materialism further in her essay (op. cit., p. 9).

understanding that the information the organism unravels in its dealings with the world has a concrete and practical character. On the contrary, the schemes that the exact sciences apply to the lifeworld are mathematical formulae designed to apprehend and represent structures, functions, and behavior formally and statistically. These formulae are, in every respect, an abstract reconstruction of concrete life.

The communication between a biological system and its environment is not determined by physical interaction laws alone (Pattee and Rączaszek-Leonardi 2012). Besides, living beings' study seems never to reveal general laws applicable outside the biosphere (Monod 1988). However, a unified conceptual framework encompassing both physics and biology could allow for a proper understanding and specification of life phenomena within a strictly scientific worldview (Bailly and Longo 2006/2011).

In principle, elementary physical laws are symmetric in time, whereas hereditary propagation requires a direction in time: "the temporal relation between the memory of a trait and a trait itself is not symmetric" (Bailly and Longo 2006/2011, p. 39). However, thermodynamics is a classic example of a physical theory dealing with irreversible phenomena (a falling egg, a cooling hot body, an expanding gas, a water jet, some chemical reactions). It thus makes it possible to understand the principles of environmental conditioning through which two physical systems exchange energy as heat or work, anticipating certain transformations and reactions crucial to explain fundamental life processes. Some statistical mechanics theories have also decisively contributed to understanding some complex biological systems' underlying physics, such as proteins, nucleic acids, neural networks, regulatory mechanisms, and cell proliferation. Schrödinger (1992) discussed the central problem of how the quantum mechanical stationary states in single molecules (understood as quantum systems) are transferred into a macroscopic system—for example, the phenotypic expression of heredity traits in the individual's interactions with the environment. Today, this problem remains a matter of dispute, especially regarding its theoretical implications. Another much disputed issue of a more strictly epistemological nature is the role of chance in molecular and cellular biology that has profound consequences to conceiving the relationship between heredity and the molecular mechanisms of genetic, morphogenetic, and physiological expression (Monod 1988; Kupiec et al. 2011; Pavé 2007).

The origin of life, reproduction, and the genetic code's transmissibility entail the relationship between symbolic information and physical structures submitted to laws. They pose the problem of understanding how "the gene's symbolic information controls material construction" (Pattee and Rączaszek-Leonardi 2012, p. 6) or, more specifically, how "a molecular structure could function as symbolic information that controlled a very complicated lawful enzyme dynamic" (ibid., p. 12). El-Hani et al. (2008, p. 77) remark that "the question concerning the evolution of different ways of storing, transmitting and interpreting information can be treated as a major theme in the history of life" and has been a central issue in evolutionary biology since the 1950s and 1960s.

This question is complex and challenging to come to grips with because of the incommensurability between genetic and human language due both to structural and functional reasons:

> (...) the subject-object relation first arises with the emergence of records of events. Records require some form of material symbols that represent the event and an agent that interprets the symbols. This largely arbitrary symbol-matter relation first appears with evolvable self-replication, which I define as the origin of life. The emergence of symbolic function arises only when useful information is recorded by an agent. (...) evolution requires a symbolic language that physical laws do not construct and cannot interpret. This is the basis of the new field of biosemiotics. The genetic language appears to be the only language necessary for all evolution, while human language, including mathematics, is the only language we have to express, communicate and interpret the genetic language, evolution, and physical laws, including quantum theory. (Pattee and Rączaszek-Leonardi 2012, p. 5)

Human language's unsuitability evokes the age-old philosophical problem mentioned above: how to establish a conceivable ostensive relation between *physis* and *logos?* Pattee and Rączaszek-Leonardi (2012) emphasize two of the most significant functional differences between genetic and brain-based languages. The first one is related to the distinction between discrete (unit based) and continuous (based on complex structure and function) syntaxes; the second one concerns the relationship between variation and selection. According to the authors, there is a noticeable inversion in the continuous/discrete relationship when one compares genetic and brain-based languages:

> Genetic language began with discrete strings of a small number of molecular symbol vehicles in single cells and ended up controlling the complex dynamical metabolic network of an entire multicellular organism. Conversely, human language began with a complex dynamical neural network and ended up producing discrete strings of a small number of written symbols. (...) We know that recognizing even a single written symbol requires a coordinated dynamic pattern of millions of neurons. The primeval genetic symbol-matter problem is how non-dynamic discrete molecules become symbol vehicles that control chemical dynamics. (...) The symbol-matter problems in brains is the converse of the problem in genes: How does a complex dynamic neural network produce the concept of a discrete symbol that can end up written as simple arbitrary mark? (Pattee and Rączaszek-Leonardi 2012, p. 26)

In this respect, Keller (2005) also suggests the possibility of reconceptualizing genes as verbs and brings to the foreground the problem of a real and ostensible transmission of biology's theoretical contents through human language. In turn, according to El-Hani et al. (2008), biosemiotics can provide the tools for building up a conceptual framework to deal with communication, information processing, and meaning-making in the life sciences and "to formulate the notion of genetic information in a manner which does not lend support to genetic determinism" (p. 19).

The other significant difference between genetic and brain-based languages concerns another functional inversion, this time in the relationship between variation and selection:

> (...) variation in genetic information must be expressed before selection can begin. Natural selection operates through the phenotype, not directly on the genotype. Animal brains, in

> contrast, have the enormous advantage of being able to acquire, evaluate, and select information before expression. In other words, using models, animal can predict before they act. That is why brains evolved. It is at the human thoughtful planning level that the concepts of *choice* and *purpose* acquired their conventional literal meaning. That is why the concept of purpose to cells, even as a metaphor, produces more polemics than enlightenment. (Pattee and Rączaszek-Leonardi 2012, pp. 26–27)

However, current physical theories seem insufficient to fully predict evolutionary phenomena taking place in living matter, let alone explain the entire biosphere. These theories "do not lead us to expect evolution," so that "the elementary laws of physics do not seem directly suitable for describing hereditary behavior" (Pattee and Rączaszek-Leonardi 2012, p. 39). As Longo argues:

> One of the major challenges of any historical science is the role of forecasting. Biology is a historical science, because organisms can only be understood in a temporal, phylogenetic and ontogenetic, perspective. In particular, the time of biology, and therefore of correlated sciences (ecology etc.), is a time of change in the "space of possibilities" (of "phases" as we say in physics, of ecosystems and species in biology), punctuated by rare events—evolutionary novelties, speciation. In physics, the space (of phases) is fixed as "a priori" of knowledge, as "condition of possibility" for "writing equations," explain Newton and Kant: it a priori contains all possible trajectories—unpredictability is within these trajectories (the randomness of a die concerns six possibilities, no more, no less). In biology, to physical randomness is added the unpredictability of changes of the space of possibilities and of rare events, to which one cannot even assign probability values. (Longo 2020a, p. 113)

The information metaphor emerged as an attempt to circumvent this problem of the biological object's indeterminate nature (Prochiantz 2012; Pagni 2015) through "the so-called automata description of molecular biology" (Pattee and Rączaszek-Leonardi 2012, p. 39). One can attribute this view to the cells' idea as "a classic machine that behaves very much like a modern large-scale computer" (ibid., p. 39). In brief, it is still the idea of cells as Cartesian mechanisms.[19]

---

[19]The current pandemic has been disclosing the ecosystemic crisis the *Homo sapiens* species has created based to a certain extent on the myth of an "adjustable" nature working as a "programmable machine," as Longo (2020a, p. 114) puts it: "Since the 1990s, many epidemiologists have been warning us: the notion of 'epidemic of epidemics' dates back to 1993 (...) some cases may be due to a synthetic biology that claims to be all-powerful and let us believe that we can fully control living organisms by modifying ('editing') their DNA/RNA, but more than 70% of these emerging infectious diseases come from animals, at new interfaces with the environment. Deforestation for agricultural settlement accompanied by intensive livestock farming enables the passage of bacteria and viruses from wild animals to livestock and then to humans. None of these cases and microorganisms were individually predictable, and none will be in the future: they and their causes are known a posteriori. The self-serving denial of the history of life, of the evolutionary construction of ecosystems, of their specificity and diversity, is the main cause of the activities that destroy them. *Often this denial finds its justification in a new scientism, which rules our science: on the one hand, the spontaneity of the man/economy/nature dynamics would choose the best possible path—a misuse of the mathematics for the equilibrium physics of the 19th century; on the other hand, nature itself would be an adjustable, even programmable machine, with disposable material and biological 'resources.'* A new awareness and a science of these phenomena is being built, change is possible: the knowledge of a history (...) and a vision of organisms in their autonomy and their dependence on the ecosystem make it possible to act" (our emphases).

Expanding the sign-theoretical view of biological processes is an unavoidable task to deepen the scientific understanding of biocomplexity. As El-Hani et al. (2008, p. 5) argue, "the major tenet of biosemiotics is probably an attempt to take biological sciences, including systems biology, a step further by expanding its theoretical domain by means of a semiotic (sign-theoretic) understanding of biological processes at many levels of complexity. The gene is not simply the same as DNA understood as a chemical substance."

4 Conclusion

Asma and Gabriel (2019) synthesize phenomenology's philosophical contribution to expanding biology's theoretical horizons, both historically and in the context of current scientific inquiry:

> Drawing from the phenomenology of philosophers Merleau-Ponty and Husserl in the last several decades, the direct perception approach has matured into a robust picture of *how perception works*, claiming that *animals are active in the world, moving their heads and bodies to make information available to their sense and to direct their actions*. This *mutual dependence between body and environment* is part and parcel with *the relational nature of meaning* we have been portraying as *central to the emotional mind*. (Asma and Gabriel 2019, p. 157, our emphases)

This general spirit animates this book and the authors that have contributed to making it possible. It offers an opportunity to re-evaluate through different disciplines and methodologies the current impact of biosemiotics and evolutionary theory on searching for the objects and objectives of human knowledge. Since evolution is both a natural and a historical fact, its study brings to the foreground the limits of assuming a clear-cut separation between nature (the systems of laws governing natural events) and nurture (human individual and collective history). In turn, biosemiotics improves our awareness of the symbolic operativity at work in nature, and not only in culture. It also emphasizes nature's role in building up the symbolic order itself. The three sections comprising this volume—"Life, meaning, and information," "Semiosis and evolution," and "Physics, medicine, and bioenergetics"—and the contributions they contain offer an opportunity to revisit some of the main reference models in biosemiotic and evolutionary research. These essays shed new light on many issues underlying the problem of the origins and evolution of meaning and knowledge. Among them, one can mention the boundaries between teleonomic and teleological evolutionary paradigms and their possible integration; the relationship of linguistics and biological sciences, especially regarding the concept of causality; biological information and their transmission mechanisms; the difference between physical and biosemiotic intentionality; the results of applying biosemiotic models for cultural transmission in the economics and engineering of design and digitalization.

These topics do not exhaust the field, of course, but provide elements to problematize and contextualize some of the fundamental questions raised by contemporary life sciences. Hence, this publication contemplates scientific and philosophical interests and can hopefully open dialogue possibilities about the theoretical and methodological pluralism manifested in the following essays. Another guiding principle in organizing this collection was its international nature, represented by the presence of researchers from all over the world with different national and cultural backgrounds. This diversity intends to stimulate the development of multilateral academic relations at the service of a more inclusive scientific and humanistic interchange.

References

Asma S, Gabriel R (2019) The emotional mind: the affective roots of culture and cognition. Harvard University Press, Cambridge, MA; London, England

Bachelard G (1971) Le nouvel esprit scientifique [The new scientific spirit]. Presses Universitaires de France, Paris (Original work published in 1934)

Bailly F, Longo G (2011) Mathematics and the natural sciences [The physical singularity of life]. Imperial College Press, London (Original work published in 2006)

Barale M et al (2001) Materiali per un lessico della ragione. ETS, Pisa

Barbieri M (2007) Introduction to biosemiotics. The new biological synthesis. Springer, Dordrecht, The Netherland

Beiser FC (2014) After Hegel: German philosophy 1840–1900. Princeton University Press, Princeton, NJ

Bergson H (2013) L'évolution creatrice. Presses Universitaires de France, Paris (Original work published in 1907)

Bergson H (2016) Histoire de l'idée de temps: cours au Collège de France, 1902–1903. Presses Universitaires de France, Paris

Berlocher SH (1998) Origins: a brief history of research on speciation. In: Howard DJ, Berlocher SH (eds) Endless forms: species and speciation. Oxford University Press, Oxford, UK, pp 3–17

Black M (1962) Models and metaphors: studies in language and philosophy. Cornell University Press, Ithaca, NY

Bowler PJ (2003) Evolution: the history of an idea. University of California Press, Berkeley and Los Angeles (Original work published in 1983)

Burtt EA (2003) The metaphysical foundations of modern science. Dover Publications, Mineola, NY (Original work published in 1925)

Cacciari M (2009) The unpolitical. On the radical critique of political reason. Fordham University Press, New York

Cacciari M (2014) Labirinto filosofico. Adelphi, Milano

Cacciari M (2019) La mente inquieta. Saggio sull'Umanesimo. Giulio Einaudi Editore, Torino

Cassirer E (1953) Substance and function and Einstein's theory of relativity. Dover Publications, Mineola, NY (Original work published in 1910)

Cobley P et al (2011) Semiotics continues to astonish. Thomas A. Sebeok and the doctrine of signs. De Gruyter Mouton, Berlin

Collingwood RG (2014) The idea of nature. Martino Fine Books, Eastford, CT (Original work published in 1945)

Darwin Ch (2009) The origin of species. Cambridge University Press, New York (Original work published in 1859)

Dawkins R (2016) The selfish gene. Oxford University Press, Oxford, UK (Original work published in 1976)

Deacon TW (2011) Incomplete nature. How mind emerged from matter. W. W. Norton & Company, New York, London
Dilworth C (2007) The metaphysics of science: an account of modern science in terms of principles, laws, and theories. Springer, Dordrecht
du Noüy L (1975) L'avvenire dello spirito. Giulio Einaudi editore, Torino (Original work published in 1941)
Eco U (2016) La struttura assente. La ricerca semiotica e il metodo strutturale. La nave di Teseo, Milano
El-Hani CN, Queiroz J, Emmeche C (2008) Genes, information, and semiosis. Tartu University Press, Tartu
Emmeche C et al (2002) Reading Hoffmeyer, rethinking biology. Tartu University Press, Tartu
Favareau D (2008) Joining sign science and life science: introduction to The American Journal of Semiotics' Special Issue on Biosemiotics. Am J Semiot 24(1–3):iii–xv
Flew A (1986) David Hume: philosopher of moral science. Blackwell, Oxford, UK
Galilei G (1967) Dialogue concerning the two chief world systems. University of California Press, Berkeley and Los Angeles (Original work published in 1632)
Galleni L (2010) Darwin, Teilhard de Chardin e gli altri. Le tre teorie dell'evoluzione. Felici Editore, Ghezzano, Italia
Gare A (2019) Biosemiosis and causation: defending biosemiotics through Rosen's theoretical biology or integrating biosemiotics and anticipatory systems theory. Cosm Hist: J Natural Soc Philos 15(1):31–90
Ghiselin MT (1997) Metaphysics and the origin of species. State University of New York Press, Albany, NY
Goodman AH, Heath D, Lindee MS (eds) (2003) Genetic nature/culture: anthropology and science beyond the two-culture divide. University of California Press, Berkeley, CA
Gould SJ (2011) The hedgehog, the fox, and the magister's pox: mending the gap between science and the humanities. Harvard University Press, Cambridge, MA (Original work published in 2003)
Hanson NR (1962) The dematerialization of matter. Philos Sci 29(1):27–38
Harney M (2007) Merleau-Ponty, ecology, and biosemiotics. In: Cataldi SL, Hamrick WS (eds) Merleau-Ponty and environmental philosophy: dwelling on the landscapes of thought. State University of New York Press, Albany, NY, pp 133–146
Henning BG, Scarfe AC (2013) Beyond mechanism [Putting life back into biology]. Lexington Books, Lanham, MD
Hoffmeyer, J. (1996). Signs of meaning in the universe. Indiana University Press, Bloomington, Indianapolis (Original work published in 1993)
Hoffmeyer J (1997) Biosemiotics: towards a new synthesis in biology. Eur J Semiotic Stud 9 (2):355–376
Hoffmeyer J (2001) Life and reference. Biosystems 60:123–130
Husserl E (2002).La crisi delle scienze europee e la fenomenologia trascendentale. Net, Milano (Original work published in 1959)
Keller EF (2000) The century of the gene. Harvard University Press, Cambridge
Keller EF (2005) The century beyond the gene. J Biosci 30:3–10
Köhnke KC (1991) The rise of neo-Kantianism: German academic philosophy between idealism and positivism. Cambridge University Press, Cambridge, UK
Kupiec J-J et al (2011) Le hasard au cœur de la cellule. Probabilité, déterminisme, génétique. Éditions Matériologiques, Paris
Lakoff G, Johnson M (1980) Metaphors we live by. The University of Chicago Press, Illinois, USA
Longo G (2020a) Thinking beyond the "epidemic of epidemics". Org: J Biol Sci 4(1):113–114
Longo G (2020b) Organisms and ecosystems like programmable machines: some consequences of the mechanistic bias on nature. Discussion paper at the ENSSER, Paris, October 2020.
Longo G (2020c) E-mail correspondence in October 23, 2020.

Longo G, Montévil M (2014) Perspectives on organisms. Biological time, symmetries and singularities. Springer, Berlin, Heidelberg

Longo G, Mossio M (2020) Geocentrism vs genocentrism: theories without metaphors, metaphors without theories. Interdiscip Sci Rev 45(3):380–405

Maran T (2017) Mimicry and meaning: structure and semiotics of biological meaning. Springer International Publishing, Cham, Switzerland

Mayr E (1985) The growth of biological thought: diversity, evolution, and inheritance. Belknap Press, Cambridge, MA (Original work published in 1982)

Merleau-Ponty M (1963) The structure of behavior. Beacon Press, Boston (Original work published in 1942)

Merleau-Ponty M (1968) The visible and the invisible. Northwestern University Press, Evanston (Original work published in 1964)

Merleau-Ponty M (2000) Parcours deux 1951–1961. Éditions Verdier, Lagrasse

Merleau-Ponty M (2002) Conversazioni. Se Srl, Milano (Original work published in 1948)

Merleau-Ponty M (2003) Nature [Course Notes from the Collège de France]. Northwestern University Press, Evanston, IL (Original work published in 1995)

Merleau-Ponty M (2004) Senso e non senso. Il Saggiatore/Net, Milano (Original work published in 1996)

Merleau-Ponty M (2005) Phenomenology of perception. Routledge, Taylor & Francis e-Library (Original work published in 1945)

Monod J (1988) Il caso e la necessità. Mondadori, Milano (Original work published in 1970)

Newton I (1965) Principi matematici della filosofia naturale. Tipografia Torinese S.p.A, Torino (Original work published in 1687)

Nuki S (2020) Temporality and historicity: phenomenology of history beyond narratology. In: Brough JB, Embree L (eds) The many faces of time. Springer, Dordrecht, pp 149–165

Pagni E (2015) I fondamenti biologici dell'indeterminatezza ontologica: l'organismo vivo tra identità e negazione. In: Lanfredini R (ed) Materia. Mimesis, Milano, pp 133–153

Patočka J (1988) Le point de départ subjectif et la biologie objective de l'homme. In: Patočka J (ed) Le monde naturel et le mouvement de l'existence humaine. Kluwer Academic Publishers, Dordrecht, pp 155–179

Patočka J (1992) Introduction à la phénoménologie de Husserl. Éditions Jérôme Millon, Grenoble

Patočka J (1995a) Corps, possibilites, monde, champ d'apparition. In: Abrams E (ed) Papiers phénoménologiques. Éditions Jérôme Millon, Grenoble, pp 117–129

Patočka J (1995b) Qu'est-ce que l'apparition? In: Abrams E (ed) Papiers phénoménologiques. Éditions Jérôme Millon, Grenoble, pp 252–257

Pattee H (2015) The physics of symbols evolved before consciousness. Cosmos Hist: J Nat Soc Philos 11(2):269–277

Pattee H, Rączaszek-Leonardi J (2012) Laws, language and life. Springer Science, Dordrecht

Pavé A (2007) La nécessité du hasard. Vers une théorie synthétiques de la biodiversité. EDP Sciences, Les Ulis

Peirce CS (2015) Semiótica. Editora Perspectiva S.A., São Paulo (Original work published in 1931)

Pievani T (2012) Introduzione a Darwin. Laterza, Roma; Bari

Prochiantz A (2012) Qu'est-ce que le vivant? Éditions Seuil, Paris

Richardson SS, Hallam S (eds) (2015) Postgenomics. Perspectives on biology after the genome. Duke University Press, Durham and London

Rickert H (2014) Kulturwissenschaft und Naturwissenschaft [Cultural sciences and natural sciences]. Celtis Verlag, Berlin (Original work published in 1899)

Ricoeur P (1983) Temps et récit I: l'intrigue et le récit historique. Seuil, Paris (Original work published in 1983)

Schaeffer J-M (2007) La fin de l'exception humaine [The end of human exception]. Gallimard, Paris

Schrenk M (2016) Metaphysics of science: a systematic and historical introduction. Routledge, London

Schrödinger E (1992) What is life? Cambridge University Press, Cambridge (Original work published in 1944)
Sharov A et al (2016) Comprehending the semiosis of evolution. Biosemiotics 9:1–6
Simanke RT (2011) Freudian psychoanalysis as a model for overcoming the duality between natural sciences and human sciences. In: Krause D, Videira AAP (eds) Brazilian studies in philosophy and history of science: an account of recent works. Springer Science, Dordrecht, pp 211–221
Simanke RT (2014) O humanismo e a tese da excepcionalidade humana [Humanism and the thesis of human exception]. In: Araújo SF, Caropreso F, Castañon GA, Simanke RT (eds) Fundamentos filosóficos da psicologia contemporânea [Philosophical foundations of contemporary psychology]. Editora da UFJF, Juiz de Fora, pp 199–216
Simanke RT (2016) Flesh and matter: Merleau-Ponty's late ontology as a materialist philosophy. Humana Mente: J Philos Stud 9(31):117–133
Stamos DN (2003) The species problem: biological species, ontology, and the metaphysics of biology. Lexington Books, Lanham, MD
Wilson EO (1999) Consilience: the unity of knowledge. Vintage Books, New York
Zachos FE (2016) Species concepts in biology: historical development, theoretical foundations and practical relevance. Springer, Dordrecht

Life Sciences and the Natural History of Signs: Can the Origin of Life Processes Coincide with the Emergence of Semiosis?

Franco Giorgi

Abstract Life has been defined as the quality that distinguishes functional beings from corpses. This distinction is primarily based on properties that living systems have expressed during evolution. On this ground, to be alive appears to depend on the system's capacity to persist invariant on the condition of holding mutually dependent such magnitudes as form, function, energy, and information through the organization of an internal hierarchy dynamically coupled with the environment. The scientific explanation of this complex relationship reduces the living hierarchy to the composite interactions of all constituting elements expressed in isolation. Thus, the resulting functional teleonomies are explained as due to the diachronic nature of the underlying mechanisms without having to account for the origin of their emergent relationships. In this study, I will argue that living systems could be more adequately signified, and not merely explained if the role of signs in the emergence of new forms and functions could be understood as instrumental in channeling natural selection toward the establishment of specific evolutionary trends.

Keywords Evolution · Sign · Complexity · Processes · Information

1 Introduction

Any attempt to treat the origin and evolution of life in relation to the history of the sign would, in principle, require the independent definition of the two terms and the subsequent conceptual elaboration of their mutual interdependence. However, of all different life definitions offered so far, none can be regarded as objectively satisfactory. Life has been variously defined as the quality that distinguishes a functional being from a dead body. The organism condition is characterized by the ability to

F. Giorgi (✉)
University of Pisa, Pisa, Italy
e-mail: giorgif@biomed.unipi.it

© Springer Nature Switzerland AG 2021

E. Pagni, R. Theisen Simanke (eds.), *Biosemiotics and Evolution*, Interdisciplinary Evolution Research 6, https://doi.org/10.1007/978-3-030-85265-8_3

grow and reproduce or alternatively as the set of experiences that makes it possible for every individual organism to exist (see Nicholson 1870; Mayr 1997).

As extensively discussed by Gayon (2010), life has been historically viewed as either a process of animation installed by the soul in a body, as a mechanism governing living beings in much the same way as humans construct their own artifacts, or eventually as a goal-oriented organization equipped with both mechanistic and animistic principles. In more modern terms, life has been reduced to the expression of specific macromolecular functions due to the complex expression of a pre-existing genetic program (Morange 2008).[1] Rather than offering a definition, very often, textbooks provide a list of overlapping properties that living organisms are expected to express or manifest under different environmental or experimental conditions. As a result, many of the taxonomic criteria employed for this scope are blurred with counterfactuals, invalidated by shared classification properties or simply made unusable for practical reasons.[2]

Defining a sign may also be equally problematic. As widely discussed in the semiotic literature, signs may be interpreted in prescriptive terms (Morris 1971) and understood as general dispositions to be shared by a community, or alternatively as something experienced meaningfully by its capacity to stand for and relate cognitively to something else. According to Peirce, a sign only signifies if interpreted, for it is not merely related to the object it is referred to in a dyadic manner, but rather triadically by the living being's capacity to use it meaningfully. As such, the sign is comprised of the *signifier*, the object as the *signified* and the *interpretant* as the effect (s) induced in a living being perceiving their relationship (Atkin 2013). It thus follows that perceiving something as a sign does not simply entail to be related to the world as it is, in its immediate physicality, but interacting with it in relation to the opportunities it may offer, i.e., for its *affordances* (Pickering 2007). Objects are not perceived for what they are made of, but for the use that every cognizant subject can make of them based on his/her previous experience and objective knowledge of the surrounding world (Deely 2001a).[3]

---

[1]Weber (2018) has defined living systems as open systems in which genetically informed autocatalytic cycles extract energy and build complex internal structures capable of evolving over a multigenerational time. To account for this teleonomic tendency, life is assumed to be endowed with the capacity to self-reproduce and self-maintain by constructing an internal organization that adapts itself to the surrounding medium (Damiano and Luisi 2010; Capra and Luisi 2014).

[2]In an attempt to overcome these obstacles, Ferreira Ruiz and Umerez (2018) have recently proposed to classify living systems based on *natural kind clusters* whose members share several common properties. Under these conditions, the co-occurring of these properties in the cluster makes the system alive, even though no single property may be said to be individually necessary for their membership. This strategy circumvents the difficulty of providing a proper life definition as something to be categorized based on the properties shared with other members of the same class (Emmeche 1998).

[3]The interpretant is defined as the third party that plays the crucial role of cognitively associating the object to the sign it stands for. Consequently, by extending their irreducible relationship to any future usage, it may become potentially available. Overall, this makes semiosis related, not just to signs as such, but to the actual actions that signs may prove feasible for (Deely 2017).

2 The Complexity of Living Processes

Living beings may only endure on condition of holding causally connected such magnitudes as *form* and *function* (Banavar et al. 2014), on the one hand, and *information* and *energy*, on the other (Benner 2010).[4] This correlation does not necessarily imply any reciprocal dependence between them since similar functions may be performed by phylogenetically unrelated structures (Wainwright 2009). On a similar ground, the information should not be seen as being deterministically constrained by specific matter/energy flows since various biological activities can actually be implemented by simply varying the spatial and/or temporal scales of their informational processing (Crnkovic 2012). The observation that several physiological processes may unfold on different time scales and be similarly shaped by natural selection indicates that a close temporal correlation between these magnitudes does not necessarily imply the existence of a causal dependence (Karmon and Pilpel 2016). This is especially true if one considers that explaining how form and function, along with energy and information, are mutually dependent on each other, does not provide us with the exclusive possibility of accounting for their causal connection (Glennan 2017).[5]

In my view, bringing the role of signs on the fore does, in principle, allow us to rephrase the question *of what life is* in *what it means to be alive*. By adopting this paradigmatic change, the role played by signs in the emergence of new living forms and functions can thus be more adequately understood as being instrumental for channelling natural selection toward the establishment of specific evolutionary trends (Thomas 2019). In this perspective, life may be more adequately signified, rather than merely explained, as an emergent property of a catalytic closure behaving as a self-organized agent (Kauffman 1995). This definition entails that life has no evolutionarily pre-stated phase space and needs only to adapt to indefinable contextual conditions before any genetic template could even evolve (Longo et al. 2012).

Bertalanffy (1950) defined living beings as complex open systems capable of persisting invariant by their capacity to entertain a continuous exchange of energy and materials with the external environment. In this view, living beings are seen as permanently incomplete and continuously striving toward more favorable environmental conditions to satisfy their energy needs, as envisioned initially by Jonas (1966) (see also Trnka 2015). This exchange occurs because living beings are provided with a boundary separating their internal space (inside) from their surroundings (outside). Due to this *topological closure*, the internal space may grow in

---

[4]In other words, to be alive, organisms are expected to make these magnitudes mutually dependent on each other through the organization of an internal hierarchy dynamically coupled with the environment. This condition is obviously at variance from what happens in human-made devices where these magnitudes are independently controlled and aprioristically connected on the basis of some principle of external causality (Balzani et al. 2008).

[5]Even though mechanisms may adequately explain how things interact and produce the effects one may observe upon system manipulation (Machamer et al. 2000), they do not account for their coming together in a productive relationship.

complexity at the expense of a continuous flow of energy and allow the environment to increase in entropic content by accumulating metabolic waste.[6]

The existence of an inside/outside relationship helps to define the spatio-temporal conditions of the subject's self-identity requirements (Atmanspacher and Dalenoort 1994; Easthope 2009). Given this condition, how should then evolutionary processes be conceptually reconciled with their entropic decay? Are these two processes opposed to one another both temporally and causally, one leading to a progressive increase in complexity and the other inexorably bound to degradation, or could the former be causally dependent on the latter? If entropy is the only real tendency in nature, the correct way of thinking about its relationship with evolution would be to assume living organisms as continuously threatened in their stability (Schreiber and Gimbel 2010). They do not tend naturally to complexity, but rather fall into complexity (Deacon 2011) due to their capacity to respond adaptively to any entropic threat that jeopardizes their stability.[7] While living beings are open to the outside for guaranteeing a constant energy supply, at the same time, they ought to be semantically closed. According to Maturana and Varela (1980), this semantic closure forces them to organize as autopoietic systems composed of a network of processes continuously producing all components that regenerate the same processes. In turn, this internal organization makes them naturally self-referential and essentially invariant in the presence of any disturbance from the outside (Hempel et al. 2011). In the end, it is through these processes that living systems learn how to construe their autonomy and develop the capacity to self-produce and self-maintain.[8]

A system can be regarded as self-referential if it has enough information to account for its self-description, along with a series of instructions for making this description operative. Under these conditions, the system is prompt to act in a contextually compatible way, i.e., to use information that includes both data and instructions. This situation entails that the system must be capable of specifying how many parts it should be comprised of and how these parts should be used to make it functionally active above a minimum complexity threshold (Pattee 1995). For a complex system to be semantically closed, two different levels of interpretation have to bind together: a first level defining the relationship between the actions to be performed and the symbols encoding these actions, and a second level to control the

---

[6]Depending on the context and on the level of complexity attained by different living systems, this boundary may be envisioned as the plasma membrane for cells, as the skin and its accessory structures for organisms or as languages or cultures for societies (Smith 1997; Breysse and De Glas 2007).

[7]In practice, living beings behave according to the so-called Red Queen Principle nicely expressed by Lewis Carroll (1871) in his book *Through the looking glass* whereby *one has to run a lot to remain in the same place* or, as said otherwise, they can only persist on condition of being capable of evolving toward higher levels of complexity.

[8]Damiano and Luisi (2010) see the resulting adaptive interaction with the environment as concerning the system's metabolic requirements and a prerequisite for selecting any sort of environmental information that may prove cognitively relevant for its autonomy and self-referential closure.

sequential realization of the actions specified in this description (Pattee 2013). Under these conditions, the system is capable of reproducing and evolving in a potentially open-ended manner. Any organism so engaged with the environment explores the significance to be part of the context that defines its referentiality (Cárdenas-García and Ireland 2019). In this conceptual approach, the system itself remains essentially autonomous, for no instructive role is expected to be played by the environment.[9] By this semantic closure, a living system can actually learn how to behave as an autonomous agent and adapt selectively to the environment (Pattee 2006, 2013).[10]

3 DNA as Information

These last considerations call into question what one should mean by information, what role it should play in cell/cell and organism/environment interactions, and, ultimately, how their overall information content could increase in complexity during evolution. Speaking of information in biology today means tackling a crucial topic for all the scientific, philosophical, and biosemiotic implications the term raises (see Cartwright et al. 2016 for an extensive discussion). It is common in lay parlance to attribute a causal role to DNA concerning the morphological adaptations that living organisms have experienced during evolution (Stegmann 2012).

Here, it suffices to emphasize only two major aspects of the DNA structure: the nucleotide sequence along one single DNA strand and the complementarity of the two paired strands. While the first condition defines the maximum DNA coding capacity, due to the absence of any sequence constraint, the second defines the possibility for information to be handed unchanged from generation to generation due to the Watson and Crick pairing rules (Lodish et al. 2000). On this ground, given the nucleotide sequence on a single strand, the complementary strand can always be replicated at every new generation. Thanks to these findings, DNA has been ever since and undisputedly considered the repository of genetic information (Godfrey-Smith 2007). However, despite its heuristic value as a molecular tool, we still need to understand what information means in this context and what causal value is to be given to its molecular structure (Travers et al. 2012). On a conceptual ground, the term information in biology is generally referred to the information theory originally elaborated by Shannon and Weaver (1949).[11] This theory understands information

[9] Any gain in autonomy is strictly correlated with the system's capacity to free itself from the external influence and, in the end, determine its own goals by moving towards higher degrees of freedom (Cariani 2011).

[10] One should note, however, that for an autopoietic system to be truly autonomous in an adaptive manner, its internal organization also has to be buffered to such an extent as to make lower-level fluctuations somehow compatible with more extensive adaptive changes, and, therefore, constrained within the limits of its semiotic threshold (Lemke 2000).

[11] One must notice that the theory was elaborated considering a communication process between a source and a receiver. Under these conditions, the message transmitted along a channel correlates

syntactically and makes it more related to the unexpected than to the anticipated, in that the higher the noise contained in the message, the more information it actually carries.[12]

By contrast, the existence of constraints in a sequence of letters reduces the actual information capacity of the message to be transmitted (Carson 1961). However, it is precisely through the presence of these constraints that words are readable and made significant for those who partake of the same language, i.e., share the same coding rules with the message source. When the same reading criteria are applied to the DNA structure interpretation, we cannot fail to notice that bases are not linked directly but joined through phosphodiester bonds. While this randomness confers a high information capacity to the DNA syntactic structure, it says nothing about its semantic content. Therefore, when correctly interpreted, the concept of information transfer adopted by the central dogma refers exclusively to the syntactic aspect of the DNA structure (Godfrey-Smith 2007). However, this would not necessarily imply that the expression of an encoded blueprint for structural or enzymatic proteins should reflect the DNA's capacity to implement and control its own decoding. In reality, nothing is actually transferred, neither matter nor energy and much less information (Hoffmeyer and Emmeche 1991). According to Bateson, information should not be related to material causality, but rather to the pattern which connects and allows the context to be perceived as *a difference that makes a difference* (Bateson 1972).

In more general terms, a message may be considered informative if written in compliance with a shared code and capable of holding a semantic relationship with the recipient system. Taking information in its causative value would imply assuming DNA replication as an entirely self-referential process and consider DNA itself in its exclusive agentive role. An additional implication would be that any digital phase should always have to precede the corresponding analogical phase, both temporally and causally, as if codes had to be necessarily worked out before the encoded products or functions (Hoffmeyer 2001).[13] Reading information as in-formation would make it equivalent to something being formed, rather than something instructing how to make it (Hoffmeyer and Emmeche 1991). In this transition, information has shifted from indicating "something happening" to the cause of "something that has already happened" or even worse, to "something that has yet to happen." Perhaps this transition justifies why DNA has gradually acquired

---

exclusively with the reduction of uncertainty perceived by the receiver, and therefore with the syntactic rules used to construct the message itself (Pattee and Rączaszek-Leonardi 2012).

[12]Thus, communication in this context is completely uprooted from the message significance, for the more random the sequence of letters in a sentence, the more informative the message content for the receiver. For instance, due to the constraints by which messages are built in any language, there might be letters that fully determine the following ones in the same row. In this case, their informative value is practically nil due to the receiver's capacity to anticipate it.

[13]The recognition that an exclusively analogical phase has historically preceded the analog/digital correspondence in any human-made device should induce us to think of this assumption as inappropriate (Indiveri and Horiuchi 2011).

an agentive role in people's minds and therefore taken as causally responsible for everything happening downstream of the transcription process. In other words, during evolution, mechanisms have anticipated the establishment of relationships. As will be clarified later on in this paper, biosemiotics holds the opposite view that relationships are first to be explored in their potential fitness and then eventually selected to develop more efficient mechanisms. In the absence of such recognition, living organisms must be conceived in computational terms, i.e., equipped since birth with all information necessary to realize and control both developmental and differentiation processes (Emmeche 1994). As already stated, this concept should be understood exclusively in operative terms, i.e., verifiable on proximate temporal scales for lack of any historical vision of their becoming on remote causal terms.

4 The Origin of Information

The objects of the inorganic world are structurally additive because the functions they express at the macroscopic level can be traced back to the symmetries of their microscopic elements (Longo 2008). If one applies this stringent logic to biology, living organisms are inevitably interpreted in reductionist terms, for any form or function they might express during evolution would also appear as reducible to the mechanisms that sustain their reproducibility. However, the living world's complexity is such as to impede any such reductionist analysis of the properties expressed at physiological and behavioral levels. The recognition of this structural and functional complexity requires the biologist to include, among the properties to be causally explained, such dimensions of reality as *individuality* and *temporality* that are not experienced in the inorganic world (Herron et al. 2018; Beaulieu 2012).

If functions are simply causally correlated with the effects that organisms produce in their acting, temporally subsequent events may appear as having a causal value for preceding events (Pettit 1996). Functional explanations always center on the question of how organisms come to express such and such property or behavior. The answer is always in the form: because it contributes to the realization of its function. To bring the functional explanation back to the canons of the nomological-deductive interpretation for making the property in question causally necessary for the production of the relative function, or merely assuming it as sufficient, even if not necessary, has not solved the problem of finalistic or goal-oriented interpretations. One can never justify any structure as evolutionarily necessary to account for the origin of the related functional activities. Structurally similar organs may develop functionally divergent strategies, in much the same way as equivalent functions may develop by structurally different organs (Alcocer-Cuarón et al. 2014).

One of the greatest difficulties in explaining these functional teleonomies stems directly from the diachronic nature of the underlying mechanisms since living systems evolve through processes that are essentially historical (Weisblat 1998). To attribute a causal value to the genetic program does not resolve the difficulty of explaining these teleonomies since it leaves out the most critical question: *the origin of the information*. We have already seen that information is significant only if expressed in conformity with the syntactic and semantic constraints imposed by a proper reading code. In structural terms, this equals saying that a message is perceived as meaningful if comprised of elements constrained by specific sequencing criteria (Barthes 1986).

In physically closed systems, the constituting elements have comparable degrees of freedom and equally probable configurations. Under these conditions of equilibrium, the system reaches an entropic maximum and the least capacity to perform work and expresses maximum symmetry.[14] According to Collier (1996), it is precisely the breaking of symmetry that gives the system the possibility of increasing its information content. More importantly, the contingency with which the asymmetry arises confers a unique historical character to its content. On a physical ground, symmetry-breaking creates some local discontinuities in the form of energy gradients in much the same way as succession constraints restrict the degree of freedom in a written message of equiprobable elements. If the change in the state of a physical system is energetically determined, any potential outcome is necessarily preordained. However, unlike symmetry breaking in physical systems, the introduction of succession constraints in a coding sequence does not change its energy content and, consequently, makes it possible for the system to choose amongst more alternative outcomes (Kull 2018). Under these conditions, any message perceived by the system may become semantically significant because of its temporal coincidence with one of the alternatives explored within the range of the accessible options. The significance of this explorative strategy lies in the different way information is handled by teleomatic and teleonomic systems (Mayr 1974). While in the first case, every variation in information content is necessarily associated with a change in the energy content, in the second, every variation occurs in an energetically free form. In physical terms, this entails that the system has degenerated to such an extent to make all possible configurations equally probable. However, from a semantic point of view, it is precisely the gratuity of this configurability that provides the system with the possibility of making significant choices. Because of this exploration's free nature, any alternative explored by the system may be perceived as contingently significant, and therefore fixed in its responding repertoire as meaningful information (Sterelny et al. 2013).

[14]By symmetry in this context, one should mean the system's ability to remain invariant when its basic configurations are subjected to all possible transformations (Brading and Castellani 2003). In principle, it is physically inconceivable that a system with a uniform symmetry could contain local discontinuities or energy gradients. Should such discontinuities arise, the system would inevitably become asymmetric and have greater information content.

5 A World of Relationships

Ever since man has started to question the world and the cognitive tools he/she could use to understand it, reality has appeared as either made of things or relationships (Marmodoro and Yates 2016). Two different theses have contended the interpretation of reality since antiquity. According to Heraclitus, reality comprises a form of energy in a dynamic and perpetual transformation, whereas, for Democritus, the reality is fundamentally made of discrete atoms dispersed in a vacuum. In essence, the world may be thought of as composed of immutable objects combined differently to give rise to the variety of multiform things around us or as a universal flux where the present moment instantiates a temporal section of a continuous variation (Dupré and Nicholson 2018).

If objects are taken as the basic stuff of the world around us, the reality is necessarily conceived in reductionistic terms, i.e., explicable by its reducibility to the simplest elements. In this view, relationships are taken as resulting from the composite interactions of the properties that all constituting elements express in isolation. As far as we can tell, most, if not all, scientific disciplines base their research programs on this paradigm. In particular, chemistry owes much of its prestigious success to the predictive value derived from the reductive analysis of many natural and artificial compounds (Villani 2017). For the reductionist mind, knowing the properties expressed in isolation by these elements suffices to predict the functions of the whole of which they are part.

However, one should note that conceiving causality this way applies pretty well to the so-called inorganic or inanimate world made of objects like molecules, atoms, and so on.[15] In this world, processes are teleomatic because they attain a state of equilibrium upon exhausting their energy potential at the maximum of the entropy content. After all, the practical outcome of any newly acquired scientific knowledge could actually be valued by the ability to predict any future effect of the observed events. In this sense, determinism and predictability come to coincide, at least in the world of inanimate things (Penfield et al. 2014). Living organisms could also be reduced to their constituting elements and, consequently, explained according to the inorganic world's physical and chemical principles. As a result, structure and function would be interpreted as causally correlated and controlled by pre-existing genetic information, and every significant change would be interpreted as guided by the DNA's sole nucleotide sequence.

However, there is plenty of evidence in nature to indicate that genetically unrelated structures may eventually perform similar functions in much the same

[15] Whatever trend these objects manifest, whether toward higher temperature, increase in entropy, lowering pH, or others, can still be explained mechanically and therefore devoid, so to speak, of any intentionality. In this sense, atomic and molecular phenomena can be labelled as teleomatic and, as such, interpreted in the full respect of the temporal asymmetry between causes and effects (Mayr 1982).

way as homologous structures may ultimately realize different functions.[16] These observations suggest that the same basic structure is potentially apt to perform different functions depending on how it interacts with and responds to the world demands. At the same time, they indicate that the DNA plays no major agentive role in accomplishing these functions and that information is not predetermined, but developmentally acquired. Not being a primary cause, DNA is, therefore, more likely to act as a bimodal archive of all explorative processes experienced by the organism in the course of development (Ochab-Marcinek and Tabaka 2010).

Based on these developmental and evolutionary considerations, the real question is how mechanisms and relationships causally relate to one another in living organisms. According to a mechanistic philosophy, many, if not all, phenomena of the living world ontologically correlate with mechanisms, rather than with the laws of nature (Bechtel and Abrahamsen 2013; Craver and Darden 2013). While laws are conceived as theoretical principles accounting for the general dependency of living processes on intrinsic causal interactions, mechanisms are taken to depict a more locally recurrent dependency under particular space and time conditions.[17]

It is difficult to challenge this approach and argue about the greater or lesser appropriateness of mechanisms with respect to laws to account for the life sciences. However, the point to be stressed is whether the mechanistic explanation of causality suffices to account for the origin of structures and their relationship with proper functions, and why some mechanisms, and not others, have come to define specific structure–function relationships over time. Mechanisms are interactions regulated exclusively by factors intrinsic to the system itself and causally determined by the interacting partners' properties. Should living processes be conceived as causally determined by mechanisms, their recurrence would always be interpreted as void of any alternative and entirely predictable in their developmental path (Kauffman 2016). On the contrary, if the specificity of the structure–function relationship could be approached in semiotic terms, their emergence could be initially attributed to recurrences of conventional type and interpreted in various ways depending on the experiential contexts (Cariani 1998; Rocha 2001).

[16]It may be of interest to note, at this juncture, that the vertebrate anterior limb develops a leg as long as functionally linked to locomotion, even though the limb bud itself has the potential to develop as either wing, fin or even hand in different vertebrates (Capdevila and Izpisua Belmonte 2001). Thus, depending on the developmental path undertaken in evolution, organisms do actually interact with the environment by running, flying, or even handling objects.

[17]Glennan (2017) sees the passage from laws to mechanisms as conceptually equivalent to the transition from theories to models.

6 Relationships versus Mechanisms

To favor the role of relationships with respect to that of mechanisms entails justifying evolutionary or developmental processes on the basis of the primacy of relationships. In the living world, it is the relationship itself that causes the interacting partners to persist. The same partners could not even exist, if not as parts of the relationship, and, in isolation, they would have no selective advantage as when included in the membership set (Nicholson and Dupre 2018). New organizational patterns can only emerge if the system's properties and processes work as an expression of the whole and not of any of its composing parts. These properties cannot be justified in terms of earlier interactions or attributed to preexisting functions but considered emergent as respect to any ancestral character (Queiroz and El-Hani 2006). This feature suggests, in turn, that the emergence of new organizational patterns does not depend on the implementation of new mechanisms but, on the contrary, on the co-option of earlier mechanisms selected from features adapted originally to different contexts. The idea that mechanisms may be co-opted rather than implemented for every new functional requirement stems directly from the modular character of all molecular and cellular interactions of the living hierarchy. In fact, modularity is a crucial biological concept that confers the possibility of using preexisting modules in novel ways to all developing and evolutionary systems. Using old resources for molecular and regulatory innovations, they avoid the necessity to embark on the costly elaboration of new elements.[18] From a biosemiotic standpoint, the emergence of new relationships can be taken to work as a sign of the system's capacity to self-organize and deal autonomously along with the necessity to maintain well balanced both redundancy and robustness (Atã and Queiroz 2019). On this ground, sign processes work by instantiating real relationships before any physical interaction could be construed and selectively stabilized by the whole system (Bains 2006).

The confrontation between mechanisms and relationships is not merely a matter of formal ontology, but a substantial distinction between what it is and what it could possibly be, between the *actual* and the *potential*. While mechanisms describe how the interacting partners actually produce effects, relationships are opportunities that the system explores to deal dynamically with newly emerging patterns. Kauffman (2000) refers to this expansion of the possibility space as the *adjacent possible* that, in his words, acts by increasing the diversity of what can happen next. Since what it can actually be done depends on the subject's capacity to choose amongst the available alternatives, the extent to which future possibilities are realized cannot be defined a priori. This makes it possible, in turn, for more complex semiotic capacities to emerge and expand exponentially in the *tendency to take habit* (Romanini and Fernandez 2014). In this sense, exploring the adjacent possible allows one to grasp

---

[18] Schlosser and Wagner (2004) have extensively discussed the role played by modularity in evolution. They take that this process primarily addresses increasing the living network's efficiency and stabilizing the effects of natural selection.

both the limits of change and the creative potential of innovation. Having said that, one can then understand how the semiotic exploration of the *adjacent possible* of interpreted relationships can ultimately result in the perception of what Favareau (2015) has defined the *relevant next,* i.e., the emergence of a meaningful and unpredictable world perceived as potentially evolvable toward indefinite goals.

Physical interaction is only explained by causal propositions and interpreted as pre-determined with no chance of being otherwise. Instead, in the world of living organisms, relationships may become meaningful whenever experienced as a selective choice between nonequivalent alternatives (Kull 2018). It follows that relationships are explained whenever tied to a single causal path, due to the absence of any alternative, while they are perceived as significant if considered as choices emerging from exploratory behavior.

7 Life as Emerging Semiosis

At the juncture between mechanisms and relationships, biosemiotics finds its own conceptual space as an interdisciplinary research program aimed at understanding how living systems communicate and interact significantly with the environment. In these terms, relationships can be interpreted as meaningful processes spanning from the genetic code to intercellular signaling, from animal behavior to such human artifacts as language and symbolic thought (Giorgi 2017). The key concepts of biosemiotics may therefore be epitomized as (1) the primacy of relationships as opposed to the properties of mechanistic interactions; (2) the selection of variants expressing more efficient relationships among the ones explored by any developing or evolving system; (3) the environment perceived as a meaningful Umwelt by living systems endowed with different sensory-motor systems; (4) and, finally, organisms understood as intrinsically teleological and sense-making subjects playing the role of agents (Kull et al. 2008, 2009; Favareau 2010).

This biosemiotic interpretation stems directly from the semiotic and pragmatic vision of Charles Sanders Peirce (1998), who considered relationships as conceptually equivalent to perceived signs. For Peirce, a sign should be understood as something that is determined by something else (an object), such as determining a corresponding effect in a perceiving subject. The effect produced in this way is referred to as the *interpretant,* which is ultimately determined by the sign in its mediated form (Atkin 2013). Thus, the sign is conceived as intrinsically triadic, that is to say, constitutively not reducible to the dyadic relations that comprised it (sign-object, sign-interpretant, object-interpretant) (Deely 2017). Removing the interpretant from this triad would reduce the resulting relationships again into a series of unrelated cause and effect dyads. Moreover, the lack of any interpretant would make the observer external to the observation and the relative effects interpreted as void of any alternative with a semiotically pending significance (Gudwin and Querioz 2007).

Unlike Saussure for whom the sign is only a cultural creation (Bergendorff 2009), Peirce conceives the sign as a pre-categorical element that belongs to both the natural and cultural worlds.[19] In this conception, signs stand objectively for something other than themselves and therefore obey the same logic in both intellectually and naturally founded worlds. In this sense, Peirce's semiotic proposal is much more revolutionary than Saussure's in assuming that semiotic mediation among living beings begun before any nature/culture divide appeared in the world's history. Since mental representations in humans are cognitively biased by experience and not merely biologically determined, sign interpretation cannot be naturally reduced to the perceived objects. It thus follows that the inner human world could not be interpreted as simply representing the world as it is, but as the world that could be used or could have been used (Deely 2001b).

Gregory Bateson's (1979) reflection is a wise example of how number zero may affect things or relationships differently. In the world of things, zero defines a simple absence of effects in an invariant context. Instead, number zero in the living world means a lack of relationship in a redundant and flexible context. This condition entails a deprivation and the establishment of an adaptive dysfunction that living organisms enact in the attempt to compensate for the missed function. It is clear that the meaning of any absence depends on how the context is perceived and, in turn, how this affects the recipient's response repertoire. In other words, context matters for meaning-making (Tønnessen et al. 2018).

Let us now return to the question of meaning in relation to the concept of information as initially elaborated by the theory of Shannon. It makes sense to talk about information transfer from a source to a receiver as long as communication is restricted to the molecular domain's complex interactions. As already mentioned, at this level, relationships are energetically balanced and, as such, necessarily determined by the underlying mechanisms. On the other hand, information is not simply transferred along a channel in the living world but interpreted by an agentive subject. Interpreting any incoming message as a sign makes it possible for the subject to restructuring its capacity to learn and act (Bateson 1972). In conclusion, while information conveyed through signals is received and eventually processed to trigger stereotyped responses, information perceived by an agentive subject is mediated by a sign and interpreted as a meaningful relationship.

Still, the question remains at which level of the living hierarchy does information processing give rise to meaningful information through interpretation. For instance, should we consider a single cell enough complex to be a semiotic system? Despite the relevance of the question, there is no unanimous consent yet as to where the semiotic threshold should be placed in the living hierarchy. Umberto Eco (2018) contends that biological processes are causally determined by lock-and-key

[19] As wisely expressed by Deely (2001a), the sign does not arise at the junction between a mind-dependent and a mind-independent reality, for no inner mental image can ever reflect the outer world. Interpreting the sign as the mental counterpart of a sensory experience would make it equivalent to an idea and, as such, it would exist as an object per se.

interactions and are therefore void of any interpretation. Thus, he excludes any possibility for biological processes to be semiotically grounded and sees social and cultural conventions as the only context in which signs and codes could be appropriately interpreted. Nöth (1994) postulates, instead, that the semiotic threshold should be placed on the transition from dyadic to triadic processes, and that, therefore, signs can only signify if properly mediated by a semiotic triad.[20]

Contrary to the above interpretations, I contend that, unlike the physical domain where molecular processes are quantitative changes abiding by a sole hierarchical scale, living systems do indeed experience alternative options in their developmental dynamics (Bruni and Giorgi 2015). The possibility for living systems to choose among various alternatives is made directly available by the cell capacity to perceive different instances of *cross-talk*, *redundancy*, and *categorical sensing* as meaningful signs of environmental signalling (Bruni 2007). By this intrinsic logic, cells may capture the very essence of ubiquitous signals, avoid undesired cross-talk, and integrate any system redundancy in their responding repertoire (Giorgi et al. 2013).

8 Mechanistic Signals or Semiotic Signs

The following lines provide some examples of how mechanistic processes may be perceived as meaningful events when interpreted within a semiotically oriented framework or, as otherwise said by Pattee, when a molecule becomes a message (Pattee 1969). G protein-coupled (GPC) receptors form the largest class of plasma membrane-linked receptors in eukaryotic cells (Aparicio and Powell 2004). By acting as signal transducers across the plasma membrane, they play critical roles in controlling cell communication with the extracellular environment. In recent years, it has become increasingly clear that cell signaling uses GPC receptors in the form of oligomeric complexes (Gurevich and Gurevich 2008). Since cell signaling is not exclusively determined by ligand availability or receptor assembly, it is only in their emerging relationship on the cell membrane that resides the selective barrier for all incoming signals. This relationship makes it possible for the receiving cell to interpret the establishment of an oligomeric receptor phase as a meaningful sign, rather than as a predetermined signal (Giorgi et al. 2010). In case monomeric receptors prove incapable of self-assembling in the oligomeric phase, they are gradually removed from the plasma membrane and, in the long run, eliminated from the cell responding repertoire.

Another example of how mechanistic relationships can be interpreted as meaningful signs has come from the analysis of olfactory receptors. That olfactory receptors realize an olfactive function may seem rather obvious. However, the

[20]At the other end of the spectrum, Barbieri (2007) maintains that there is no interpretation in biology. He argues that molecular and cell interactions are defined solely by the genetic code's rules that have remained invariant throughout evolution for all organisms.

question is logically legitimated by the observation that they are also ectopically expressed in other tissues such as the testis and the distal renal nephron (Feldmesser et al. 2006). The hypothesis to justify their functional diversity has been that homologous olfactory receptors may share the same fundamental role of conferring a functional cell identity under different environmental conditions. By adopting this point of view, the functional attribution as olfactory or chemotactic sensors to these receptors should not be seen as simply resulting from a genetically predetermined program. Instead, it should be regarded as a direct consequence of the environmental conditions to which receptor genes have been exposed during evolution. Exposure to odorant or chemotactic patterns under specific contextual situations may, in fact, confer them the capacity to perceive their relationships with the environment as an emergent meaningful sign (Giorgi et al. 2011).

A final example of mechanistic relationships can be found in the world of cell communication. As well known, cells communicate by receptor signaling, junctional coupling, or microvesicle release. The interactions between membrane receptors and ligands are regulated by the ligand's conformational flexibility in the extracellular milieu and the receptor capacity to self-assemble on the cell surface (Chen et al. 2017). As such, it is an exclusive molecular process that acts on target cells via the integration of highly complex transduction mechanisms in their responding repertoire (Slavoff and Saghatelian 2012). Cells joined by gap junctions communicate directly with each other by exchanging signals through specific coupling channels. In so doing, all connected cells come to share the same information content and behave as highly coordinated clusters (Goldberg et al. 2004). Vesicular traffic instantiates a substantially different type of communication. Interacting cells are not required to adjust their structural coupling or develop high-affinity binding sites but to deliver their vesicles regardless of their target specificity. The information transferred by extracellular vesicles occurs independently of the cell capacity to match the delivery mechanism with their signaling contents. This independence requires the cells to simply monitor the efficiency of their delivery mechanism regardless of the repertoire expressed by the target cells, along with the possibility of capturing any newly emergent relationship with the environment as a meaningful sign (Giorgi and Auletta 2016).

9 Conclusions

Living beings comprise hierarchical structures at different levels of complexity. The extent levels interact with one another within this hierarchy entails a principle of reciprocal adaptation led by some selective choices of either cooperative or competitive nature. These alternatives are always semiotically mediated by the subject's capacity to interpret the context adequacy of the perceived message and, as such, causally linked to his/her ability to render the choice experientially meaningful (Favareau 2015). The subject has then to adapt to any newly emerged complexity such that, during transition along the hierarchy, each level is selected as a part of an interacting network of relationships satisfying both heredity and fitness conditions. New individualities can thus emerge as a whole if, and only if, they present greater selective advantages concerning the levels of the singularities that have preceded them in the same hierarchy (Bruni and Giorgi 2016). The persistence of these intricate relationships or, at least, their long-term stability, presupposes the existence of several turnover processes within and among different complexity levels. At every level of this hierarchy, each fundamental component can be replaced, and their function restored autopoietically through the insertion of newly emergent elements.

Life has solved the problem of entropic uncertainty by rendering energetically simpler and more efficient all those living architectures that evolution has rendered hierarchically too complex. At any developmental stage, the living system faces the alternative of either elemental substitution or elimination, and this is obviously a semiotic interpretation that calls into question the energy requirement and the functional efficiency of every choice (Berthoz 2012). According to a strictly deterministic logic, living organisms should attain this goal through dyadic relationships proper to any stimulus–response interaction. On the contrary, relationships become significantly advantageous for the system's survival if interpreted biosemiotically as signs of triadic relationships. Thus, signs can be recognized as successful choices of alternative antecedents and suitable reminders of their historically constrained adaptability.

References

Alcocer-Cuarón C, Rivera AL, Castaño VM (2014) Hierarchical structure of biological systems: a bioengineering approach. Bioengineered 5(2):73–79

Aparicio SA, Powell J (2004) Genetic approaches to unravelling G protein-coupled receptor biology. Curr Opin Drug Discov Dev 7:658–664

Atã P, Queiroz J (2019) Emergent sign-action classical ballet as a self-organized and temporally distributed semiotic process. Eur J Pragmatism Am Philos 11(2):1–19

Atkin A (2013) Peirce's theory of signs. In: Zalta EN (ed) The Stanford encyclopedia of philosophy. University Stanford Press, Stanford

Atmanspacher H, Dalenoort GJ (1994) Inside versus outside. Springer Verlag, Berlin Heildeberg

Bains P (2006) The primacy of semiosis: an ontology of relations. University of Toronto Press, Toronto

Balzani V, Credi A, Venturi M (2008) Molecular devices and machines: concepts and perspectives for the nanoworld. Wiley-VCH Verlag GmbH and Co.

Banavar JR, Cooke TJ, Rinaldo A, Maritan A (2014) Form, function, and evolution of living organisms. PNAS 111(9):3332–3337

Barbieri M (2007) Is the cell a semiotic system? In: Barbieri M (ed) Introduction to biosemiotics. The new biological synthesis. Dordrecht, Springer, pp 179–207

Barthes R (1986) Elements of semiology. Hills & Wang, New York

Bateson G (1972) Steps to an ecology of mind. University of Chicago Press, Chicago, Illinois

Bateson G (1979) Mind and nature. A necessary unity. Bantam Books, New York

Beaulieu A (2012) Alfred North Whitehead précurseur des théories de l'auto-création. Revued'histoire des sciences 65:81–101

Bechtel W, Abrahamsen AA (2013) Thinking dynamically about biological mechanisms: networks of coupled oscillators. Found Sci 18:707–723

Benner SA (2010) Defining life. Astrobiology 10(10):1021–1030
Bergendorff S (2009) Simple lives, cultural complexity. Lexington Books, UK
Berthoz A (2012) Simplexity: simplifying principles for a complex world. Yale University Press, New Haven
Brading K, Castellani E (2003) Symmetries in physics: philosophical reflections. Cambridge University Press, Cambridge
Breysse O, De Glas M (2007) A new approach to the concepts of boundary and contact: toward an alternative to mereotopology. Fundamenta Informaticae 78:217–238
Bruni LE (2007) Cellular semiotics and signal transduction. In: Barbieri M (ed) Introduction to biosemiotics. Springer, Dordrecht, pp 365–408
Bruni LE, Giorgi F (2015) Towards a heterarchical approach to biology and cognition. Prog Biophys Mol Biol 119(3):481–492
Bruni LE, Giorgi F (2016) Multilevel semiosis. A paradigm of emergent innovation. Biosemiotics 9:1–12
Capdevila J, Izpisua Belmonte JC (2001) Patterning mechanisms controlling vertebrate limb development. Annu Rev Cell Dev Biol 17:87–132
Capra F, Luisi PL (2014) The systems view of life. Cambridge University Press, Cambridge
Cárdenas-García JF, Ireland T (2019) The fundamental problem of the science of information. Biosemiotics 12:213–244
Cariani P (1998) Towards an evolutionary semiotics: the emergence of new sign-functions in organisms and devices. In: Van de Vijver G, Salthe S, Delpos M (eds) Evolutionary systems. Kluwer, Dordrecht, pp 359–377
Cariani P (2011) The Semiotics of cybernetic percept-action systems. Int J Signs Semiot Syst 1 (1):1–17
Carroll L (1871) Through the looking-glass. Macmillan and Co., London
Carson DH (1961) Letter constraints within words in printed English. Kybernetik 1:46–54
Cartwright JHE, Giannerini S, González DL (2016) DNA as information: at the crossroads between biology, mathematics, physics and chemistry. Philos Trans R Soc A374:20150071
Chen J, Almo SC, Wu Y (2017) General principles of binding between cell surface receptors and multi-specific ligands: a computational study. PLoS Comput Biol 13(10):e1005805
Collier J (1996) Information originates in symmetry breaking. Symmetry: Cult Soc 7:247–256
Craver CF, Darden L (2013) In search of mechanisms: discoveries across the life sciences. University of Chicago Press, Chicago
Crnkovic GD (2012) Information and energy/matter. Information 3:751–755
Damiano L, Luisi PL (2010) Towards an autopoietic redefinition of life. Orig Life Evol Biosph 40:145–149
Deacon TW (2011) Incomplete nature: how mind emerged from matter. W.W. Norton, New York
Deely J (2001a) Four ages of understanding. University of Toronto Press, Toronto
Deely J (2001b) A sign is what? A dialogue between a semiotist and a would-be realist. Sign Syst Stud 29(2):705–743
Deely J (2017) Academic culture and the science of signs. In: Semetsky I (ed) Edusemiotics—A handbook. Springer Science+Business Media Singapore
Dupré J, Nicholson DJ (2018) A manifesto for a processual philosophy of biology. In: Dupré J, Nicholson DJ (eds) Everything flows. towards a processual philosophy of biology. Oxford University Press, Oxford
Easthope H (2009) Fixed identities in a mobile world? The relationship between mobility, place, and identity. Identities 16(1):61–82
Eco U (2018) Giorgio Prodi and the lower threshold of semiotics. Sign Syst Stud 46(2/3):343–351
Emmeche C (1994) The computational notion of life. Theoria—Segunda Epoca 9(21):1–30
Emmeche C (1998) Defining life as a semiotic phenomenon. Cybernetics Human Knowing 5:3–17
Favareau D (2010) Essential readings in biosemiotics. Springer, Dordrecht
Favareau DF (2015) Creation of the relevant next: how living systems capture the power of the adjacent possible through sign use. Prog Biophys Mol Biol 119(3):588–660

Feldmesser E, Olender T, Khen M, Yanai I, Ophir R, Lancet D (2006) Widespread ectopic expression of olfactory receptor genes. BMC Genomics 7:121
Ferreira Ruiz MJ, Umerez J (2018) Dealing with the changeable and blurry edges of living things: a modified version of property-cluster kinds. Eur J Philos Sci 8:493–518
Gayon J (2010) Defining life: synthesis and conclusions. Orig Life Evol Biosph 40:231–244
Giorgi F (2017) Un mondo di relazioni. Tangram Edizioni Scientifiche, Trento
Giorgi F, Auletta G (2016) Semiotic tools for multilevel cell communication. Biosemiotics 9:365–382
Giorgi F, Bruni LE, Maggio R (2010) Receptor oligomerization as a process modulating cellular semiotics. Biosemiotics 3:157–176
Giorgi F, Maggio R, Bruni LE (2011) Are olfactory receptors really olfactive? Biosemiotics 4:331–347
Giorgi F, Bruni LE, Maggio R (2013) Semiotic selection of mutated or misfolded receptor proteins. Biosemiotics 6:177–190
Glennan S (2017) The new mechanical philosophy. Oxford University Press, Oxford
Godfrey-Smith P (2007) Information in biology. In: Hull D, Ruse M (eds) The Cambridge companion to the philosophy of biology. Cambridge University Press, Cambridge, pp 103–119
Goldberg GS, Valiunas V, Brink PR (2004) Selective permeability of gap junction channels. Biochim Biophys Acta 1662:96–101
Gudwin R, Querioz J (2007) Semiotics and intelligent systems development. Idea Group Publishing, London
Gurevich VV, Gurevich EV (2008) GPCR monomers and oligomers: it takes all kinds. Trends Neurosci 31:74–81
Hempel S, Pineda R, Smith E (2011) Self-reference as a principal indicator of complexity. Procedia Comput Sci 6:22–27
Herron MD, Zamani-Dahaj SA, Ratcliff WC (2018) Trait heritability in major transitions. BMC Biol 16(1):145
Hoffmeyer J (2001) Life and reference. Biosystems 60:123–130
Hoffmeyer J, Emmeche C (1991) Code-duality and the semiotics of nature. In: Anderson M, Merrell F (eds) On semiotic modeling. Mouton de Gruyter, Berlin, pp 117–166
Indiveri G, Horiuchi TK (2011) Frontiers in neuromorphic engineering. Front Neurosci 5:118
Jonas H (1966) The phenomenon of life. Harper and Row, New York
Karmon A, Pilpel Y (2016) Biological causal links on physiological and evolutionary time scales. elife 5:e14424
Kauffman SA (1995) At home in the universe. The search for laws of self-organization and complexity. Oxford University Press, Oxford
Kauffman SA (2000) Investigations. Oxford University Press, Oxford
Kauffman SA (2016) Humanity in a creative universe. Oxford University Press, Oxford
Kull K (2018) Choosing and learning: semiosis means choice. Sign Syst Stud 46(4):452–466
Kull K, Emmeche C, Favareau D (2008) Biosemiotic questions. Biosemiotics 1:41–55
Kull K, Deacon T, Emmeche C, Hoffmeyer J, Stjernfelt F (2009) Theses on biosemiotics: prolegomena to a theoretical biology. Biol Theory 4(2):167–173
Lemke JL (2000) Opening up closure: semiotics across scales. Ann N Y Acad Sci 901:100–111
Lodish H, Berk A, Zipursky SL, Matsudaira P, Baltimore D, Darnell J (2000) Molecular cell biology, 4th edn. W. H. Freeman, New York
Longo G (2008) Symmetries and Symmetry-breakings: the fabric of physical interactions and the flow of time. In: Vidal C (ed) The evolution and development of the universe. Ecole Normale Supérieure, Paris
Longo G, Montévil M, Kauffman S (2012) No entailing laws, but enablement in the evolution of the biosphere. arXiv:1201.2069
Machamer P, Darden L, Craver CF (2000) Thinking about mechanisms. Philos Sci 67:1–25
Marmodoro A, Yates D (2016) The metaphysics of relations. Oxford University Press, Oxford. UK

Maturana HR, Varela FJ (1980) Autopoiesis and cognition: the realization of the living. Cohen R, Wartofsky M (ed). Reidel Publishing, Boston
Mayr E (1974) Teleological and teleonomic: a new analysis. Boston Stud Philos Sci 14:91–117
Mayr E (1982) The growth of biological thought. Harvard University Press, Cambridge, MA
Mayr E (1997) This is Biology: the science of the living world. Harvard University Press, Cambridge, MA
Morange M (2008) Life explained. Yale University Press, New Haven
Morris C (1971) Writings on the general theory of signs. Mouton, Paris; The Hauge
Nicholson HA (1870) A manual of zoology. Blackwood & Sons, Edinburgh
Nicholson DJ, Dupre J (2018) Everything flows: towards a processual philosophy of biology. Oxford University Press, Oxford
Nöth W (1994) Oppositions at the root of semiosis. In: Nöth W (ed) Origin of semiosis: signs evolution in nature and culture. Mouton de Gruyter, Berlin, pp 37–60
Ochab-Marcinek A, Tabaka M (2010) Bimodal gene expression in noncooperative regulatory systems. PNAS 107:22096–22101
Pattee HH (1969) How does a molecule become a message? Dev Biol Suppl 3:1–16
Pattee HH (1995) Evolving self-reference: matter, symbols, and semantic closure. Commun Cognit Artif Intell 12:9–28
Pattee HH (2006) The physics of autonomous biological information. Biol Theory 1(3):224–226
Pattee HH (2013) Epistemic, Evolutionary, and physical conditions for biological information. Biosemiotics 6:9–31
Pattee HH, Rączaszek-Leonardi J (2012) Laws, language and life. Springer, Dordrecht
Peirce CS (1998) The essential Peirce, selected philosophical writings, Volume 2 (1893–1913): the Peirce edition project. Indiana University Press, Bloomington, IN
Penfield T, Baker MJ, Scoble R, Wykes MC (2014) Assessment, evaluations, and definitions of research impact: a review. Res Eval 23:21–32
Pettit P (1996) Functional explanation and virtual selection. Br J Philos Sci 47:291–302
Pickering J (2007) Affordances are signs. tripleC 5(2):64–74
Queiroz J, El-Hani CN (2006) Semiosis as an emergent process. Trans Charles S Peirce Soc 42 (1):78–116
Rocha LM (2001) Evolution with the material symbol systems. Biosystems 60:95–121
Romanini V, Fernandez E (2014) Peirce and biosemiotics: a guess at the riddle of life. Springer, Dordrecht
Schlosser G, Wagner GP (2004) Modularity in development and evolution. University of Chicago Press, Chicago
Schreiber A, Gimbel S (2010) Evolution and the second law of thermodynamics: effectively communicating to non-technicians. Evol Edu Outreach 3:99–106
Shannon CE, Weaver W (1949) The mathematical theory of communication. The University of Illinois Press, Urbana, IL
Slavoff SA, Saghatelian A (2012) Discovering ligand-receptor interactions. Nat Biotechnol 30:959–961
Smith B (1997) Boundaries: an essay in mereotopology. In: Hahn LH (ed) The philosophy of Roderick Chisholm. Open Court, pp 534–561
Stegmann U (2012) Causal control and genetic causation. Noûs 48(3):450–465
Sterelny K, Joyce R, Calcott B, Fraser B (2013) Cooperation and its evolution. The MIT Press, Cambridge, MA
Thomas JL (2019) Meaningfulness as sensefulness. Philosophia. https://doi.org/10.1007/s11406-019-00063-x
Tønnessen M, Maran T, Sharov A (2018) Phenomenology and biosemiotics. Biosemiotics 11:323–330
Travers AA, Muskhelishvili G, Thompson JM (2012) DNA information: from digital code to analogue structure. Philos Trans Math Phys Eng Sci 370:2960–2986

Trnka J (2015) Hans Jonas on teleology in biology. The relevance of existential philosophy for biological thinking. Lìmite Revista Interdisciplinaria de Filosofía y Psicología 10(34):41–49
Villani G (2017) Chemistry: a systemic complexity science. Pisa University Press, Pisa
von Bertalanffy L (1950) The theory of open systems in physics and biology. Science 111:23–29
Wainwright P (2009) Innovation and diversity in functional morphology. In: Laubichler M, Maienschein J (eds) Form and function in developmental evolution. Cambridge University Press, Cambridge, pp 132–152
Weber B (2018) Life. In: Zalta EN (ed) The Stanford encyclopedia of philosophy. Stanford University Press, Stanford
Weisblat DA (1998) Embryonic development as a quasi-historical process. Int J Dev Biol 42:475–478

A Proposal for a Biosemiotic Approach to Digitalization: Literacy as Modeling Competence

Alin Olteanu

Abstract The advantages of a biosemiotic approach to the social effects of digitalization are explicated. Biosemiotics is a semiotic modeling theory that is inherent to a phenomenology of the body. Even though it offers an encompassing and comprehensive account of meaning as modeling, it has only recently been employed in analyzing cultural and social matters, which are the traditional foci of semiotic theories that tended to overlook the role of the body and its relation to the environment in meaning-making. Here, it is argued that biosemiotics offers a theoretical framework that captures the multimodal and fast communication dynamics of digital societies as affordances of the human body. As such, the chapter explores a biosemiotic perspective on sustainable development as fostered by digitalization.

Keywords Digitalization · Literacy · Embodiment · Text · Mediality

1 Introduction: Semiotics, Embodiment, and Digitalization

This chapter explores new pathways that semiotic theories can take in approaching the specific representational possibilities and constraints that human societies beget by expanding onto the digital (digitalization). The insights that a semiotic approach offers are here revealed in light of Jeremy Rifkin's (2011) proposal that merging digital communication networks and the renewable energy grid will produce an industrial revolution that can foster sustainable development. The implications for the literacy of digitally modeled societies and environments are explored as a critical development for ushering this third industrial revolution. I argue that, as noticeable in several areas of the humanities and social sciences, semiotic theories that assume a phenomenological account of modeling present certain advantages in approaching

A. Olteanu (✉)
University of Tartu, Tartu, Estonia

Kaunas University of Technology, Kaunas, Lithuania
e-mail: Alin.Olteanu@rwth-aachen.de

© Springer Nature Switzerland AG 2021

E. Pagni, R. Theisen Simanke (eds.), *Biosemiotics and Evolution*, Interdisciplinary Evolution Research 6, https://doi.org/10.1007/978-3-030-85265-8_4

digitalization in comparison to discursive theories. The latter tend to draw on a glottocentric (read *language-centered*) notion of meaning as *text*, inherited from literary theory. Construals of representation and text have profound consequences for understanding contemporary societies, which are inherent in human–machine interactions. As Ihde remarks, "our existence is *technologically textured*" (1990: 1). Among semiotic modeling theories, biosemiotics can particularly serve as a framework for understanding the social and cultural implications of digitalization because of a coincidence between its theoretical focus and specific characteristics of representations in digital media. Notably, biosemiotics offers comprehensive theoretical tools to grasp (1) multimodality, as contrasted to glottocentrism, and (2) simultaneous (public) participativity, as contrasted to *feedback* as resulting in situations where interlocutors are distinguishable and take turns in their dialogue. These two specific aspects of digital media are discussed in detail in Sect. 4.

Biosemiotics has not been employed yet in the study of mediality. The prospect for a biosemiotic understanding of mediality, however, was clearly stated by Hoffmeyer, claiming that "[k]nowledge is a bodily phenomenon" (2018: 5), that it is always contextual and that it is a matter of *becoming*, "not at all something I already *have* but something *created* in this very moment" (p. 4). From the biosemiotic perspective, modeling is embodied, wherein the affinity of this theory with cognitive linguistics (e.g., Lakoff and Johnson 1999) and, of course, cognitive semiotics (Brandt 2011) occurs. Unsurprisingly, this is aligned with the conclusions that semiotic theory observes for education, where educational processes are understood as pertaining to the self's continuous becoming (Stables 2012). Taking this as a starting point, an approach to mediality grounds competences for engaging with texts (literacy) in the competences for meaning-making of the human body (on embodied semiotic competences see Stjernfelt 2006; Hoffmeyer 2008; Hoffmeyer and Stjernfelt 2016). It can underpin the view in media semiotics that mediality is primarily evoked by corporeality (Elleström 2018, 2019) and that technological media are evoked and developed in light of the basic human competences for meaning-making. The phenomenological account of meaning in biosemiotics endorses a view of the environment (*Umwelt*) as an "open field of actions, where organisms exchange meaningful and interpretative communications" (Pagni 2016: 60), revealing a mutuality between environment and mediality.

Further, biosemiotic theory implies conceptualizing technological media development as a process of scaffolding (Hoffmeyer 2007) ever more comprehensive representations, attuned to circumstantial requirements. This perspective can be developed to complete the argument, as suggested by McLuhan (1964) and further explored by Danesi (2002), that societies shaped by electronic media do not accommodate modern ontological dualism, bringing about the reintegration of body and mind, instead. Representations specific to one type of technological media and unavailable otherwise must still share some media characteristics with representations as afforded by human corporeality. Digital representations bring to the fore aspects of human modeling (multimodality, collective participation) that previous technological media left aside, given the technological possibilities of the time, in favor of other aspects (e.g., recording, in the case of print, narrativity, in the case of cinema) that served as scaffoldings, circumstantially and pragmatically. In light of the affordances that digital media share with those that natural landscapes present to the human embodiment, Hopkins (2020: 49) semiotic argues that "digital media is a uniquely human technology—perhaps the most human of all technologies—that can be considered alongside other abstract human creations such as language or art."

2 The Disembodiment of *Text*

Mediality cannot be separated from meaning: media has meaning effects and, thus, interpretation takes different forms in different medialities.[1] As such, mediation of meaning (communication) is not identical in, for instance, face-to-face communication, television, and Facebook. This difference implies that the emergence of new media results in the emergence of new types of specific (digital) textualities unavailable in societies shaped by previous technological media. For the same reason, approaching mediality from a biosemiotic perspective also contributes to biosemiotic theory, not only to conceptualizing mediality more comprehensively.

The concept of *text* mainly refers to artifacts that humans engage with by an act of interpretation, as an instantiation of language (Halliday and Webster 2009; Lotman 1990). This concept came from literary studies' tradition and became a cornerstone tool in cultural studies, designating anything that can be interpreted (e.g., Eco 1979). It is the instrumental concept that generated a semiotic criticism of processual models of communication (Fiske 1990; Eco 1976, 1979) by arguing that since acts of communication always involve an operation of interpretation, they cannot be fully grasped in terms of inputs and outputs.

Etymologically originating in the Latin word for *cloth* (*textus*), the term *text* implies the act of reading or, so to say, engaging with a code. The metaphor that facilitated this semantic development supposes that a writer masters craftsmanship: codifying a message in writing is like weaving a cloth (Bringhurst, 2004: 25). Arguably, this metaphor, emerging in the early pre-print stages of the spreading of reading and writing competences, involved an embodied aspect: the act of writing involves visual and tactile interaction, like with textiles. Bringhurst (2004) explains that, in its early conception, *the text* was understood as inherent in its material incarnation, namely in typographic style.

The modern philosophical and educational frame of mind produced a split between *the text* as content and its materiality. As particularly championed in Ferdinand de Saussure's (1959 [1916]) semiology, the dichotomization of meaning into form and content inspired the linguistic turn and its corresponding glottocentric epistemology (e.g., Barthes 1991 [1957]; Rorty 1967). This split blurs the awareness

[1] As consensually accepted in semiotics, I take *the meaning* to be the product of *interpretation*. As *interpretation* continuously unfolds throughout an organism's life, *the meaning* is never a complete and finalized product.

of the meaning effects of mediality, obscuring what McLuhan (1997: 148) expressed, around the same time, by stating that "the medium is the message." Also, I argue that the form/content dichotomy, termed double articulation in linguistics (Martinet 1962), disregards technology by presenting it as the production of forms either devoid of content or imposing fixed, rigid content that does not allow for the interpretation. In either version, technology is deemed to bring hermeneutical *laziness*. Following de Saussure, if sign systems' operationality resides in the arbitrariness of the articulation of form and content into meaning, then mediality cannot be deemed to have meaning effects. From this perspective, proper *semiotic work* consists in establishing and using relations of abstract conventionality. By ignoring how a (material) form may accommodate some meanings better than others, media changes produced by technological progress are not deemed to offer possibilities for creativity. More fundamental than disregarding media affordances for interpretation, and the production of meaning as design, the problem with this view is that it misses that meaning is embodied. It illustrates the tenets of what Lakoff and Johnson (1980) criticized as the fallacious myth of subjectivism, in contrast to the fallacious myth of objectivism: in brief, that mind-independent objects do not impose any constraints on how we might construct reality. The pejorative perspective on the technology of glottocentrism is also displayed in classical phenomenology, such as coming from Heidegger and Jaspers, as Verbeek (2005 [1992]: 100) explains:

> Heidegger's hermeneutical approach attempted to understand technology as an alienating way of disclosing reality, reducing concrete technological artifacts to the fruits of such disclosing. Jaspers's existential philosophy of technology attempted to understand technology in terms of bureaucracy, mass production, and the "limits of technology," and then likewise reduced it to what this made possible or to what imposed limits on it. Both philosophies appear to be governed by what one might call a "transcendental fix."

According to Verbeek, classical phenomenology also failed to acknowledge the hermeneutical possibilities of technology because of a dualist assumption, namely the classical philosophical dichotomy of idealism and realism, reflecting the opposition of subjectivism to objectivism, respectively, regarding meaning-making. In light of this, Verbeek proposes a postphenomenological approach that collapses the subject/object dichotomy in favor of an environmental and relational perspective on (human) consciousness. By acknowledging that the environment with which human beings are involved "always codetermines in which ways they can be present to the world and each other" (Verbeek 2005: 112), postphenomenology offers a philosophy of technological artifacts as *things* that neither determine nor bypass the human–world relation, but participate to its environmental shaping. In brief, technological artifacts are active mediators in this relation. This postphenomenological perspective is compatible with the semiotic-inclined account that I develop here, particularly in light of the notion of *literacy*. This position is also supported by social semiotics recently, where the construal of meaning articulation as multimodal design reveals a view of technology "both as the product of social work and [...] a means of furthering social arrangements" (Kress and Selander 2012: 266).

As the set of skills required for surviving and thriving in society, the concept of *literacy* has been understood in terms of the competences for coding and decoding texts. Literacy is a concept specific for modernity, emerging, as a socially observable phenomenon, with the spread of print technology (e.g., Anderson 2006: 37) and, consequently, of public schooling (Rifkin 2011: 35). As such, the crux of the revolution of modernity consists in identifying skills for surviving and thriving in society, on the one hand, and for reading and writing, on the other. The understanding of literacy expanded in the humanities tradition, from where the scholarly concept originates, to possibilities of operating with symbolic systems more broadly, including not only the alphabet (letters) but also numbers. The evolution of the *literacy* concept is tied to the evolution of the notion of *text*, particularly in semiotic scholarship (see Stables and Bishop 2001: 90–92). As modern science, philosophy and education, most importantly, have been carried out in the medium of print, *the text* became associated with the linear modality of representation best afforded by this medium (see Kress and van Leeuwen 2001: 1), exclusively drawing on one code (system of signs). It prototypically consists of combinations of (ideally black) letters, belonging to one finite, clearly defined set of symbols (conventionally chosen, socially approved signs), against an (ideally white) background (as exhibited by the present paper). The preference for what in contemporary terms is called monomodality, which print modernity cultivated, underpinned a philosophy of disembodiment, as monomodal representations correspond to the thesis shared by rationalism (Descartes 2008 [1643]) and empiricism (Locke 1836 [1690]) that knowledge consists in ideas. Ideas, in the modern view, are purely mental, understood as in opposition to corporeal. Modern philosophy dissociates between the mind, which is responsible for knowledge and operates with ideas, and the body. In this view, ideas are not embodied. Because of its linearity and monomodality, print is supposed an excellent medium for communicating knowledge, then, as it can create materialized vehicles for ideas (printed books) that do not distract the mind by any innovation of design.

Any material embellishment would be, from this perspective, a distraction from the mind's work of making abstractions and comprehending ideas. It would engage perception senses, which are bodily and, while connected to, distinct from the mind. Literacy founded on the notion of text as monomodal implies a theory of knowledge and, consequently, an educational philosophy that exhibits transcendentalism, either as objectivism or subjectivism. In laying out the foundation of a cognitive approach to meaning, George Lakoff described this traditional and disembodied view of meaning and reason as "the mechanical manipulation of abstract symbols which are meaningless in themselves, but can be given meaning by virtue of their capacity to refer to things either in the actual world or in possible states of the world" (Lakoff 1987: 7). By drawing attention to the multimodal constitution of meaning, social semiotics implicitly brings to the fore the role of the body's phenomenology in knowledge acquisition. Herein, cognitive- and social-oriented approaches to meaning find a bridge.

Texts of the digital age have a very different appearance than those of the Gutenberg Galaxy, to use Marshall McLuhan's (1962) celebrated idiom for the age of print. Arguably, the representation modalities employed in digital texts are more similar to those employed in preliterate human societies than to those of previous technological media. This has profound environmental consequences. If this is indeed the case, then representation theories that consider the body's role in meaning-making are better suited for understanding digital textuality than glottocentric ones. This similarity between digital and directly corporeal medialities is most clearly revealed by an affordance prism (Gibson 1979; Majchrzak et al. 2013; Hopkins 2020) on media, which suggests that there must be some transferable skills between digital (Mills 2016; Scolari et al. 2018) and environmental literacies (Roth 1992; Stables and Bishop 2001). Digital literacies refer to skills of using digital technology by which individuals thrive in contemporary societies. The notion of environmental literacy, while extensively debated, remains vague in its scholarly uses, but it mostly refers to the competences of engaging with texts that allow an environmentally sustainable lifestyle (Stables 2001). The ambiguity comes from the various ways of understanding the relationship between text and environment: to what extent is the environment interpretable as *text*?

Indeed, something that cannot be interpreted cannot be comprehended at all. The difficulty comes from the modern understanding of *the text* as conventional, as a code that consists of symbols and socio-cultural rules. The environment, on the other hand, is not entirely conventional. There are cues for interpreting a river, a mountain, or a tree that have nothing to do with social and cultural conventions. An organism has a reliable understanding, independently of cultural considerations, of what might happen to it if, for instance, it dives into a river. The understanding is based on the organism's cognitive capacity of mentally simulating possibilities of events, and it is critical for its survival. It uses basic semiotic competences that belong to the organism's primary modeling system (Sebeok 2001: 148–149) and can be employed regardless of linguistic or other more sophisticated modeling capacities. For example, Campbell (2017) points out that the gradual sophistication of Eco's (1979, 1999) notion of text, progressing along with Eco's partial but growing accommodation of Peirce's semiotics with his own thought, where the former's "cognitive turn" was critical, and consisted in the process of "dismantling of the wall between nature and culture" (Campbell 2017: 136–137). This process led, in the case of Eco, from a strict anthropocentric and conventional notion of the text to a more broadly encompassing exploration of the claim that a sign is "an inchoative text" (Eco 1979: 184; Campbell 2017: 137).

In humans and other animals' everyday experience, neither are social texts entirely conventional nor is the environment understood in an entirely non-conventional way. In the biosemiotic approach to ecology, which was termed ecosemiotics (Nöth 2001), the impossibility of purely cultural or purely non-cultural objects in the human world has been taken as a premise (see Maran and Kull 2014: 42–43; Maran 2014: 298). However, this does not cancel the possibility of pragmatically bypassing linguistic and cultural elements in modeling. If a human must swim across a river to survive, she can abstract her cultural representations of rivers and swimming. On the other hand, bypassing primary modeling operations is not possible. Cultural artifacts that are connotations of swimming, such as swimming gear, the architecture of a swimming pool, it being an Olympic sport make no sense in the abstraction of the human competence to swim, given the morphology and physiology of the human body. Embodied competences, such as that of swimming, before any culturalization, are semiotic competences. For swimming to be performed, the (human) performer must develop a model, having the form of an internalized representation of it. An even more straightforward example is eating. Arguably, a fork is only a fork culturally, just like chopsticks are chopsticks culturally. Outside their cultural context, they are both nothing more than *things*, as material objects. The postphenomenological notion of *things* (Verbeek 2005) accounts for that a fork can be used as a comb and chopsticks can be used as drumsticks, or they might both be abstract art. In the abstraction of how humans consume organic matter to survive by eating through a buccal cavity, which can chew and swallow, and aided by hands and the other limbs and the entirety of the human body, forks and chopsticks are not forks and chopsticks and neither can they be pragmatically reinterpreted as, for instance, combs or drumsticks. *Things* make sense as *things* because of the semiotic competences of the human body. Semiotic competences and modeling, in general, must be understood in view of the possibilities and constraints that the organism-environment relation evokes, which is to say, in light of the affordance perspective (see also Olteanu and Stables 2018; Campbell et al. 2019). This view opens an understanding of technology that does not disconnect it from the body or previous cultural modeling. Particularly remarkable, the affordance approach, in a semiotic key, has been adopted in artificial intelligence. Roli and Kauffman (2020) consider that the emergence of artificial organisms depends on the possibility of endowing artificial systems with evolutionary possibilities by creating "mutually consistent affordances" (p. 1170).

As suggested in the ecosemiotic vein, a key to tackling questions regarding interpreting the environment resides in acknowledging that while *landscape* constitutes the origin of possibilities and constraints for meaning-making, it is, in turn, modeled by media representations. This view calls for an investigation of the potential implications that digitalization has for the natural environment. Understanding the mediality that environments provide to organisms complements the mirrored approach to media as environments in media ecology (Strate 2008; Scolari Carlos 2012).

3 Mediality from an Affordance Perspective

Affordance is a concept in ecology, referring to the possibilities and constraints for an organism's action, stemming from the morphological body–environment relation. The affordances of the environment, as coined by James Gibson, consist in "what [the environment] offers the animal, what it provides or furnishes, either for good or ill. [...] It implies the complementarity of the animal and the environment" (Gibson 1979: 127).

Human beings' environmental affordances change with the development of technologies, which, by evoking new medialities, allow for new representations of society and the environment. It is the relationship between organisms and the environment, thus, which evokes mediality. This relationship is implied by Sebeok's biosemiotic concept of model as interchangeable with the environment because:

> All organisms communicate by use of models (*umwelts*, or self-worlds, each according to its species-specific sense organs), from the simplest representations of manoeuvres of approach and withdrawal to the most sophisticated cosmic theories of Newton and Einstein. (Sebeok 2001: 23)

The position in biosemiotics that communicational competencies are means for modeling justifies the adoption of the concept of *affordance* in approaches to technology, media, and communication, as proposed by Treem and Leonardi (2013) independently of semiotic theories. Thus, a more specific concept of *media affordances* has been coined to refer to *technology affordances* as displayed in the use of specific media (Majchrzak et al. 2013). *Technology affordances* are defined as "the mutuality of actor intentions and technology capabilities that provide the potential for a particular action" (Majchrzak et al. 2013: 39, see also Faraj and Azad 2012). This mutuality leads to a prism that renders "the symbiotic relationship between the action to be taken in the context and the capability of the technology" (Majchrzak et al. 2013: 39) unavoidable, I argue, in cultural studies broadly. Text, from this point of view, is shaped within media affordances.

For this reason, a terminological shift from *text* to *the model* can be insightful for understanding the cultural dynamics entailed by digitalization and, also, for scrutinizing the environmental (ecological) implications of digitalization. A modeling perspective on meaning, as exhibited by biosemiotics, avoids the loopholes of conventionalism often implied by the literary notion of text. As I argued elsewhere (Olteanu 2019), when committed to the implications of the traditional notion of *text*, semiotic theory risks culturalism, namely the ideological theory of culture as a rigidly determining worldview (see Stjernfelt 2012; Eriksen Stjernfelt 2012). On the other hand, a semiotic theory, such as biosemiotics, that revolves around a phenomenological understanding of modeling, easily avoids this by acknowledging the embodied and environmental conditions of meaning-making (cf. Cobley 2016). It follows the consideration in cognitive semiotics, as well as in cognitive linguistics, that:

> (...) meaning and meaning production in thinking or communication are seen as based, grounded, anchored in human facts, that is, approached as emerging from personal and interpersonal conscious experience in the actual life world of humans. This evident but nevertheless most often neglected principle could be called the *phenomenological prerequisite*. We cannot just grasp meaning patterns as ideas combined with other ideas out of the blue. The mind is embodied and situated; the 'body' in question is not only a physiological entity but also a social and emotional entity equipped with consciousness—inner life, if you will. Experience takes place in domains of experience, or semantic realms, so in any situation, a human subject will be interacting with specific physical circumstances, with other subjects, collectively or individually, and with its own thinking and feelings. We live in situations that have to be represented while we live them, that is, represented in certain

domains. The present moment of a human being is also in itself a representation! (Brandt 2011: 53–54)

Thus, the situatedness of the embodied mind is the phenomenological prerequisite for mediality as it evokes representational and communicational possibilities. The body's presence in time-space implies representation. By developing technological media, humans attune representational possibilities to specific needs and goals for modeling and communicating by clearing away the landscape's noise, which contains elements that are unnecessary for a focused message.

For a similar realization, albeit in very different terms, Charles Peirce considered that the best medium for representations of logical operations is a blank page. His view is not a strive for monomodality but has to do with iconicity, which supports multimodality. He found that logic is spatial and, hence, best represented graphically, which led him to explain why a blank page is a perfect medium for graphic representations of logic (see Stjernfelt 2015: 38). He termed the blank sheet *the sheet of assertion* or *phemic sheet* (CP 4.553, Stjernfelt 2015: 43), as anything inscribed on a blank sheet makes, in virtue of its contrast to the sheet, an existential assertion. The phemic sheet, according to Peirce, iconically represents the Universe of Discourse (CP 4.396, 4.561). By Universe of Discourse, following Augustus de Morgan, Peirce (CP 2.536) meant the "circumstances of [a proposition's] enunciation," which "show that it refers to some collection of individuals or of possibilities, which cannot be adequately described, but can only be indicated as something familiar to both speaker and auditor." Because it is blank, the phemic sheet is a graph that iconically represents the universe of discourse. To represent an assertion on the phemic sheet is to *scribe* a token of a graph on the sheet. The blank page works very well as a medium for graphic representations of logic because it presents this particular affordance: anything inscribed on it appears to the human eye as an assertion of an existential quantifier within a Universe of Discourse. How graphs can represent logical operations in virtue of a phemic sheet's affordances exemplifies how texts are afforded (made possible and constrained) by media.

Because a medium frames and represents a Universe of Discourse, each medium must be considered to have specific affordances and, hence, particular implications for modeling. Indeed, no medium can have entirely different affordances to all other media. All technological media relate to the human body in ways in which the natural environment also does. The means of developing a new technological medium consist of the affordances of existing ones. An older technological medium serves as the semiotic scaffolding (see Cobley and Stjernfelt 2015) that supports the next one: handwriting constituted the scaffolding for print, print for the telegraph, telegraph for radio, radio for television, and all of these for the Internet. All technological media are based on the affordances for meaning-making that the human body–environment relation entails at the bottom line. Thus, a basic set of potential affordances is deemed to stem from human embodied situatedness, in agreement with Brandt (2011), within the landscape in which human embodied morphology, sense perception channels, and cognitive capacities evolved, following Gibson's (1979) affordance theory. This position is at the confluence of many recent semiotic accounts, where an "embodiment turn" (Stjernfelt 2007: 257) is observed in the study of meaning. In a recent but already crucial study in media semiotics, Elleström explains that mediality is evoked not only by technology and premediated devices but also by corporeality (2018: 270–271). Assuming a basic mediality that stems from human corporeality is the same as Brandt's (2011) observation that the present moment of a human being, or, it may be added, of any living organism, already is representation. This assumption allows for a medium-centered study of meaning and communication, which positions a medium's concept at the heart of cultural studies. It implies that *text* must be understood regarding mediality. Because media model texts, the concept of the model also comes to the fore.

From this perspective, both specific and shared representation possibilities of digital media can be observed not only in contradistinction to those of print but also to those of other electronic media, such as radio and television. For instance, in the context of organizational communication, Treem and Leonardi (2013) explain that social media affordances "are of important consequence [. . .] because they afford behaviors that were difficult or impossible to achieve in combination before these new technologies entered the workplace" (2013: 178). While Treem and Leonardi focus on organizational communication, their observation can be extended to many aspects of human societies where behavior is dependent on media affordances. Also, the implications of such affordances for organizational communication are deemed to further impact society broadly, as the latter is modeled by the organizations that constitute it. Industry and the lifestyles that result from it are an essential ingredient of modeling society. Many novel approaches, such as Jeremy Rifkin's (2011) theory of industrial revolution, semiotic approaches to the creative industries (Hartley 2015), and the recent emergence of the anthropology of experts (Boyer 2008), among others, point out, each in its own terminology. Also, following Rifkin, media consist of one of the two elements that merge to constitute an industry. The other one is energy infrastructure.

4 Digital Media and Natural Modeling

In at least two critical regards, digital media are more similar to the basic mediality evoked by humans' natural environment than previous media technologies that shaped human societies. These two are multimodality and simultaneity. As the digital and natural environments share media characteristics (a concept explored in detail in Elleström 2014), they also share affordances. This condition is arguably due to the network structure of digital media, which resembles ecological dynamics. Unlike previous technological media, digital media are not unidirectionally linear or nucleic but spread the flow of meaningful content (semantic information) in network structures. This network character is also the primary motivation of the biosemiotic approach to ecology, given that the focus here is on schematic representation (Kull 2003). The shift from ladder-like or tree-like to web-like models that ecology produced in the natural sciences is similar to semiotics' effects in the humanities, revealing the insightfulness of a semiotic approach to ecology (i.e., ecosemiotics).

4.1 *Biosemiotics and Multimodality*

The concept of multimodality refers to the involvement of various sense-perception channels and representational modalities in meaning-making (Kress and van Leeuwen 2001). It was mostly explored within the field of social semiotics (see van Leeuwen 2005), which, arguably, developed as a semiotic offshoot of systemic functional linguistics (Halliday 1978; Halliday and Webster 2009). The development of functional linguistics into social semiotics is, in part, due to the difficulties that linguistics encountered in analyzing meaning in societies where texts rely more and more on the merging together of varieties of modalities. For the same reason, a semiotic turn is recently observed in translation studies. For instance, reviewing the existing research on multimodality in translation, Pérez-González remarks that translation and interpretation "often interact with the semiotics of the human body" (2014: 122). This interaction reveals the mutuality between the emergence of the concept of multimodality and the embodiment turn. Also, these are inextricably related to the iconic turn (Boehm and Mitchell 2009; Moxey 2008; Stjernfelt 2007: 53, 69), namely the recent shift of focus in the humanities and cognitive sciences from language, as the primary vehicle of knowledge, to embodied image schemata (e.g., Lakoff 1987; Lakoff and Johnson 1999).

Instead of discussing texts as multimodal, I consider that discussing culture in "the interrogation of modeling" (Cobley 2016: 28) offers a better understanding of human embodiment and situatedness within a landscape. This interrogation is the cornerstone of the biosemiotic approach to culture, as Cobley proposed (2010, 2016). In general, the semiotic take on cultural studies or, more precisely, *cultural semiotics*, analyzed cultural phenomena as *texts*. The biosemiotic alternative of *modeling* has its root in Thomas Sebeok's criticism of cultural semiotics (e.g., 1991, 2001), which justified his biosemiotic project as a humanities endeavor. Sebeok's criticism was not pointed at cultural semiotics as much as it targeted the prevalent tendency toward glottocentrism in cultural studies. Relying on the state-of-the-art evolutionary anthropology of the time, Sebeok argued that linguistic modeling systems are based on more basic, non-verbal, internalized modeling systems of organisms (2001: 43, 136). As such, modeling systems that are supralinguistic, such as human culture, do not draw only on the affordances of (one) language. More than merely *non-verbal* structures of meaning, in an organism's knowledge of its world (*Umwelt*), there *are non-verbalizable* (elements of) meaning. Thus, what is non-verbalizable is modeled and communicated by medialities other than language strictly. The models that are carriers (von Uexküll 2010 [1934, 1940]: 140) of such meaning, it is safe to assume, have a schematic (or diagrammatic) structure. The iconic turn implies more than merely accepting that non-linguistic structures are meaningful and can be used in modeling, but also that language, as well, is functional because it is based on schematic structures. By introducing a primary, non-verbal modeling system as a hypothesis of biosemiotics, Sebeok not only brought semiotic theory up to date with the evolutionary theories of his time (e.g., Gould and Vrba 1982) but also delivered a theory that was to be aligned with more recent views in, among others, philosophy of embodiment, cognitive sciences, and visual arts. In light of the iconic turn on meaning-making, cultural studies are ripe to fully adopt biosemiotics as an explanatory theory for cultural phenomena. Until recently, an avant-garde, biosemiotics now has the opportunity to take the foreground in media and communication studies on account of the possibility to approach mediated material as the model(ing).

As already implied, this perspective also favors a phenomenological rather than literary approach to culture. Unlike *text*, *a model* is not identifiable with an objective artifact in a moment in spacetime. Instead, it is a phenomenal interpretation produced by an organism's embodied competencies for meaning-making. While mostly a scholarly development, the iconic turn was enhanced, as made evident in social contexts, by digitalization (Boehm and Mitchell 2009: 107–108, 113). The specific media affordances of the digital reveal some important aspects of modeling in general, which are the focus of the present chapter.

As entailed by the development of electronic media, the diversification of representation modalities calls for an account of meaning and an approach to communication that go beyond the affordances of verbal (or verbalizable) language, as also argued by Elleström (2018, 2019). In this light, his semiotic, medium-centered communication model aims to encompass the broad spectrum of medial affordances. In this same regard, the answer coming from social semiotics consists in the multimodality framework (e.g., Kress 2010). A cornerstone of this framework is the criticism of what is generally referred to as the double articulation hypothesis:

> Where traditional linguistics had defined language as a system that worked through double articulation, where a message was an articulation as a form and as a meaning, we see multimodal texts as making meaning in multiple articulations. (Kress and van Leeuwen 2001: 4)

Like any animal, humans make sense of their environment by using all their sense perception channels together. A wide variety of modalities is always involved in meaning-making. This variety means that what humans receive as a "message" always consists of multiple articulations. The dominance of the medium of print during the modern age and its implicit coincidence with literacy's popularization inculcated a narrow notion of meaning (or text) as monomodal. This understanding is linked to modern mind/body dualism that construed knowledge as consisting of ideas, namely purely mental, disembodied entities. From this point of view, linearity and monomodality are desirable in communicating content that, presumably, requires an intellectual effort because any expressivity would disturb the idea's mental purity. Thus, education (apart from teaching art, perhaps), which has the role of delivering the required literacy skills, has been carried out monomodally. Because monomodality is intrinsic to modern educational curricula, it also framed media affordances, resisting multimodal media's adoption as canonical in various genres. It is not a surprise that academic discourse, a modern discourse *par excellence*, strongly resists adopting forms of modeling other than afforded by printed language. While, in some sciences, non-linguistic schematic representations are customarily used (e.g., diagrams in mathematics and logic, sketches of organs and the body in biology), the humanities have been particularly conservative in their monomodalism. This attitude explains the current crisis that the humanities curriculum is undergoing (Nussbaum 2010; Jay 2014), which has recently been tackled from semiotic perspectives (Martinelli 2016; Cobley 2017). The humanities are the scholarly tradition from where the notion of *literacy* originates. This tradition used to be the most influential in shaping educational curricula, until its historically recent marginalization, in the face of short-term profit-driven capitalism (see Cobley and Stjernfelt 2015; Cobley 2017). Arguably, the conservative perseverance of rigid concepts of *text* and *literacy* is the main reason for the current lack of popularity of the humanities. However, the recent adoption of digital methods in humanities research and the emergence of a digital humanities framework (McCarty 2005) is changing the stubborn preference for monomodality. Notably, the emerging scholarship on *digital literacy* (Mills 2016, Scolari et al. 2018) is revealing the social value of both critical thinking, in the traditional sense, as a discursive, argumentative competence, and of competences for using technology, previously regarded as aside the scope of the humanities.

Moreover, the interconnectedness of these two is particularly remarkable, as revealed by the understanding that discourse is medial (see Machin 2013). The mutuality of critical thinking and technology skills is implied in Rifkin's (2011:257) proposal of sustainability as merging media networks and energy infrastructure. Herein could lie the solution to the humanities' impasse, which requires, as argued above, the adoption of an embodied theory of meaning in cultural studies. Thus, biosemiotics is a suitable platform for advancing the state-of-the-art in the humanities in the age of digitalization. This view is supported by Cobley's (2017: 12) claim that: "In identifying the semiotic basis of natural processes, biosemiotics has fundamentally challenged the mechanist worldview that is routinely promulgated by school teaching's reliance on a Newtonian model of science."

4.2 Biosemiotics and Simultaneous Participativity

The second criterion that renders the digital similar to the natural environment is simultaneous participativity. This participativity is one of McLuhan's foundational realizations for the development of media studies, resulting in his coining of the celebrated "global village" idiom. The global human community is made possible by electronic media:

> [...] the electro-magnetic discoveries have recreated the simultaneous "field" in all human affairs so that the human family now exists under conditions of a "global village.' We live in a single constricted space resonant with tribal drums. (McLuhan 1962: 31)

Not only does electronic media network the whole planet, but it so networks it in the structure of a village, where public communication happens, so to say, *on the spot*. The main point here is not that humanity, at the global level, socially functions as a village of preliterate times, but that it has the capacity and habit to communicate *on the spot*. It is mostly in this regard that the global community of digital times resembles the *village*. In the preliterate human tribe or village, public communication is a possibility for each member, not only for a specific *literate* group with access to broadcasting technology. This is to say that the village has the medial affordance of simultaneous participativity. Immediate public exposure is a media affordance of societies that do not require literacy, understood as a competence for dealing with codified texts. Just like speaking out in a village makes the communicated content available to the entire community, so does a Facebook post or a tweet, potentially. Each member of a village knows every other member, in contradistinction to a national community's requirements. According to Benedict Anderson's celebrated theory (2006), a nation is an imagined community because it requires that none of its members knows all her co-nationals. Thus, the national community is as real as imaginably affordable. Imagining the nation, the argument goes, is possible because the medium of print offers the illusion of monolingualism (Lähteenmäki 2010: 18, Lähteenmäki et al. 2011), namely the lack of linguistic variation over vast territories (see also Anderson 2006: 90, Said 1994: 299, Cobley 2014: 37–39). Anderson referred to this media effect as print-capitalism (1983: 36, 37–46). McLuhan stressed that electronic and typographic medialities differ critically, because the electronic "shapes and structures of human interdependence and of expression [...] are 'oral' in form even when the components of the situation may be non-verbal." (McLuhan 1962: 3) The modeling entailed by electronic media, this is to say, is somewhat similar to the modeling of the environment through media that do not rely on heavily symbolic and technical codes, which are commonly termed *texts*. Also, the participation of the entire global population to one network structure, eventually makes imagining the *nation* obsolete. Even though the entire human population is too numerous for one individual to know all of the others, in a personal way, their networked interaction and interdependence collapses the affordance of imagining nations. Rather, imaginative competencies are now engaged in conceiving the audience of social media communicative instances, termed *imagined audience* (Marwick and Boyd 2010), often potentially subsuming interlocutors localized anywhere on the Globe. The imagined audience transcends groups that can be conceived monolingual.

Media shifts, McLuhan considered resonantly to Anderson's thesis, require a "reorganization of imaginative life" (1962: 3). That nationalism has been historically possible only in societies medially dominated by print is displayed in McLuhan's argument that electronic media re-tribalize (1997: 78–79) human psychic and social awareness. He explicitly stated that "printing evoked both individualism and nationalism in the sixteenth century" (1997: 10). From a media semiotic perspective, Danesi argues that the retribalization effect of electronic media further led to the reintegration of body and mind in the Digital Galaxy, as he terms the age of digital media, echoing McLuhan's term:

> Rather than take the mind out of the human body and transfer it to cybersystems or humanoid machines, the whole technology that undergirds the Digital Galaxy will [...] bring about a reintegration of the body and the mind [...]. This process of re-embodiment is a result of what McLuhan (1964) called "re-tribalization". As digital technologies continue to advance the possibility of global communication "on the spot" [...] people want the protection and emotional shelter of the "tribe" more and more. This is because, like most other species, humans have always lived in groups. The tribe remains the type of collectivity to which human beings instinctively relate even in modern times. (Danesi 2002: 177)

This is manifest in the polarizing tendency of contemporary political discourse (e.g., Wodak 2017). A stringent dualism is manifest between, on the one hand, culturalism, nationalism, or regionalism, as the preference for the shelter of tribes of old media and, on the other hand, cosmopolitanism, as the preference for the shelter of the global tribe, afforded by new media. Accordingly, Rifkin argues that the industrial revolution generated by digitalization restructures human organization in ways that eventually dissipate nation-states, leaving no room for conceiving a nation's idea as a principle of social organization (Rifkin 2011: 188–189). Just like medieval scholars could not imagine a political organization driven by the notion of citizenship, we currently find it (at least) difficult to imagine a nationless world (Rifkin 2011: 234). By affording and facilitating cooperation in networks structured regardless of nation-state boundaries, digitalization fosters a "biosphere consciousness" (2011: 143). As supported both by Hartley (2015) and Rifkin, a biosphere consciousness, seeing "the evolution of life and the evolution of the planet's geochemistry as a co-creative process in which each adapts to the other, assuring the continuation of life within the Earth's envelope" (Rifkin 2011: 188), is fostered by the urban organization of human life. In this sense, "urban" does not refer to large geographic agglomerations of the human population but to a lifestyle that supposes an awareness of global inclusion, citizenship, and industrious creativity (Hartley 2015: 82, 90). While modern communicational media supported this lifestyle in dense demographic agglomerations, digital media networks allow for such cosmopolitanism without the absolute need for vast populations' geographic colocalization. As the main attribute of biosphere consciousness, sharing among vast populations started to be a practice of the "new energy highway" (Rifkin 2011: 143), consisting of the merged digital media networks and renewable energy grid into one infrastructure, currently in its incipient phase. Thus, the resulting sense of collectivity at a global level further endorses the affinity between digital and environmental literacies. Environmental issues are always global issues and require globally minded solutions.

Despite digitalization adding medial strata to humans' representation of environment and society, it undoes the print's effect of dichotomizing body and mind. It does so because of these two critical affinities with the human modeling's natural affordances: multimodality and simultaneous participativity ("on the spot"). As such, individuals' cognitive models of the world, intimately internalized as they might be, are easily translated onto digital representations. Digital media afford to transpose one's cognitive schemata with less effort than print or audio, or kinematic (narrative) media. This transposition presents the opportunity for an ecological turn:

contrary to the mind-body dichotomization and segregational effects of print, a digitalized society is semiotically akin to the natural human environment. The environment of humans is not a literary production, such as the term *text* came to designate. It is a media product (Elleström 2014, 2019), which is modeled within media affordances.

The same argument is evident in the semiotic turn observed in recent translation studies. For encompassing the wide variety of modalities involved in the emergence of media products, the need for a concept of translation as not strictly linguistic was noticed in scholarship. For instance, the mainstreaming of audiovisuality resulted in construing translation in the interrogation of semiotic resources:

> Recent research [. . .] marked a shift away from purely linguistic or verbal only approaches, opening up new insights into the intersemiotic and multimodal nature of advertising texts [. . .]. (Torresi 2009: 6)

This is a *post-factum* reiteration of McLuhan's anticipation of the re-tribalizing effect of electronic media, albeit with a translation concern. It is one of the many movements away from glottocentric theories revolving around the notion of text toward an embodied and environmental account of human knowledge.

5 Concluding

The case for a biosemiotic approach to digitalization was developed in views of the affinities between digital media communication and natural environmental affordances. The theoretical conception of modeling in biosemiotics meets the requirements to nest digital media communication models, particularly given the re-embodiment implied by the digital. The underpinning principle of modeling as schematic cross-modal translation, inherited in biosemiotics from Peirce's logic, answers to state-of-the-art positions in current media theory, where semiotic theory is invoked because, as Elleström explains, "[a] model should be understood as a clearly outlined cognitive scheme that is both described with the aid of language, and depicted as a diagram" (2018: 270). This directly indicates Peirce's system of graphic logical representations, termed existential graphs, which has the purpose "to afford a method (1) as simple as possible (that is to say, with as small a number of arbitrary conventions as possible), for representing propositions (2) as iconically, or diagrammatically and (3) as analytically as possible." (CP 4.561) Existential graphs are logical notations of evident iconicity by their juxtaposing onto a *sheet of assertion*, namely a blank page where any inscription implies an assertion of an individual existent. At the time, Peirce could not explicate that the idea of a sheet of assertion supposes a basic mediality that results from the body's interaction with the environment. However, his choice for the verb "afford" in describing the purpose of logical notations suggests or, at least, justifies an affordance approach to media. Thus, his semiotics can now underpin, particularly in its biosemiotic development, this currently emerging theory. As a method, existential graphs, or, in general, schematic logic based on the same rationale as existential graphs, can underpin a notion of literacy that accommodates digital skills. The transposition of digitalized content in the form of schematic propositions onto a *blank* sheet of assertion allows for the re-contextualization of the content into a knowing subject's environmental situatedness. Traditional construals of *text* miss many aspects of modeling multimodal content with a network and processual (developing) structure. As a voluntary modeling effort, such a cross-modal diagrammatization avoids the illusion that digital media delivers "naked information" (Hoffmeyer 2018: 5) but re-contextualizes online content in a critical thinking manner. Considering biosemiotics, Hoffmeyer warned about the danger of not being aware of the implicit modeling that our situatedness implies, which can be considered a lack of criticism in engaging with media:

> The only reason why we can so easily retrieve any thinkable piece of information, is that Google already has digitised it. But in the very process of digitizing it, the information necessarily loses its natural connection to the rest of the world. Information is converted to fact. But this implies that the information must be given a new context—or otherwise remain incomprehensible to the human mind. Most of the time we manage this without giving it another thought, and experience the information as meaningful. This means that the information has now—unknowingly perhaps—become re-contextualised. Without this re-contextualisation, we would have only the "naked information", and no matter how many pieces of such "naked information" we add, we will never reach any kind of understandable knowledge. (2018: 5)

If, as a result of digital modeling, the content might lose its "natural connection to the rest of the world," then a modeling method is needed to re-establish these connections and map anew, by appropriating the content in the knowing subject's context in critical awareness. Arguing that contemporary societies require an educational philosophy that acknowledges literacy forms other than language literacy, Lacković (2020a, 2020b) exhibits a first and compelling step in this direction. Consequently, she proposes a media literacy method, termed inquiry graphics, based on Peirce's icon-based semiotics. When extrapolated to a modeling theory, Peirce's logic brings these connections to the fore (see also Lacković and Olteanu 2020) by its commitment to the iconicity between representation and represented object, as supported, among others, by Kralemann and Lattmann (2013), Elleström (2013) and Stjernfelt (2015). This is already visible in the adoption of Peircean existential graphs in digital computing (Sowa 2014) and his iconicity concept as a cornerstone for modeling in the digital humanities (Ciula and Eide 2017; Ciula and Marras 2016, 2019). However, despite Sebeok's (2001) founding of biosemiotics as a Peircean-based modeling theory, this path has not been explored in a digital or, more broadly, media regard at all. The biosemiotic perspective fits the multimodal affordances of schematic modeling by its account of meaning according to which mediality is a matter of embodied environmental situatedness. This argument has not been explored explicitly so far but is an implication of the basic tenets of the biosemiotic concept of the environment as a model (*Umwelt*). This approach particularly allows for bridging environmental and digital literacies, in view of a comprehensive and phenomenological account of modeling as scaffolding, instead of the traditional notion of literacy as based on textual competencies. This account, I propose, is an insightful pathway to exploring the educational dynamics of the third industrial revolution, which will stem from the merging of digital media and renewable energy networks.

Acknowledgment This research was supported by the Estonian Research Council (grant MOBJD346 "Towards a joint environmental and digital literacy: An ecosemiotic approach to digitalization").

References

Anderson, B ([2006] 1983) Imagined communities: reflections on the origin and spread of nationalism. Verso, London, New York

Barthes, R (1991 [1957]) Mythologies. Trans. Cape, Jonathan. The Noonday Press, New York

Boehm G, Mitchell WJT (2009) Pictorial versus iconic turn: two letters. Culture, Theory and Critique 50(2–3):103–121

Boyer D (2008) Thinking through the anthropology of experts. Anthropology in Action 15 (2):38–46

Brandt PA (2011) What is cognitive semiotics? A new paradigm in the study of meaning. Signata: Annales des sémiotique/Annals of Semiotics 2:49–60

Bringhurst R (2004 [1992]) *The elements of typographic style.* Version 3.0. Hartley & Marks, Vancouver

Campbell C (2017) Exploring the textual woods: Umberto Eco's growing concept of text. In: Torkild T, Bent S (eds) Umberto Eco in his own words. De Gruyter Mouton, Berlin, pp 134–142

Campbell C, Olteanu A, Kalevi K (2019) Learning and knowing as semiosis: extending the cultural apparatus of semiotics. Sign Systems Studies 47(3/4):325–381

Ciula A, Eide Ø (2017) Modelling in digital humanities: signs in context. Digital Scholarship in the Humanities 32(1):i33–i46

Ciula A, Marras C (2016) Circling around text and language: towards "pragmatic modelling". Digital humanities. *DHQ: Digital Humanities Quarterly* 10(3):17–29

Ciula A, Marras C (2019) Exploring a semiotic conceptualisation of modelling in digital humanities practices. In: Olteanu A, Stables A, Borţun D (eds) Meanings & co.: the Interdisciplinarity of communication, semiotics and multimodality. Springer, Cham, pp 33–52

Cobley P (2010) The cultural implications of biosemiotics. *Biosemiotics* 3(2):225–244

Cobley, P ([2014] 2001). *Narrative*, Second Edition. Routledge, New York

Cobley P (2016) Cultural implications of biosemiotics. Springer, Dordrecht

Cobley P (2017) What the humanities are for – a semiotic perspective. In: Kristian B, Cobley P (eds) Semiotics and its masters. de Gruyter Mouton, Berlin, pp 3–23

Cobley P, Stjernfelt F (2015) Scaffolding development and the human condition. Biosemiotics 8 (2):291–304

Danesi M (2002) Understanding media semiotics. Arnold, London

Descartes R (2008 [1643]) Meditations on first philosophy with selections from objections and replies. Oxford University Press, Trans. Michael Moriarty. Oxford

Eco U (1976) A theory of semiotics. Indiana University Press, Bloomington

Eco U (1979) The role of the reader: explorations in the semiotics of texts. Indiana University Press, Bloomington

Eco U (1999 [1997]) Kant and the platypus: essays on language and cognition. A Harvest Book, Harcourt, Inc, San Diego

Elleström L (2013) Spatiotemporal aspects of iconicity. In: Elleström L, Fischer O, Christina L (eds) Iconic investigations. John Benjamins, Amsterdam, pp 95–117
Elleström L (2014) Media transformation: the transfer of media characteristics among media. Palgrave Macmillan, London
Elleström L (2018) A medium-centered model of communication. Semiotica 224:269–293
Elleström L (2019) Modelling human communication: Mediality and semiotics. In: Olteanu A, Stables A, Borţun D (eds) Meanings & co.: the Interdisciplinarity of communication, semiotics and multimodality. Springer, Cham, pp 7–32
Eriksen J-M, Stjernfelt F (2012) The democratic contradictions of multiculturalism. Telos, New York
Faraj S, Azad B (2012) The materiality of technology: an affordance perspective. In: Leonardi PM, Nardi B, Kallinikos J (eds) Materiality and organizing: social interaction in a technological world. Oxford University Press, New York
Fiske, J (1990 [1982]). Introduction to communication studies. Second Edition. Routledge, London
Gibson JJ (1979) The ecological approach to visual perception. Houghton Mifflin, Boston
Gould SJ, Vrba ES (1982) Exaptation – a missing term in the science of form. Paleobiology 8 (1):4–15
Halliday MAK (1978) Language as social semiotic: the social interpretation of language and meaning. Arnold, London
Halliday MAK, Webster JJ (eds) (2009) Continuum companion to systemic functional linguistics. Continuum, New York
Hartley J (2015) Urban semiosis: creative industries and the clash of systems. Int J Cult Stud 18 (1):79–101
Hoffmeyer J (2007) Semiotic scaffolding of living systems. In: Barbieri M (ed) Introduction to biosemiotics: the new biological synthesis. Springer, Dordrecht, pp 49–166
Hoffmeyer J (2008) The semiotic body. Biosemiotics 1(2):169–190
Hoffmeyer J, Stjernfelt F (2016) The great chain of semiosis. Investigating the steps in the evolution of semiotic competence. Biosemiotics 9(2):7–29
Hopkins J (2020) The concept of affordances in digital media. In: Friese H, Nolden M, Rebane G, Schrieter M (eds) Handbuch Soziale Praktiken und Digitale Alltagswelten. Springer, Wiesbaden, pp 47–54
Ihde D (1990) Technology and the lifeworld: from garden to earth. Indiana University Press, Bloomington
Jay P (2014) The humanities "crisis" and the future of literary studies. Palgrave Macmillan, New York
Jesper H (2018) Knowledge is never just there. Biosemiotics 11(1):1–5. https://doi.org/10.1007/s12304-018-9320-4
Kralemann B, Lattmann C (2013) Models as icons: modelling models in the semiotic framework of Peirce's theory of signs. Synthese 190:3397–3420
Kress G (2010) Multimodality: a social semiotic approach to contemporary communication. Routledge, London
Kress G, Selander S (2012) Multimodal design, learning and cultures of recognition. Interned and Higher Education 15(4):265–268
Kress G, van Leeuwen T (2001) Multimodal discourse: the modes and Media of Contemporary Communication. Arnold, London
Kull K (2003) Ladder, tree, web: the ages of biological understanding. Sign Systems Studies 31 (2):589–603
Lacković N (2020a) Inquiry graphics in higher education: new approaches to knowledge, learning and methods with images. Palgrave Macmillan, Cham
Lacković N (2020b) Thinking with digital images in the post-truth era: a method in critical media literacy. Postdigital sciences and education 2:442–462

Lacković N, Olteanu A (2020) Rethinking educational theory and practice in times of visual media: learning as image-concept integration. Educ Philos Theory. https://doi.org/10.1080/00131857.2020.1799783
Lähteenmäki M (2010) Heteroglossia and voice: conceptualizing linguistic diversity from a Bakhtinian perspective. In: Lähteenmäki M, Vanhala-Aniszewski M (eds) Language ideologies in transition: multilingualism in Russia and Finland. Peter Lang, Frankfurt am Main, pp 17–34
Lähteenmäki M, Varis P, Leppänen S (2011) Editorial: the shifting paradigm: towards a re-conceptualisation of multilingualism. Apples – Journal of Applied Linguistics 5(1):1–11
Lakoff G, Johnson M (1980) Metaphors we live by. The University of Chicago Press, Chicago and London
Lakoff G, Johnson M (1999) Philosophy in the flesh: the embodied mind and its challenge to Western thought. Basic Books, New York
Lakoff G (1987) Women, fire, and dangerous things: what categories reveal about the mind. University of Chicago Press, Chicago, IL
Locke J (1836 [1690]) An essay concerning human understanding. T. Tegg and Son, London
Lotman Y (1990) Universe of the mind: a semiotic theory of culture. Tauris, London
Machin D (2013) Introduction: what is multimodal critical discourse studies? Crit Discourse Stud 10(4):347–355
Majchrzak A, Faraj S, Kane GC, Azad B (2013) The contradictory influence of social media affordances on online communal knowledge sharing. J Comput-Mediat Commun 19:38–55
Maran T (2014) Biosemiotic criticism: modelling the environment in literature. Green Letters 18 (3):297–311
Maran T, Kull K (2014) Ecosemiotics: Main principles and current developments. Geografiska Annaler: Series B, Human Geography 96(1):41–50
Martinelli D (2016) Arts and humanities in Progress: a manifesto of Numanities. Springer, Cham
Martinet A (1962) A functional view of language. Oxford University Press, Oxford
Marwick AE, boyd d. (2010) I tweet honestly, I tweet passionately: twitter users, context collapse, and the imagined audience. New Media Soc 13(1):114–133
McCarty W (2005) Humanities computing. Palgrave Macmillan, New York
McLuhan M (1962) The Gutenberg galaxy: the making of the typographic man. University of Toronto Press, Toronto
McLuhan M (1964) Understanding media. Routledge & Kegan Paul, London
McLuhan M (1997) In: McLuhan E, Zingrone F (eds) Essential McLuhan. Routledge, London
Mills KA (2016) Literacy theories of the digital age: social, critical, multimodal, spatial material and sensory lenses. Multilingual Matters, Bristol
Moxey K (2008) Visual studies and the iconic turn. J Vis Cult 7(2):131–146
Nöth W (2001) Ecosemiotics and the semiotics of nature. Sign Syst Stud 29(1):71–81. https://doi.org/10.12697/SSS.2001.29.1.06
Nussbaum MC (2010) Not for profit – why democracy needs the humanities. Princeton University Press, Princeton
Olteanu A (2019) Multiculturalism as multimodal communication: a semiotic perspective. Springer, Cham
Olteanu A, Stables A (2018) Learning and adaptation from a semiotic perspective. Sign Systems Studies 46(4):409–434
Pagni E (2016) Why explicit semiotic grounding is essential to biology as science? The point of view in biosemiotics. HumanaMente Journal of Philosophical Studies 9(31):57–72
Peirce, CS (1931–1935, 1958) The Collected Papers of Charles Sanders Peirce. Ed. Charles Hartshorne, Paul Weiss and A. W. Burks. Belknap, Cambridge. [In-text references are to CP, followed by paragraph number]
Pérez-González L (2014) Multimodality in translation and interpreting studies: theoretical and methodological perspectives. In: Bermann S, Porter C (eds) A companion to translation studies. Wiley Blackwell, Oxford, pp 119–131

Rifkin J (2011) The third industrial revolution: how lateral power is transforming energy, the economy, and the world. Palgrave Macmillan, New York
Roli A, Kauffman SA (2020) Emergence of organisms. Entropy 22(10):1163. https://doi.org/10.3390/e22101163
Rorty RM (ed) (1967) The linguistic turn: essays in philosophical method with two retrospective essays. University of Chicago Press, Chicago
Roth CE (1992) Environmental literacy: its roots, evolution, and directions in the 1990s. ERIC, Columbus, OH
Said EW (1994 [1993]) Culture and imperialism. Vintage Books, New York
Saussure F (1959 [1916]) Course in general linguistics [*Course de linguistique générale*]. Eds. Bally, C., Sechehaye, A. Trans. Baskin, W. The Philosophical Library, New York
Scolari Carlos A (2012) Media ecology: exploring the metaphor to expand the theory. Commun Theory 22:204–225
Scolari CA, Masanet M-J, Guerrero-Pico M, Establés M-J (2018) Transmedia literacy in the new media ecology: teens' transmedia skills and informal learning strategies. El Profesional de la Informacion 27(4):801–812
Sebeok T (1991) A sign is just a sign: advances in semiotics. Indiana University Press, Bloomington and Indianapolis
Sebeok T (2001 [1994]) Signs: an introduction to semiotics. University of Toronto Press, Toronto
Sowa J (2014) From existential graphs to conceptual graphs. In: Management Association, I (ed) Computational linguistics: concepts, methodologies, tools, and applications. IGI Global, Hershey, pp 439–472
Stables A (2001) Who drew the sky? Conflicting assumptions in environmental education. Educ Philos Theory 33(2):245–256
Stables A (2012) Be(com)ing human: semiosis and the myth of reason. Sense Publishers, Rotterdam
Stables A, Bishop K (2001) Weak and strong conceptions of environmental literacy: implications for environmental education. Environ Educ Res 7(1):89–97
Stjernfelt F (2006) The semiotic body. A semiotic concept of embodiment? In: Nöth W (ed) Semiotic bodies, aesthetic embodiments and Cyberbodies. Kassel University Press, Kassel, pp 13–48
Stjernfelt F (2007) Diagrammatology: and investigation on the borderlines of phenomenology, ontology and semiotics. Springer, Dordrecht
Stjernfelt F (2012) Liberal multiculturalism as political philosophy: will Kymlicka. Monist 95 (1):49–71
Stjernfelt F (2015) Iconicity of logic – and the roots of the "iconicity" concept. In: Hiraga MK, Herlofsky WJ, Shinohara K, Kimi A (eds) Iconicity: east meets west. John Benjamins, Amsterdam, pp 35–53
Strate L (2008) Studying media as media: McLuhan and the media ecology approach. MediaTropes 1:127–142
Torresi I (1998) 2009. Advertising. In: Baker M, Saldanha G (eds) Routledge encyclopedia of translation studies, 2nd edn. Routledge, Oxford, pp 6–10
Treem JW, Leonardi PM (2013) Social media use in organizations: exploring affordances of visibility, Editability, persistence, and association. Ann Int Commun Assoc 36(1):143–189
van Leeuwen T (2005) Introducing social semiotics. Routledge, London
von Uexküll, J (2010 [1934, 1940]) A foray into the worlds of animals and humans with a theory of meaning. University of Minnesota Press, Minneapolis
Verbeek, P-P (2005 [1992]) *What things do: Philosophical reflections on technology, agency, and design*. Trans. Crease, Robert P. The Pennsylvania State University Press, University Park, PA
Wodak RE (2017) The "establishment", the "Élites", and the "people": Who's who? Journal of Language and Politics 16(4):551–565

Threshold, Meaning, and Life

Arthur Araujo

Abstract In this chapter, essentially, I argue that meaning is an activity that distinguishes the life process from inanimate ones. Since meaning and life are overlapping processes, methodologically, the notion of a threshold zone is an important tool to understand where the meaning process (or *semiosis*) begins and makes sense of life in nature. In arguing that meaning and life are overlapping processes, I endorse a view of continuity in nature. Taking into account Peirce's view of cognition as an indeterminate process, in particular, I suggest an understanding of meaning as a mosaic or web, which begins by a process of beginning. In using the image of the web, notably, I contrast this with Darwin's tree of life. The idea that I have in mind is to put forward the hypothesis that *where there is meaning, there is life*.

Keywords Threshold · Meaning · Life · Continuity · Indeterminateness

1 Introduction

As a hypothesis for dealing with the issue of a threshold to differentiate meaning and life in nature, I will assume that *where there is meaning, there is life—no meaning, no life!* (and vice versa).[1] The idea is that the relation between meaning and life is

This chapter is one of the results of a visiting period at the Department of Semiotics at the University of Tartu (Estonia) in 2018.

[1] In general terms, I agree with Evan Thompson in *Mind in Life* (2007) and his understanding of continuity between mind and life when he says, "The theme of this book is the deep continuity of life and mind. *Where there is life, there is mind* [my italics], and mind in its most articulated forms belongs to life. . . . From this perspective, mental life is also bodily life and is situated in the world. The roots of mental life lie not simply in the brain but ramify through the body and environment.

A. Araujo (✉)
Department of Philosophy, Federal University of Espírito Santo, Vitória, Brazil

© Springer Nature Switzerland AG 2021

E. Pagni, R. Theisen Simanke (eds.), *Biosemiotics and Evolution*, Interdisciplinary Evolution Research 6, https://doi.org/10.1007/978-3-030-85265-8_5

based on a sign inference process or semiosis whose logical form is "since p, q": the first, therefore the second (Manetti 2002: 285). As Umberto Eco (1976: 17) notes, it is not by chance that ancient philosophy has so frequently associated meaning and inference.[2] Although my hypothesis may take the logical form of material implication, what I actually have in mind is to stress the overlapping of meaning and life—that is to say: meaning and life cannot, and do not, exist independently.[3]

To illustrate my hypothesis (*where there is meaning, there is life*), I was inspired by William James's thesis on experience and life from his radical empiricism: experience is the immediate flux of life in the form of radical eventfulness. The ideas of flux and eventfulness stand for "asubstantialism" in James's view of experience (Weber 2013: 96): as the world is either an experiencing or an experienced, experience holds the world relationally together. Summarizing James's radical empiricism, as experience is the immediate flux of life, one can say that *where there is experience, there is life—no experience, no life!* It is important to note that insofar as experience is the immediate flux of life, it means "sensation" or "feeling." Therefore, it is not exclusively *conditio humana* (a human condition). In his radical empiricism, James describes experience as a mosaic structured in constructivist terms. In comparison with James's empiricism, I espouse the idea that life as a whole consists of a dynamic web of meaning relations. Using the image of the web, I will highlight this notion in contrast with Darwin's tree of life.[4] So I will refer to life as a web in which meaning is engendered. Taking into account my hypothesis that life and meaning overlap, and following Thomas Sebeok (2001: 3), it seems fair to assert that meaning is the activity that distinguishes the life process from inanimate ones.[5] As I see it, meaning could not have existed before the evolution of life and vice versa (Sebeok 2001; Hoffmeyer 2008).

Our mental life . . . cannot be reduced simply to brain processes inside the brain" (Thompson 2007: IX). Being philosophically sympathetic to Thompson, however, I have developed my hypothesis independently. If "mind" is replaced with "meaning," this is perfectly in accordance with my hypothesis; that is, *where there is meaning, there is life*. What seems to be my point of distinction from Thompson's understanding of continuity between life and mind is that I regard continuity of meaning and life as extending in the universe as a whole (and not restricted to Earth).

[2]Peirce adopted the designation "semiosis" (in a variant transcription) from Philodemus's fragmentary Herculanean papyrus *On Signs*, where the Greek equivalent appears at least 30 times to represent a type of reasoning or inference from signs (Sebeok 2001: 74).

[3]As I am assuming here a (bio)semiotic understanding of meaning, it comprises non-human and human forms of life. In line with Floyd Merrell (1997: x), I take meaning to be an activity "flowing along within the semiosic process." The result is plurality and continuity more than singularity and discreteness. As a general idea in this chapter, then, I have in mind to merge meaning in the core of process metaphysics.

[4]In *Theoretical Biology* (1926), interestingly, Jakob von Uexküll introduces the expression "web of life." In his understanding, the living world is much more a web than a ladder.

[5]Instead of using "capacity of" with a meaning that may suggest a human-like competence, I think that "activity" seems more adequate to process thought. Once meaning is assumed to be an activity more than *conditio humana*, it is characteristic of life-forms at multiple levels in organic nature. The idea suggests that meaning is an activity of *making sense*.

As presented by Umberto Eco (1976) with regard to the issue of the *missing link*, the challenge here will be to consider whether there must actually be a threshold zone from which one can discriminate meaning and life in nature. As a consequence, the challenge will also be to inquire whether the notion of a threshold zone can be applied in our understanding of where the meaning process or semiosis begins. In this chapter, thus, in accepting that the notion of the threshold zone makes sense of meaning and life in nature as a process of transition from the inorganic to the organic, I will look into how one can understand the differentiation of non-living and living entities, as well as the differentiation of meaningless (or non-semiotic) and meaningful (or semiotic). For that, I will present and review different conceptions on the issue of a threshold. As a particular case for discussion, I will rival David Bohm's conception of meaning (1985). To him, since meaning is taken to be a fundamental physical property of reality, the notion of the threshold does not seem to be epistemologically and ontologically relevant. In my contrary view, without consideration of the notion of the threshold zone, the ideas of meaning and life make no sense in the world at all.

In conclusion, I will argue that the missing link can be seen much more as a *metaphor* than an ontological claim. I will also argue that one needs the notion of the threshold zone to make sense of meaning and life as a process of transition from the inorganic to the organic. Having in mind Peirce's view of cognition as an indeterminate process, finally, I will suggest a mosaic understanding of meaning and life in that the web of life begins by a process of beginning; holistically speaking, as an indeterminate process, the web of life is without a center or periphery. In short, the idea is to put forward my hypothesis that *where there is meaning, there is life—no meaning, no life!* In speaking of life, I am not speaking of the origin and forms of life on earth. In line with Whitehead's cosmology, I have in mind the ideas of meaning and life as processes extending and overlapping in the universe as a whole. Having different contexts in natural sciences, epistemology, and biosemiotics as a background, in short, this chapter exploits a conceptual unity between the threshold, meaning, and life.

2 Threshold and Missing Link

In *A Theory of Semiotics* (1976), Umberto Eco introduces the notion of "natural boundaries" to delineate the borders of the semiotic approach. For both semiotics and biosemiotics, "natural boundaries" is a notion that epistemologically acquires a distinctive value. By "natural boundaries," accordingly, Eco means the point of transition from the non-semiotic to the semiotic as a sort of missing link. Having in mind such a transition, he speaks of a "lower threshold":

> By natural boundaries, I mean principally those beyond which a semiotic approach cannot go; for there is non-semiotic territory since there are phenomena that cannot be taken as sign-functions. . . . The phenomena on the lower threshold should rather be isolated as indicating the point where semiotic phenomena arise from something non-semiotic, as a sort of

> "*missing link*" [my italics] between the universe of signals and the universe of signs. (Eco 1976: 6, 21)

It is evident that Eco assumes the notion of a lower threshold as indicating the point of inflection (in the form of a missing link) from which semiotic phenomena arise. In regarding phenomena such as function, sign, value, and meaning, something seems to be absent in nature, and the door is opened for claiming that there must be a missing link. In delineating an epistemological border for understanding the extent of the semiotic approach, however, Eco's ideas of the missing link and the threshold have motivated interesting questions in semiotics and biosemiotics:

> One important question that divides people in semiotics is the question often referred to as the "semiotic threshold," i.e., the problem of defining the simplest system capable of semiosic activity. (Hoffmeyer 2011: 282)

> It is thus an open and crucial issue of research to determine, empirically and conceptually, the different thresholds in this zone between such simple reproducing and evolving systems and contemporary terrestrial organisms that appear to depend unambiguously on semiotic processes. (Cobley et al. 2011: 27–28)

> Semiotic threshold ... defines "a boundary between semiotic and non-semiotic areas." (Higuera and Kull 2017: 109)

> The concept has helped delimit and shape the whole area of semiotic studies. Theorizing of the so-called "lower semiotic threshold" has also provided biosemiotics with a way to set (and break) some of its boundaries, specifying the level where one can refer to sign action in opposition to non-semiotic activity. (Higuera and Kull 2017: 110)

Not only the notion of the semiotic threshold divides opinions, but also the meaning of the missing link does. "[Even though] the discovery of hominid fossils in Africa is a good example of transitional morphologies, for instance, it is not a missing link" (www.livescience.com/32530-what-is-the-missing-link.html). Also, as Atmanspacher (2020) states, "introducing mental states as the essential *missing link*" [my italics] in relation to the brain's physical states "is highly speculative from a contemporary perspective" in terms of physical theory. In this latter case, many terms have been created to resolve the supposed missing link between the mental and the physical. Orphans of the Cartesian pineal gland, many philosophers of mind and cognitive neuroscientists have created a sort of mythology by attributing psychic additions to the brain—the explanatory gap, the hard problem, supervenience, and so on—as a sort of missing link between the mental and the physical.

As was very well illustrated by Whitehead in *The Concept of Nature* (1948: 29, 43), for instance, the proponents of the theory of psychic additions treat "the greenness as a psychic addition into nature." As a consequence, they split up nature into real and additional properties. The forms of such a bifurcation are historically known as the theories of primary and secondary properties. All of these theories seem to have in common the belief that there must be a missing link between primary and secondary properties, as well as between mental and physical properties. That is why many philosophers make qualia a sort of disastrously homeless property in their naturalistic explanation of mind.

Like Darwin, who denied any missing link in the evolutionary process, many believe that the notion of a missing link is to be taken much more as a "metaphor" (Donaldson 2015) than as a claim of an evolutionary explanation of behavior and macroscopic order:

> The reluctance of many social scientists to appreciate or take advantage of the richness of the evolutionary approach is a direct consequence of a widespread tendency to overlook a crucial link in the causal chain from evolution to behavior: the level of innate psychological mechanisms, described as information processing systems. (Cosmides and Tooby 1987: 277)

> The fact that there are spontaneous inorganic processes that generate macroscopic order is seen by many as a missing link between living and non-living processes. (Deacon 2012: 264)

In advocating the notion of the threshold, I think one should not understand the meaning of the missing link as the inflection point between non-living and living or between the non-semiotic and the semiotic. In my opinion, the missing link can be seen much more as a metaphor than an ontological claim in favor of a causal chain in nature. Instead of a missing link, it would be more productive to take the notion of a threshold as an epistemological tool for understanding the transition and *continuity* from non-living to living, as well as from meaningless to meaningful in the world. As introduced by Pattee and Rączaszek-Leonardi (2012), interestingly, the notion of an *epistemic threshold* indicates the boundary zone where matter has much more than only physical properties and includes something else such as meaning and life.

In semiotics and biosemiotics, as noted previously, many authors seem to agree that there must be a threshold zone in which meaning and life overlap (*pace* Sebeok). To make epistemologically explicit the notion of a threshold zone, I introduce Whitehead's differentiation of entities as it is presented by him in *Process and Reality*:

> In the actual world, we discern four grades of actual occasions, grades which are not to be sharply distinguished from each other. First, and lowest, there are the actual occasions in so-called "empty space"; secondly, there are the actual occasions which are moments in the life-histories of enduring non-living objects, such as electrons or other primitive organisms; thirdly, there are the actual occasions which are moments in the life-histories· of enduring living objects; fourthly, there are the actual occasions which are moments in the life-histories of enduring objects with conscious knowledge. (Whitehead [1929] 1978: 177)

It is clear that Whitehead makes of life and non-life a difference of degree rather than essence.[6] What results from such a differentiation is "the blurring of the

[6]Concerning Whitehead's ontology, additionally, as organisms are events temporally and spatially differentiated, I agree with Nicholson and Dupré (2018: 1) that "the living world is a hierarchy of processes, stabilized and actively maintained at different timescales ... molecules, cells, organs, organisms, populations, etc. ... Although the members of this hierarchy are usually thought of as things, we contend that they are more appropriately understood as processes." There is, however, a critical point on which I disagree with Nicholson and Dupré. I disagree not because they are introducing a non-Whiteheadian approach to the process thought in the philosophy of biology. As I see it, they seem to embrace a physicalist interpretation of the process thought. As a consequence of such an interpretation, they assume that organisms are merely happenings—e.g., something that

difference between inanimate and animate nature" (Jonas [1966] 2001: 96). For Whitehead, accordingly, continuity between life and non-life goes down to the elementary physical entities. Inspired by Whitehead, interestingly, Hans Jonas ([1966] 2001: 1) asserts that "the organic even in its lowest forms prefigures mind." However, unlike Whitehead, Jonas advocates a strict separation between living and non-living entities on the basis of the assumption that the former is a form organically emancipated from matter. In line with my hypothesis in this chapter, presumably, Jonas ([1966] 2001: 96) accepts the idea of a threshold to differentiate non-living and living entities.

Two conceptual clarifications are needed here to make Whitehead's grades of entities more precise. First, it must be noted that an "enduring object" is no more than a succession of entities (Emmet 1932: 173). However, as every actual entity emerges from a background, it must be more or less enduring in different periods and according to a historical route. Depending on the route that an entity takes, it takes the form of a living or biological organism.[7] As a lesson from Whitehead's grades of entities, it is evident that the differentiation of routes sorts the enduring objects (and consequently their respective life-histories) and draws a line between mechanical and organic processes. Second, unlike Descartes's distinction of the two species of substance (bodies and minds), Whitehead ([1929] 1978: 239, 244, 277) points out that an actual entity is always dipolar: while one pole is physical, the other one is mental, and they cannot be separated. The controversial consequence of such a conception may be that it opens the door to pan-psychism: the idea that mentality can be found everywhere, even in elementary atomic particles. Although I acknowledge that the activity of meaning and mind are not *conditio humana*, the idea that elementary particles are alive and capable of meaning sounds extremely unreasonable. In parallel with Whitehead's understanding of grades of entities, for me the activity of meaning (and mind) depends on the organism's bodily plan.[8]

---

happens. In this sense, Nicholson and Dupré leave untouched important philosophical topics such as agency, intentionality, consciousness, or qualia.

[7] "In the case of an animal, the mental states enter into the plan of the total organism and thus modify the plans of the successive subordinate organisms until the ultimate smallest organisms, such as electrons, are reached. Thus, an electron within a living body is different from an electron outside it, by reason of the plan of the body. The electron blindly rims either within or without the body; but it runs within the body in accordance with its character within the body; that is to say, in accordance with the general plan of the body, and this plan includes the mental state. But the principle of modification is perfectly general throughout nature, and represents no property peculiar to living bodies" (Whitehead 1948: 80).

[8] In line with Gregory Bateson, for instance, to conceive the existence of mind (and arguably the activity of meaning) in some entity, a minimum of organization and complexity is required. It is not the case of atomic particles as such: "there is a lower level of division such that the resulting parts, when considered separately, lack the complexity necessary to achieve the criteria of mind. In a word, I do not believe that single subatomic particles are 'minds' in my sense because I do believe that the mental process is always a sequence of interactions between parts. The explanation of mental phenomena must always reside in the organization and interaction of multiple parts" (Bateson 1979: 92).

Considering Whitehead's differentiation of entities, however, I think one can map a threshold zone from which there are no degrees of meaning and life below the life-histories of enduring living objects. Also, as Whitehead differentiates degrees of entities, every actual entity is a "drop of experience." Borrowing the expression "drop of experience" from William James, Whitehead ([1929] 1978: 18, 68) affirms that the actual entities are ultimately the final facts of which the world is made up. As drops of experience, and insofar as actual entities are complex and interdependent, the world results from emergent and relational processes. Having the idea of drops of experience in mind and stressing the relational nature and eventfulness of reality, in consequence, Whitehead discredits a substantialist ontology. Discrediting such an ontology, accordingly, Whitehead favors the image of life as a relational web of processes.

In following Whitehead's differentiation of degrees of entities and considering that they are drops of experience, I am convinced that one can trace a threshold zone of the transition between life-histories of enduring non-living and living objects. Using Whitehead's differentiation, accordingly, the threshold zone exemplifies the transition between degrees of entities and experience, indicating the point from which there is no activity of meaning (such as quarks, protons, or electrons). For me, it is entirely nonsense to claim that "there is something it is like to be a quark or a photon or a member of some other fundamental physical type" (Chalmers 2017: 19). Although I agree that semiosis supposes interpretation, in many cases it is also sign inference. It is hard to think that atomic particles can perform interpretation and sign inference. In my view, what determines the activity of meaning is the entity's bodily organization. Below the life-histories of enduring living objects and insofar as the entity's bodily organization is rudimentary at that level, there is *no* (activity of) *meaning* and consequently there is *no life*. The most important aspect for differentiating the activity of meaning is that there is no semiosis without sign inference or interpretation. Moreover, once the meaning is taken to be an activity of making sense, it does not seem fair to claim that atomic particles make sense of anything. As I see it, this is the reason why the notion of a threshold should be taken seriously in order to map at which level of natural processes the activity of meaning (or semiosis) is engendered. As Sebeok (2001: xiv) pointed out, incidentally, the idea is that semiosis is life.

In the broadest sense, as noted by William James in 1909 (McDermott 1977: 280), the experience of activity is synonymous with life. That is to say that there is no experience at the level of the inactive world, such as electrons and primitive organisms. Using Whitehead's terminology, the experience of activity is supposed to be found in moments of the life-histories of enduring living entities.[9] *If there is the experience of the activity, there is life—no experience, no life!* Once one

[9]In his Harvard Lectures (1924–1925), paralleling James's conception of the experience of the activity, Whitehead underlines that "the essence of the activity . . . individualizes itself in a plurality of real things" (Whitehead *apud* Ford (1984: 286)). Furthermore, insofar as James understands that experience means the immediate flux of life, it individualizes itself in a plurality of activities.

acknowledges that the life-histories of enduring living entities are engendered in nature in continuity with the life-histories of enduring non-living entities, I do think we can epistemologically accept the existence of a threshold zone as a type of inflection in nature where meaning and life overlap. The idea is that *if there is an activity of meaning, there is life—no meaning, no life!* (and vice versa). In tracing a parallel with James's notion of experience of an activity, in particular, I have it in mind to stress that the activity of meaning indicates something else more than material conditions and merely happenings in nature—that is, meaning stands for the very sense of experiencing life as a transitional process, which results in a threshold zone for creation and novelty to take place in the world.

Illustration 1 Transitional process of experience (TP) and the threshold zone (TZ).[10] According to William James's empiricism, as a process of differentiation between physical and mental contexts in the flux of experience, one can speak of thresholds as transitions.[11] Incidentally, in line with James's conception of

[10]It is interesting to note here William James's use of "threshold." Referring to Gustav Fechner, James sees the "threshold" as the discrete character of sensible experience: "Fechner's term of the threshold, which has played such a part in the psychology of perception, is only one way of naming the quantitative discreteness in the change of all our sensible experiences. They come to us in drops" (James 1909: 231–232). Incidentally, James's considerations of Fechnerian philosophy occur primarily in the *Principles of Psychology* of 1890, in his *Lecture on Human Immortality* of 1898, and, finally, in an article in the *Hibbert Journal*, which became the fourth chapter of his *Pluralistic Universe* of 1909 (Marshall 1974: 304). In *Human Immortality*, particularly, James puts forward the conception of the "threshold" from Fechner's *Psychophysik*. As held by Fechner, James (2010: 165) notes that the condition of consciousness corresponds to a kind of psychophysical movement in the sense of reaching a certain degree of activity, which is called the "threshold." In Fechner's own words, "More general and higher mental phenomena, such as the total consciousness of the people depending on sleeping and waking, the consciousness of individual thoughts, the attention in a given direction have a point of extinction and origination, we will use the term and expression the threshold … the conditional, the elevation of consciousness to the threshold or which they correspond, but it can raise the question whether we are not in favor of adopting a threshold value of the underlying psychophysical movement" (Fechner 1966: 175–176). The idea that I develop in this chapter is that the conditions for mentality depend on a certain threshold of experience in terms of a psychophysical activity. Indeed, as noted by M. E. Marshal (1974: 309), one aspect of Fechner's philosophy becomes important to James's *The Pluralistic Universe*: the constitution of reality is identical throughout. Following James's empiricism and Fechner's metaphysics, it is fair to claim that mentality is distributed in a series of levels throughout the experience.

[11]"If one and the same experience can figure twice, once in a mental and once in a physical context … one does not see why it might not figure thrice, or four times, or any number of times, by running into as many different mental contexts, just as the same point, lying at their intersection, can be continued into many different lines" (James 1977: 210).

experience as a transitional process of differentiation, "there is a primary semiotic threshold opposing physics (that which is not alive) to biology (living things, including internal biological processes, known since Sebeok as endosemiotics), and there is a secondary semiotic threshold, which opposes the latter to that which is language-like (discussed in sociology and semiotics of culture)." (Sonesson 2006: 203)

In Illustration 1, James's conception of experience is depicted as continuity of transitional processes whose image is a dynamic mosaic; indeed, James's radical empiricism is a mosaic philosophy. In perceiving a mosaic when considering the context, admittedly, one experiences a cyclic, non-linear, and continuous process of uniting discrete elements, resulting in a dynamic gestalt.[12] Using the mosaic metaphor, James seeks to show that experience consists of a dynamic field of non-linear relations and is centrifugally structured.

In the sequel to his theory of the transitive parts of the stream of consciousness (introduced in *The Principles of Psychology*), in his essays on radical empiricism, James explores the idea of continuity as designating an unbroken chain of processes in the experience. For James, in its immediate structure, experience consists in a space-time continuity of transitional processes.[13] Accordingly, insofar as experience is the immediate flux of life, life acquires an empirical sense of consisting in the parts as much as in the transition:

> I called [philosophy of pure experience] a mosaic philosophy. In actual mosaics the pieces are held together by their bedding, for which bedding the Substances, transcendental Egos, or Absolutes of other philosophies may be taken to stand. In radical empiricism there is no bedding; it is as if the pieces clung together by their edges, the transitions experienced between them forming their cement. Of course, such a metaphor is misleading, for in actual experience the more substantive and the more transitive parts run into each other continuously, there is in general no separateness needing to be overcome by an external cement . . . the metaphor serves to symbolize the fact that Experience itself, taken at large, can grow by its edges. That one moment of it proliferates into the next by transitions which, whether conjunctive or disjunctive, continue the experiential tissue, cannot, I contend, be denied. Life is in the transitions as much as in the terms connected. (James [1909] 1996: 33)

Since one cannot find gaps in experience because it is an empirical continuity of relations, the idea of a missing link arguably makes no sense. Through countless transitions in experience, accordingly, life grows *here* and *everywhere* as an empirical process.

---

[12]As was very well noted by Harry Heft, influenced by James's understanding of psychological experience as an extended flow, for Gibson in his ecological approach, perceiving is a mode of activity rather than the reception of sensory stimulation. In Gibson's own words, "The act of picking up information, moreover, is a continuous act, an activity that is ceaseless and unbroken. The sea of energy in which we live flows and changes without sharp breaks. . . . Hence, perceiving is a stream, and William James's description of the stream of consciousness applies to it. Discrete percepts, like discrete ideas, are as mythical as the Jack of Spades" (Gibson 1979: 204).

[13]Interestingly, in line with Whitehead, James (1983: 227) compares the river of life or river of elementary feelings to the Heraclitean river.

James's understanding of experience as a continuous process points to the fact that it has no bottom and ultimate layer. Metaphysically speaking, such an account of experience commits James's radical empiricism to a form of anti-foundationalism: nothing that is not derived from experience acquires a sense of reality or can be known; by knowing, in particular, James means related to portions of experience. Interestingly enough, in his notes for a psychological seminary of 1895–1896, James describes the immediate data of experience as a "field":

> "fields" that "develop", under the categories of continuity with each other [categories such as]: sameness and otherness [of] things [or of] thought-streams, fulfilment of one field's meaning in another field's content, "postulation" of one field by another, cognition of one field by another, etc. (James *apud* Perry (1976: 365))

As James was well educated in the sciences, there is little doubt that he was aware of the theory of (electromagnetic) fields. In this case, as noted by Heft (2017: 118), "the electrified wire and the needle [are] not bounded, separate entities, but instead they reside in a field of continuous relations that they themselves generate." For James, in parallel with the field theory, both objects and their relations are equally experienced, rather than objects only. Accordingly, there is no relation that is not experienced, and the relations connecting experiences are themselves experienced relations.

As was also observed by Heft (2017: 128), in clear contrast to "the Newtonian–Lockean view that natural phenomena, including mind, are fundamentally composed of discrete units (e.g., ideas)," James claims that experience is essentially a continuous process described as a dynamic field of relations. In occasional moments in the flux of experience as such, and because of the process of differentiation, a field of relational transitions emerges and frames a threshold zone giving rise to meaning, life, and mind.

As depicted in Illustration 1, in occasional moments of experience, the "threshold zone" consists in a transitional process of connecting and differentiating parts of the experience and, for instance, bringing forth life/lifelessness and mind/matter distinctions. Since the parts of experience can be differentiated empirically, the threshold zone stands for a transitional process from which meaning emerges and takes place in the world. The idea here is to put forward an empiricist understanding of the threshold zone as a sort of transitional process in the flux of experience. In the threshold zone, once it connects and differentiates particular parts of the experience, meaning and life overlap: that is to say, *where there is experience, there is life*.

It must be noted that I am using activity of meaning in the sense of semiosis. So, in speaking of semiosis, I speak of the activity of meaning and life.[14] Once more, I stress that there is no meaning and life below the life-histories of enduring living objects. Using Whitehead's differentiation of entities in terms of degree, it is fair enough to say that one can epistemologically assume the transition of life-histories of enduring non-living and living entities as indicating the threshold zone in nature

[14] "We follow Sebeok (1979) in defining the emergence of life as the threshold for the semiosphere" (Hoffmeyer 2008: 5).

from which meaning and life overlap.[15] Even though we lack empirical evidence to affirm the existence of the threshold zone, I insist that it is an important epistemological tool for our understanding of meaning and life in nature.

Although for Whitehead ([1929] 1978, p. 161), every actual entity has the capacity for knowledge, I believe that the notion of the threshold zone is epistemologically needed.[16] For instance, in the third grade of the entity, according to Whitehead's terminology, one can speak of knowledge and meaning without assuming the existence of consciousness. In stressing that the activity of meaning is not *conditio humana*, I believe that there must be a threshold zone of differentiation and transition between the inorganic and the organic, as well as between the meaningless (or non-semiotic) and the meaningful (or semiotic) so that meaning and life can make sense in nature.[17] If one differentiates grades of entities and experiences, it is plausible to assume that what a living entity does is meaningful since the activity of meaning cannot be reduced to mere physical occurrences. It depends on the entity's functional organization. In this sense, I do agree with Sebeok when he claims that "semiosis is what distinguishes all that is animate from lifeless" (Seboek *apud* Cobley et al. (2011: 2)). From a minimal level of functional organization, therefore, one can speak of the activity of meaning, mentality, and life. Perhaps the most important aspect of differentiating grades of entities and experience is the fact that the threshold zone indicates how one can epistemologically understand the transition and continuity from mere happenings to animacy and activity of meaning in nature. As I see it, the idea that meaning overlaps with the activity of living entities stresses epistemologically the need for a threshold zone of differentiation and transition

[15] In *Teoria Semântica da Evolução* (*Semantic Theory of Evolution*), Marcello Barbieri adds to the Darwinian worldview the dimension of meaning in nature. Much more than just variation, adaptation, and selection processes, nature is also rich in plurality and meaning. As Barbieri notes, there is indeed meaning in nature (*pace* von Uexküll (1982)). To the extent that each organism or life-form incorporates processes of meaning, they engender a language in nature: "Life is the language that nature has learned to speak on the surface of our planet" (Barbieri 1985: 169). In the sense that I understand Merleau-Ponty's notion of "prose of the world," in particular, it means the language that nature has learned to speak.

[16] According to Whitehead, all entities are capable of knowledge. If that is so, we need to count on a way of differentiating the entities to understand the levels of knowledge. I think that what can precisely distinguish the degrees of entities is their respective functional organizations. For instance, in comparing an electron with a cell, it is evident that the former has an inferior functional organization. As a result, it is hard to assert that the electron is capable of knowledge if this is taken to be a way of meaning the world. In contemporary contexts of cognitive sciences and philosophy of mind, it is more proper to speak of cognition instead of knowledge. In comparison with Whitehead's lower grades of actual entities, for example, they instance the capacity of minimal cognition and mind: "By accepting the existence of mind in animals, we commit ourselves to answer many difficult questions. For example, where is the lower evolutionary threshold for mind? Does mind require brain or at least some kind of nervous system?" (Sharov 2013: 243).

[17] "I don't like the idea that consciousness should be present in atoms . . . I like to see semiosis as an emergent phenomenon, where the increase in semiotic freedom is indeed the one most conspicuous fact we have about organic evolution" (Hoffmeyer *apud* Pickering (2012: 197)).

between lifelessness and life in the world; in a broad sense and more than *conditio humana*, meaning acquires the sense of experiencing life.

David Bohm has a remarkable understanding of continuity and meaning in nature. On the one hand, he advocates the principle of continuity. For him, it is unhelpful to postulate arbitrary discontinuities in trying to explain reality. On the other hand, he understands meaning as a basic property:

> I want to introduce a new notion of meaning which I call soma-significance. . . . In this approach meaning is clearly being given a key role in the whole of existence. . . . The notion of soma-significance implies that soma (or the physical) and its significance (which is mental) are not in any sense separately existent, but rather that they are two aspects of one over-all reality. (Bohm 1985: 72–73)

> Meaning and matter may not have the same sort of consciousness that we have, but there is still a mental pole at every level of matter, and there is some kind of soma-significance. (Bohm 1985: 89–90)

> Modern physics has already shown that matter and energy are two aspects of one reality. . . . The energy of mind and of the material substance of the brain are also imbued with a kind of significance which gives form to their over-all activity. So quite generally, energy enfolds matter and meaning, while matter enfolds energy and meaning. (Bohm 1985: 90)

Bohm's understanding that physical and mental are not separately existent seems to be in convergence with James's radical empiricism, and I totally agree. In fact, for James, experience presents in itself physical and mental poles. But, regarding Bohm's understanding that meaning is a fundamental concept applied to the human-made world, as well as to the physical reality, I am not in line with him. Even though I assume that meaning is not *conditio humana*, it does not follow that I take it to be applied to the physical reality to a "greater or lesser extent" or that semiosis occurs at all levels of the natural and human-made world (Pickering 2012: 198). Additionally, I am not convinced that "there is a mental pole at every level of matter" as Bohm believes to be the case. Insofar as he takes for granted such a belief, he seems to espouse a form of ultra-experimentalism on meaning. Although I recognize that James's empiricism is a form of pan-experimentalism, I think that there must be a threshold zone from which the very notions of experience and meaning are differentiated in making sense of the world. I do not believe that it makes sense to speak of experience and activity of meaning for quarks, protons, or electrons. In such cases, one would presumably commit a *category mistake* (*pace* Gilbert Ryle).

I am trying to maintain here that there must be a threshold zone from which one can epistemologically understand the transition from non-living to living processes, as well as from meaningless (or non-semiotic) to meaningful (or semiotic). Considering the notion of the threshold zone, it does not seem that there is an activity of meaning below living entities. In stating that, again, I do not have in mind that meaning is *conditio humana*. On this point, I do agree with Bohm:

> Meaning, though, has nevertheless been regarded as peculiar to our own minds and not as a proper part or aspect of the objective universe. However, if there is a generalized kind of meaning intrinsic to the universe, including our own bodies and minds, then the way may be opened to understanding the whole as self-referential through its meaning for itself—in other words, by whatever reality is. And the universe as we now conceive it may not be the whole thing. (Bohm 1985: 92)

Following my hypothesis (that *where there is meaning, there is life*), and following William James's conception of experience as the flux of life, I presumably accept Bohm's ultra-experimentalism on meaning. In contrast to Bohm, however, I understand that the activity of meaning depends on an entity's bodily organization. It is not clear to me whether Bohm assumes such a conceptual distinction in his understanding of meaning. Moreover, I suppose Bohm would regard the notion of a threshold zone as an arbitrary discontinuity and unhelpful in our attempts to explain reality.

In contrast to Bohm, who regards meaning as a fundamental property, in convergence with James's empiricism and in the form of a mosaic philosophy, I consider meaning in terms of constructivism. Just as the very essence of experience, for James, is that everything is structurally related and connected, I think of meaning as resulting from relational processes and forms of life. Unlike Bohm, I have in mind that meaning results much more from a dynamic web of relational processes than from a fundamental physical property.

Despite the absence of empirical evidence for the belief of continuity in nature, interestingly, Darwin assumed the hypothesis of continuity. As part of an epistemological belief in the explanation of the origin and development of species and mental capacities, Darwin ([1859] 1979: 445) endorsed the concept that natura non facit saltum. From Darwin's point of view, continuity corresponds to the uncountable gradation of previous and intermediary stages between species whose reality is unobservable in nature. In contrast to Leibniz's view, Darwin's hypothesis of continuity has no metaphysical or immaterial significance in the explanation of the natural world. In my view, which is consistent with Darwin's hypothesis of continuity, this is also the case with a threshold zone: despite the absence of empirical evidence to affirm the existence of a threshold zone in nature, methodologically, it stands for our understanding of transition and continuity between non-living and living, as well as the emergence of meaning activity in nature.

Even though we do not have empirical evidence, the notion of a threshold zone can be likened to the belief in continuity in nature. Although the conception of continuity corresponds to an unobservable reality in nature, it acquires a strategic relevance in explaining the development of species and mental capacities in the world. In my opinion, comparatively, the notion of a threshold can also be an important epistemological tool for our understanding of the transition and continuity from non-living to living entities, as well as from meaningless to meaningful in nature; that is to say, methodologically, the threshold acquires strategic relevance in our understanding of the world as a dynamic and systematic continuity.

3 Two Paradigms of Continuity

I will use the term "paradigm" in Thomas Kuhn's sense to mean a particular type of worldview. In such a sense, a new paradigm introduces a worldview in that it rivals a previous one. In the first place, as illustrated below, it is depicted as the paradigm of the chain of being, or *Scala Naturae*. An important historical aspect in the pre-Darwinian and pre-Lamarckian period, the chain of being stands for the idea of continuity between species that presents a solution to the problem of an interval between man and the rest of the natural world.

The chain of being meant an attempt to establish order among creatures in the world by a hierarchical scale of ascendancy, whose top would be reserved for man as the work of God's creation (Lewin 2005: 4).

Wikipedia: http://en.wikipedia.org/wiki/Great_chain_of_being

Illustration 2 Introduced by Charles Bonnet (1781) as "Contemplation de la Nature," the idea of *Scala Naturae* can be found among different authors (from Plato to naturalists and philosophers of the eighteenth century) and means three general characteristics of the universe: completeness, continuity, and gradation. The chain of being is the conception according to which life is organized on an ideal and linear progression from the simplest atom to the most complex and perfect being (or human being). Since the chain of being constitutes a continuous progression, it is not broken and includes no intervals. In general terms, the chain of being indicates a fixist, essentialist, and determinist worldview.

Replacing the worldview of *Scala Naturae*, a new paradigm emerges in giving an alternative interpretation of the idea of a *continuum* in nature. As noted by Terrence Deacon, quoting Gregory Bateson:

> The turn on the logic of the classic "Chain of Being":
>
> As Gregory Bateson has described: "Before Lamarck, the organic world, the living world, was believed to be hierarchic in structure, with Mind at the top. The chain, or ladder, went down through the angels, through men, through the apes, down to the infusoria or protozoa, and below that to the plants and stones. What Lamarck did was to turn that chain upside down. When he turned the ladder upside down, what had been the explanation, namely: The Mind at the top, now became that which had to be explained." (Deacon 2012: 119)

To explain continuity in nature, the nascent paradigm rivals the ideas of fixism, essentialism, and determinism as they can be found in the worldview of *Scala Naturae*. Besides regarding the problem of the interval between man and the rest of the natural world, for instance, the idea is to interpret continuity as resulting from dynamic processes, gradual changes, and individual variation. That is to say: as an alternative to the worldview of *Scala Naturae*, the idea of continuity indicates that life is dynamically activity and process. Thus, much more than being organized on an ideal and linear progression, life has to do with non-fixism, non-essentialism, and non-determinism.

Bearing in mind that continuity can be dynamically explained in nature, the new paradigm can be divided into two moments: the "tree of life" and the "web of life." Although the notion of the web contrasts with the Darwinian view of the tree of life, they have in common the idea that life is a continuous process dynamically structured without a center or periphery.

(a) *The tree of life*

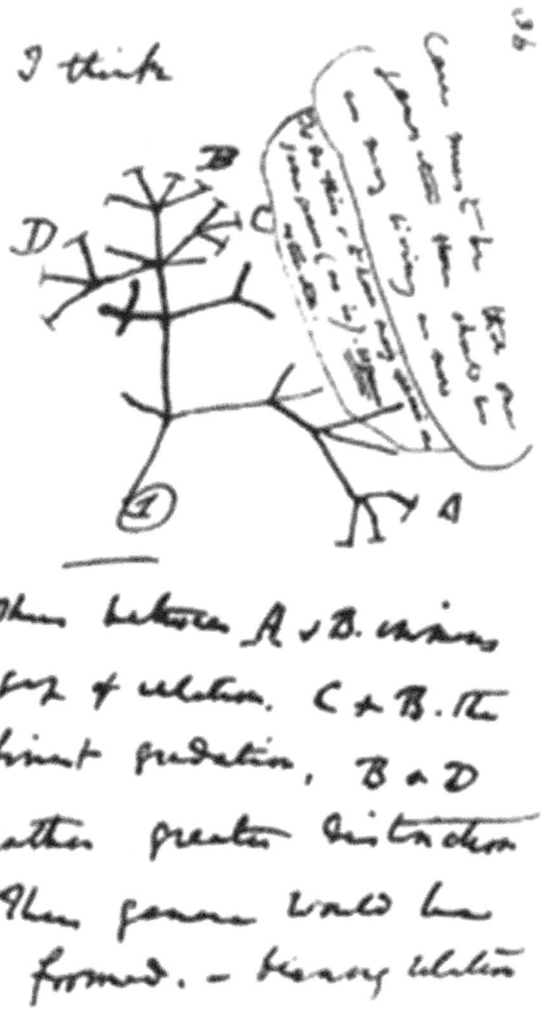

Illustration 3 Darwin's first sketch of a phylogenetic tree, from *Notebook B [Transmutation of Species (1837–1838)]* (Darwin 2008: 64).

In contrast to the essentialism of *Scala Naturae*, Darwin takes up the principle of continuity and introduces a dynamic and causal explanation of the supposed intervals between species. As such, Darwin's explanatory structure has three basic elements: (1) the individual is the matter of biological variation, (2) natural selection engenders the mechanism of efficient action on the individual, and (3) it is believed that gradual changes occur between species as a result of individual variations. Particularly on this last point, Darwin traces his maxim of gradualism:

> On the theory of natural selection, we can clearly understand the full meaning of that old canon in natural history, "Natura non facit saltum". This canon, if we look only to the present inhabitants of the world, is not strictly correct, but if we include all those of past times, it must by my theory be strictly true. (Darwin 1859: 206)

Under the maxim that natura non facit saltum, one may identify the philosophical matrix of Darwin's continuity hypothesis by reference to Leibniz's principle of continuity. In other words, the development of different species and mental characteristics corresponds to a continuous and gradual process in nature:

> Nothing takes place suddenly; one of my great and best confirmed maxims says that nature never makes leaps. I have called this maxim the Law of Continuity. . . . This law does a lot of work in natural science. It implies that any change from small to large or vice versa passes through something in between. (Leibniz [1765] 1996: 57)

> As for the gradual connection of species: we have already had something to say about that in a previous discussion, when I commented that philosophers have in the past reasoned about a vacuum among forms or among species. In nature everything happens by degrees, nothing by jumps; and this rule about change is one part of my law of continuity. (Leibniz [1765] 1996: 473)

It seems evident that Darwin and Leibniz share the belief that everything proceeds by degrees in nature and nothing by jumps. Moreover, given such an understanding of gradualism in nature and differently from the essentialism of *Scala Naturae*, for Darwin, continuity means a dynamic process in nature much more than a fixed property.

(b) *The web of life*

As a type of deflection in the new paradigm of continuity, von Uexküll (1926: 258) employs the expression "web of life" in contrast to the Darwinian view of the "tree of life." In using the image of the web in a chapter titled "The Species" in *Theoretical Biology*, von Uexküll intends to replace the concept of life expressed by the Darwinian image of the evolutionary ladder. For him, evolution results from modifications in a web of complex relations in nature. The idea is that the natural world is best depicted as a web of linked entities (Ricou and Pollock 2012):

> What happens with evolutionism when moving from Modern to Post-Modern, is that we leave behind the whole concept of life's progress as expressed in the tree of life and instead understand the evolution as modifications in the web of life. (Kull 2004: 101)

> This is a paradigm that can be best characterized by the metaphor of web, as used by Thomas A. Sebeok in the expression of "the semiotic web", and as introduced by Jakob von Uexküll. (Kull 2004: 100)

> The complex web of causal dependencies between the various levels means that we cannot fully specify the nature of an entity merely by listing the properties of its constituents and their spatial relations. It also means that we cannot pick out any level in the hierarchy as ontologically or causally primary. Whereas a substance ontology that presupposes a structural hierarchy of things only allows bottom-up causal influences, a process ontology has no trouble in recognizing that causal influences can flow in different directions. (Nicholson and Dupré 2018: 21–22)

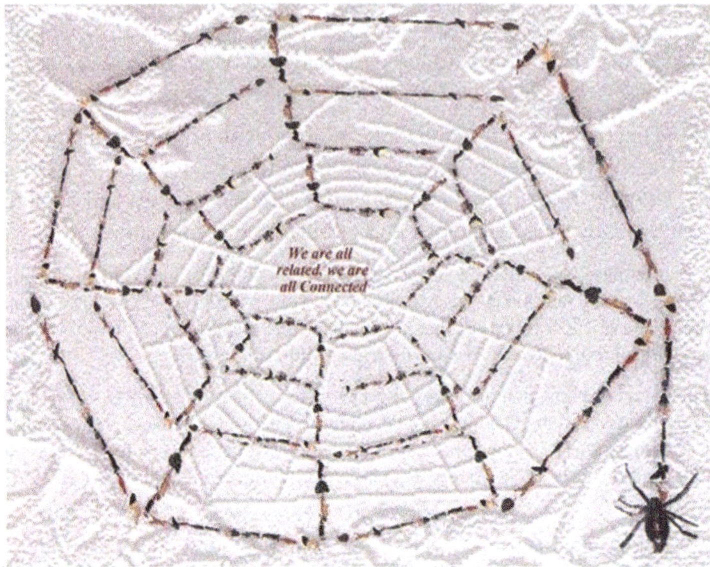

Urbaniak, 2000 (https://www. ncy.netnaturalfreque/Ray/webofLife.htm)

Illustration 4 The web of life—a complex web of causal dependencies between the various levels in the world: "Displaced was the long-held schema of nature as a static chain of being (*Scala Naturae*) comprised of a succession of material entities and culminating in spiritual entities [… in] its place there is a dynamic realm of thoroughly natural, co-evolved entities functioning in a web of environmental interdependencies" (Heft 2001: 13). Holistically speaking, as an indeterminate process, the idea is that the web of life is without a center or periphery.

My use of the image of the web of life is my tribute to the brilliant description of the spider's web by Jakob von Uexküll. As an illustration of the theory of meaning, which is the second part of his Umwelt theory, von Uexküll ([1934] 1982: 42) describes the spider's weaving of the web. For the spider, since the web *means* the fly, it functions as a way of significantly relating to the world. According to von Uexküll's theory of meaning, life takes the form of a dynamic semiotic web in the sense that *where there is meaning, there is life—no meaning, no life!* That is to say: meaning or semiosis is life (*pace* Sebeok (2001)).

As a metaphor, the notion of a web depicts an image of life in which organisms are all dynamically related and connected in the world.[18] In parallel with James's mosaic philosophy, for instance, the idea is that one understands life as resulting

[18] *Process-Relational Philosophy—An Introduction to Alfred North Whitehead*, C. Robert Mesle (2008: 9, 59) makes use of the "web" to assert that the universe is a web of relational processes. As such, the universe does not correspond to a hierarchy of things based on a substance ontology and

from a dynamic web of plural facts in the form of constructivism. Whitehead highlighted that since every actual entity is a "drop of experience," they all construct a web of experiences and reach the sense of activity together. For James (1909: 373), such a sense of activity is nothing but synonymous with the very sense of life. Insofar as the web of life takes the form of a dynamic mosaic, one must point out two points here: first, no level is taken to be ontologically or causally primary; second, the causal influences flow in different directions and not necessarily in a bottom-up sense (*pace* Nicholson and Dupré (2018)). As Illustration 4 suggests, life consists of a hierarchy of continuous processes rather than ontologically separate things.

4 Threshold and Ententionality

In retaking the threshold's notion, I will explore the ideas of transition and continuity in nature in that they give rise to life and meaning. Even though one takes for granted that life consists of a hierarchy of continuous processes, the issue seems to be from which level in nature one can differentiate life and lifelessness. As I am trying to argue, the notion of a threshold assumes a distinctive epistemological function through which one can differentiate the processes of transition and continuity from lifelessness to life in nature. Once again, I stress that the notion of a threshold does not contradict the assumption of continuity in nature. Quite the opposite is true: in assuming that there is a threshold zone in nature, one can make intelligible the processes of transition and continuity in nature from which life and meaning emerge.

As was interestingly noted by Lewis Ford (1984: 3), unlike the traditional view of pan-psychism, Whitehead's assumption is that there are no degrees of mentality below the threshold on which organisms can sustain intellectual and cognitive activity. However, for the later Whitehead, as "all actualities have some degree of mentality," mentality here means "novelty of response" (Ford 1984: 42). Nevertheless, as many organisms have a rudimentary functional organization, they lack the activity of meaning and any degree of mentality. In this sense, once again, I believe that one can epistemologically differentiate a threshold zone from which meaning is supposedly found in nature, indicating cognitive activity as a distinctive trace of living organisms. The idea of the threshold indicates a zone of deflection where much more than material conditions matter by giving rise to mentality and activity of meaning as the novelty of response in nature.

Then, as I see it, one can only speak of the activity of meaning from the life-histories of enduring living organisms. Accordingly, instead of being a fundamental property, meaning seems to indicate much more an emergent process in nature that

"the idea of ... continuously persisting entities which can be integrated into various material structures" (Koutroufinis 2014: 17).

separates living processes from merely physical–chemical processes (Deacon 2012: 144).[19] In my view, such a separation indicates the transitional character of reality and does not suggest absolutely the ideas of bifurcation or discontinuity in nature.[20] As I am developing here, the activity of meaning does not correspond to a phenomenologically discontinuous aspect of experience and must be taken to be as real as anything in nature.

To avoid undue assimilation with a substantialist ontology and the idea that reality consists of successive levels of organization as a hierarchy of things (elementary particles, atoms, molecules, cells, organisms, etc.), I understand the organism as a process and transition of processes in asymmetric forms in nature (*pace* Whitehead ([1929] 1978)). Each organism has certain temporal stability at different scales (atoms, molecules, organs, organisms, populations). Accordingly, in metaphorical terms, I agree with Pattee (Pattee and Rączaszek-Leonardi 2012: 228) that "life is matter with meaning." In this sense, it is plausible to claim that there must be a threshold as a zone of transition and continuity from which meaning, life, and mind take place in nature:

> Though subjective awareness is different from the simple functional responsiveness of organisms in general, both life and mind have crossed a threshold to a realm where more than just what is materially present matters. (Deacon 2012: 26)

---

[19] I am assuming here a weak version of emergence. The idea is that "although in emergent transitions there may be a superficially radical reorganization, the properties of the higher and lower levels form a continuity, with no new laws of causality emerging." In the sense of strong emergentism, contrarily, "emergent transitions involve a fundamental discontinuity of physical law" (Deacon 2012: 551–552).

[20] As noted by Whitehead, "What I am essentially protesting against is the bifurcation of nature into two systems of reality, which, in so far as they are real, are real in different senses. One reality would be the entities such as electrons, which are the study of speculative physics. This would be the reality which is there for knowledge; although on this theory it is never known. For what is known is the other sort of reality, which is the byplay of the mind" (Whitehead 1919: 30). In Robert K. Logan's *The Extended Mind—The Emergence of Language, the Human Mind and Culture* (2007), incidentally, one finds a defense of bifurcation and discontinuity in the origins of speech and the human mind. The argument is based on the premise of a transition from percept-based thinking to concept-based thinking (Logan 2007: 5). The whole idea is that such a transition represented the emergence of language as a major discontinuity in human thought. For Logan (2007: 18–19), indeed, the discontinuity results partially from the discontinuity between linear and non-linear dynamics in that non-linear systems exhibit emergent behavior. By assuming Terence Deacon's claim that human speech is "an evolutionary anomaly and not merely an evolutionary extreme," Logan commits himself with a form of strong emergentism. In his view, accordingly, to the extent that non-linear dynamics exhibit emergent behavior, the emergence of language represents a "discontinuity, a quantum leap, in the behavior of animal life" (Logan 2007: 18). In Ian Tattersall's *L'Emergence de l'Homme* [*The Emergence of Man*] (1999), one also finds a similar defense of a strong emergentism in human evolution: "*Homo sapiens* who eliminated the Neanderthals . . . are in one way or another linked to language. . . . From the point of view of our species, the crucial cognitive leap that was that of complex symbolic reasoning was . . . accomplished very late in human evolution" [my translation] (Tattersall 1999: 307, 309).

In the wake of Deacon, therefore, it is fair to affirm that the existence of meaning, life, and mind depends on the organism's functional organization. Once one speaks of the activity of meaning, one can speak of types of mind in nature. Also, in speaking of the activity of meaning and types of mind, the door is opened to state the existence of life. That is why I do not believe that one can speak of meaning, life, and mind below the life-histories of enduring living objects.[21]

In assuming that the activity of meaning depends on the organism's functional organization, for instance, I commit myself to a strong assumption of continuity of life and mind—"the view that the organizational structures and principles distinctive of mind are simply enriched versions of the structures and principles grounding life itself" (Ward et al. 2017: 370). In my opinion, meaning, life, and mind are spatially and temporally overlapping processes. The idea here is that the world consists of a hierarchy of levels and processes all related and connected—just like the web image, which depicts the world as a mosaic of plural facts (*pace* James). Moreover, it is not the case to look for the hierarchy level as ontologically or causally primary. As I will present in the last part of this chapter, from Peirce's conception of cognition, meaning arises by an indeterminate process of beginning (as well as life), and nothing has ontological priority in the web of life. Very briefly, the idea is that the world consists of a hierarchy of levels and processes, all of them related and connected, giving rise to the emergence of meaning, life, and mind.

As it seems plausible to speak of minimal cognition in primitive forms of life such as bacteria, the notions of meaning and value can perfectly describe when something becomes good in the organism's environment. By virtue of the dynamics of an organism's embodiment of the environment, some cognitive functions have satisfaction conditions in that they engender forms of meaning and value from a

---

[21] However, in many cases in nature, one can speak of minimal cognition as indicating types of mind. Paraphrasing Antonio Damasio (1999), one can say that the mind is based on an organism's capacity of feeling; e.g., the organism feels itself as well as it feels the environment. Moreover, considering Whitehead's conception of feeling as positive prehension, it is clearly not an anthropomorphic view of feeling. The idea is that feeling has to do with grades of feeling the world; hence, it is not *conditio humana*. In *Self Comes to Mind* (2010), Antonio Damasio devotes an analysis to qualia in which he seeks to understand how organisms' ability to sense has an origin at different scales in nature: "There are aspects of cell life that suggest the presence of forerunners of a 'feeling' function. Unicellular organisms are 'sensitive' to threatening intrusions. Poke an amoeba, and it will shrink away from the poke. Poke a paramecium, and it will swim away from the poke. We can observe such behaviors and are comfortable to describe them as 'attitudes,' knowing full well that the cells do not know what they are doing in the sense that we know what we do when we evade a threat. But what about the other side of this behavior, namely, the cell's internal state? The cell does not have a brain, let alone a mind to 'feel' the pokes, and yet it responds because something changed in its interior. Transpose the situation to neurons, and therein could reside the physical state whose modulation and amplification, via larger and larger circuits of cells, could yield a protofeeling, the honorable counterpart of the protocognition that arises at the same level" (Damasio 2010: 197). In convergence with Damasio, and in responding to the Dalai Lama's question as to whether a one-celled creature such as an amoeba is a sentient being, Varela says, "From this point of view, there is no question. There is no way for me to draw a line and distinguish my cognition from the cognition of frogs, hydras, amoebas, or bacteria" (Hayward and Varela 2001: 66).

subjective perspective. For instance, in the chemical composition of the bacterium's environment, some have positive meaning and value, and some do not. Such a process of meaning and value corresponds to what Terrance Deacon calls "ententional phenomena." Although "ententionality" has a certain degree of kinship with the traditional conception of intentionality, it has nothing to do with a human-like mental property:

> I propose that we use the term ententional as a generic adjective to describe all phenomena that are intrinsically incomplete in the sense of being in relationship to, constituted by, or organized to achieve something non-intrinsic. By combining the prefix en- (for "in" or "within") with the adjectival form meaning something like "inclined toward" . . . ententional phenomena include functions that have satisfaction conditions, adaptations that have environmental correlates, thoughts that have contents, purposes that have goals, subjective experiences that have a self/other perspective, and values that have a self that benefits or is harmed. (Deacon 2012: 30)

> Ententional: A generic adjective coined in this book for describing all phenomena that are intrinsically incomplete in the sense of being in relationship to, constituted by, or organized to achieve something non-intrinsic. This includes function, information, meaning, reference, representation, agency, purpose, sentience, and value. (Deacon 2012: 550)

To the extent that ententional phenomena achieve something non-intrinsic, they are essentially relational. As understood here, meaning, in particular, is a case of an ententional phenomenon in the sense that it is never complete in itself, depending on an actual process of achievement. Besides, according to Deacon, ententional phenomena are asymmetrically and hierarchically interrelated. For instance, representations depend on the information, and information depends on the functional organization. That is why one can speak of ententional phenomena as transitional processes and essentially incomplete in nature. Once ententional phenomena apparently depend on the organism's functional organization, they indicate a threshold zone in the natural world from which one can differentiate meaning and value in multiple degrees. As I see it, this is why one must place a threshold zone in order to differentiate ententional phenomena from the rest of natural processes (Deacon 2012: 40).

In comparison with Deacon's ententional phenomena, Wilfrid Sellars (1981), one of the most important analytical philosophers, introduces a process-based ontology and distinguishes between different normativity degrees. For Sellars, even in the lowliest primitive organisms such as bacteria, cognitive content is nothing but a function. In parallel with Peirce and James's pragmatism, Sellars has in mind to bridge the supposed bifurcation between "fact" and "norm." The idea is that many natural processes accomplish certain forms and degrees of normativity and so meaning and value are not *conditio humana*.[22] As such, I think it is fair to take

[22]Interestingly paralleling the notion of linguistic convention, Marcelo Barbieri claims that "natural conventions add meaning to information." According to him, "The processes that have created the genetic code and those that have led to the choice for dextrorotatory sugars or levorotatory amino acids are more well-known examples of natural conventions . . . the biological evolution has been produced not only through natural selection but also through natural conventions" [my translation]

meaning and value to be the very essence of the web of life in that they differentiate many forms of ententional phenomena in nature. Taking into account Sellars's process-based ontology and following Deacon's notion of ententional phenomena, I believe they strengthen the need for epistemologically defining a threshold zone in which one differentiates the activity of meaning as indicating the transition and continuity from lifelessness to life in nature.

Additionally, as noted by Whitehead ([1929] 1978: 214), the notion of a "process" has two interconnected species: microscopic and macroscopic processes. The former is efficient, and the second is teleological. That is to say: whereas microscopic processes bring about the transition from the present conditions to the future, macroscopic processes provide the ends for the transition. For Whitehead ([1929] 1978: 214–215), moreover, once an "organism" means a combination of processes in a twofold manner, "the community of actual things is an organism; but it is not a static organism." To the extent that an organism consists of dynamically inefficient and teleological processes, the structural coupling between organism and environment means process, hierarchy, and transition of processes (including microscopic and macroscopic levels). The idea is that the structural coupling results in a web of transitional processes much more than a simple arithmetic combination of organism and environment, indicating continuity in different directions. As a result, the transition and continuity of processes are gradually transformed into ententionality, activity of meaning, and life-forms in nature.

5 Meaning, Life, and Indeterminateness

To affirm the need for a threshold in the natural world from which one differentiates meaning and life, I start this last part of the chapter with the following question: Must there have been a first semiosis for making sense of meaning and life in the world? In my opinion, the answer is no. The reason is quite simple. Having in mind the image of life as a web, there is no level taken to be ontologically or causally primary, and the causal influences flow in different directions (and not necessarily in a bottom-up sense). As I will show, the image of the web discredits the idea of a first semiosis as a necessary condition for meaning and life to be accomplished in the world. Borrowing Peirce's conception of indeterminate cognition, I will argue that meaning and life consist of overlapping processes.

The series of papers in which Peirce criticizes the Cartesian foundationalism on cognition is known as the Cognition Series, and it appears in the *Journal of Speculative Philosophy* (1868). In *Questions Concerning Certain Faculties Claimed for Man* ([1866] 1958: 37), particularly, Peirce criticizes the traditional view of intuition as "a cognition not determined by a previous cognition." Once he embraces

(Barbieri 1985: 159). What Barbieri contends is, in short, that conventions can also be found in nature.

an epistemological anti-foundationalism, Peirce argues that our knowledge is by nature inferential and so any cognition is determined by a previous cognition:

> No cognition not determined by a previous cognition, then, can be known. It does not exist, then, first, because it is absolutely incognizable, and second, because cognition only exists so far as it is known. (Peirce 1966: 37)

> The reply to the argument that there must be a first is as follows: In retracing our way from conclusions to premises, or from determined cognitions to those which determine them, we finally reach, in all cases, a point beyond which the consciousness in the determined cognition is livelier than in the cognition which determines it. (Peirce [1868] 1958: 37)

For Peirce, therefore, it is not true that there must be a first cognition. The idea is that cognition consists of a dynamic web of sign inference processes. According to him, the first cognition consists in an epistemologically undecidable question and a paradox of Achilles:

> Suppose an inverted triangle ▽ to be gradually dipped into water. At any instant, the surface of the water makes a horizontal line across that triangle. This line represents a cognition. At the subsequent date, there is a sectional line so made, higher upon the triangle. This represents another cognition of the same object determined by the former. . . . The apex of the triangle represents the object external to the mind which determines both these cognitions. The state of the triangle before it reaches the water represents a state of cognition which contains nothing which determines these subsequent cognitions. . . . For any such section is at some distance above the apex, otherwise it is not a line. Let this distance be *a*. Then there have been similar sections at the distances 1/2*a*, 1/4*a*, 1/8*a*, 1/6*a*, above the apex, and so on as far as you please. So that *it is not true that there must be a first* [my italics]. Explicate the logical difficulties of the paradox (they are identical with those of Achilles) in whatever way you may. I am content with the result as long as your principles are fully applied to the particular case of cognitions determining one another. . . . The point here insisted on is not this or that logical solution of the difficulty, but merely that cognition arises by a process of beginning, as any other change comes to pass. (Peirce [1868] 1958: 37–38)

Peirce's assertion that cognition arises by a process of beginning can be illustrated as follows:

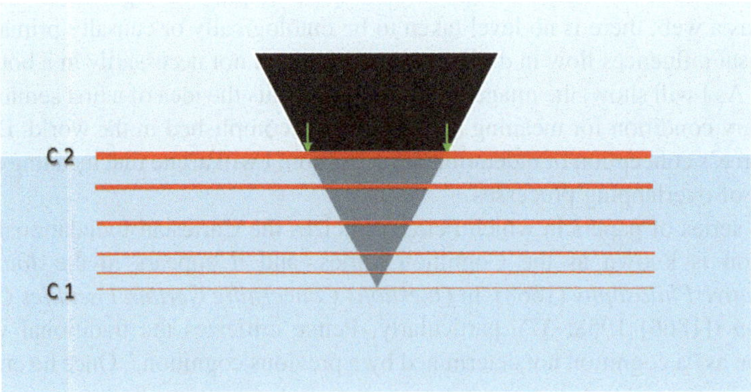

Illustration 5 In Peirce's conception of cognition, it arises by a process of beginning—C1: cognition 1; C2: cognition 2, and so on. According to Peirce, it is not the case that there is no beginning of cognition. The cognition begins gradually, and there is no cognition that is not determined by another cognition.

In referring to Peirce's Cognition Series, I basically intend to point out that meaning begins with an indeterminate process of beginning. Furthermore, in tracing a parallel with Peirce's description of the cognitive states, I have in mind to illustrate the idea that meaning and life are overlapping processes; that is, there must not have been a first semiosis in order to make sense of meaning and life. As an overlapping of processes, the web of life arises by a process of beginning in that meaning makes no sense except in the *rush* of semiosis (Merrell 1997: xi). The idea is that the overlapping of meaning and life consists in an indeterminate process; consequently, there is no first semiosis taken to be ontologically or causally primary in creating the web of life. Retaking William James's image of a mosaic, incidentally, the idea of the web of life is that of a dynamical structure without a center or periphery.[23]

In parallel with Peirce's conception that cognition arises by a process of beginning, for instance, William James insists on characterizing experience (as a whole) as a continuous process in space and time. As well as the visual field's form, for instance, James asserts that the field of experience (or world experienced) is equally fringed and has no definite boundaries. Unlike traditional empiricism, for James, the connecting relations between experiences are taken to be experienced relations. Consequently, in James's radical empiricism, experience has no bottom layer, and the structure of an experience is in the transitions of the experienced relations as a type of dynamic gestalt. Indeed, according to such an anti-foundationalist view, there is no gap in the structure of experience, and it forms a sort of continuum. Moreover, since experience consists in an immediate flux of life, "life is in the transitions as much as in the terms connected" (James [1904], in McDermott (1977: 212)). The idea of transition discredits the need for a substantialist ontology on which life is supposed to be built up.

---

[23] In many aspects, I think, the idea that the web of life begins by a process of beginning is akin to Tibetan Buddhism's picture of the "Wheel of Life" (*Bhavacakra*). In this context, incidentally, the word *nidāna* means the processes by which a being comes into existence, and it is bound to the Wheel of Life (Humphreys 2005: 152, 259). Being a "wheel," it portrays the concatenation of cause and effect; nevertheless, there is no starting point. In Buddha's teachings, indeed, it is said that our beginning is inconceivable and that its starting point cannot be indicated (Buswell 2004: 185). Additionally, in the words of the Dalai Lama (1999: 30), "what is the substantial cause of the material universe way back in the early history of the universe, we trace it back to the space particles which transform into the elements of this manifest universe. And then, we can ask whether those space particles have an ultimate beginning. The answer is no. They are beginningless. Where other philosophical systems maintain that the original cause was God, Buddha suggested the alternative that there are no ultimate causes. The world is beginningless. Then the question would be: Why is it beginningless? And the answer is: It is just nature. There is no reason. Matter is just matter." Indeed, it is interesting to note that in Buddhist cosmology, "the universe has no specific creator; the sufficient cause for its existence is to be found in the Buddhist cycle of causal conditioning known as *Pratityasamutpada* [or dependent origination]" (Gethin 2004: 183).

Once again, concerning James's anti-foundationalism, it is in parallel with Peirce's Cognition Series. Both discredit the epistemological need for a bottom layer in order to understand the cognition process. As I see it, this line of thinking can be extended to the conception of the web of life. Assuming there is no bottom layer taken to be ontologically or causally primary, it is fair to say that the web of life begins by a process of beginning as a sort of continuity. The idea of continuity does not rule out the need for a conception of a threshold zone in which one epistemologically differentiates the transition between non-living and living processes, as well as the beginning of the web of life.

As I am assuming that *where there is meaning, there is life (no meaning, no life!)*, the web of life is based on an indeterminate process. For the web of life to be coherent, however, the meanings (that make up the web) must cohere with each other. That is to say, since meaning arises by a process of beginning, such a process forms the very essence of the web of life. In line with Peirce's view of cognition, I think it is plausible to assert that the web of life consists in a dynamic web of meaning processes. Similarly to coherentism in epistemology, for instance, the idea that the web of life is based on meaning processes discredits essentialism. Indeed, to the extent that meaning and life are overlapping processes, the web of life acquires a holistic nature (see Illustration 4). In tracing a parallel with Peirce's Cognition Series in that cognition begins by a process of beginning, holistically speaking, I contend that the web of life is without a center or periphery and that it begins by a process of beginning.

Inspired by Peirce's Cognition Series, and in order to characterize the overlapping of meaning and life, I retrace Anaximander's *ἄπειρον* (apeiron): an indeterminate process engenders everything in the universe. Insofar as the web of life comprises an unlimited and indefinite activity, there is no need for a process taken to be ontologically or causally primary. As a consequence of such indeterminateness, the web of life arises from a process of beginning in overlapping with meaning. When some processes reach a threshold zone indicating a sort of deflection in nature, something else makes a difference more than only material presence. Using Whitehead's vocabulary, meaning and life overlap supposedly from the third grade of entities on which one may differentiate the transition and continuity in nature between non-living and living processes.

In comparison with Peirce's Cognition Series, it is fair to assert that everything arises from a process of beginning in the web of life without a process taken to be ontologically or causally primary. Accordingly, more than what is materially present in the web of life, what matters is its relational nature from which meaning emerges as resulting from transition and continuity with lifeless processes in the sense that "something … stands for something else by reason of a relation" (Pattee and Rączaszek-Leonardi 2012: 156). As can be seen in Illustration 4, likewise, the web of life consists in a dynamic mosaic of relational processes, which begins by a process of beginning. Having in mind the image of the web of life, in brief, I want to stress the fact that life as a relational process is matter with meaning (*pace* Pattee and Rączaszek-Leonardi (2012)). In following James's mosaic metaphor, once again, I insist that meaning should be understood in terms of constructivism more than according to Bohm's ultra-experimentalism and the idea that meaning is a fundamental property of nature.

6 Final Remarks

It is not easy to conclude a chapter in which one seeks a conceptual unity between three thorny themes in contemporary contexts in natural sciences, epistemology, and biosemiotics: the threshold, meaning, and life. The idea here may well be to outline the prospects for an investigation that justifies a conceptually coherent approach to themes and problems related to the threshold, meaning, and life. Since it is not a matter of understanding the significance and origin of life on earth, the notions of the threshold and meaning indicate an alternative to an understanding of life as a dynamic web of relations. Therefore, in considering the image of the web of life, the idea is to understand that meaning and life cannot, and do not, exist independently. And to the extent that the notion of the web of life suggests a process of dynamic relations in the form of constructivism, it does not seem to make sense to understand that meaning consists in a basic physical property.

In this chapter, I have tried to indicate in what sense one can understand meaning and life according to the notion of a threshold zone from which transition and continuity are differentiated between non-living and living, without, however, accepting the idea of a supposed missing link in nature. In considering that the notion of a missing link does not seem to justify a coherent understanding of continuity in nature, what I show is that like the very notion of meaning, the image of the web of life suggests an indeterminate process that begins by beginning as illustrated by Peirce's Cognition Series. Insofar as the web of life takes the form of a dynamic mosaic, meaning and life overlap and begin when certain processes reach a threshold zone of transition and continuity with the non-living. What derives from such a process without a bottom layer is the understanding that *where there is meaning, there is life—without meaning, without life!* (and vice versa).

References

Atmanspacher H (2020) Quantum approaches to consciousness. In: Zalta EN (ed) The Stanford encyclopedia of philosophy. Metaphysics Research Lab, Stanford. https://plato.stanford.edu/archives/sum2020/entries/qt-consciousness. Accessed 30 Oct 2020

Barbieri M (1985) Teoria semântica da evolução (trans: Pinheiro ML). Fragmentos, Lisbon

Bateson G (1979) Mind and nature—a necessary unity. Dutton, New York

Bohm D (1985) Unfolding meaning. Routledge, Oxford

Bonnet C (1781) Contemplation de la nature (Vol. 4). De l'imprimerie de Samuel Fauche, libraire du roi

Buswell RE (2004) Encyclopedia of Buddhism. Macmillan, New York

Chalmers D (2017) Panpsychism and panprotopsychism. In: Brüntrup G, Jaskolla L (eds) Panpsychism—contemporary perspectives. Oxford University Press, Oxford
Cobley P, Deely J, Kull K, Petrilli S (2011) Semiotics continues to astonish—Thomas A. Sebeok and the doctrine of sign. de Gruyter, Berlin
Cosmides L, Tooby J (1987) From evolution to behavior: evolutionary psychology as the missing link. In: Dupré J (ed) The latest on the best essays on evolution and optimality. Bradford, London
Dalai Lama (1999) Consciousness at the crossroads—conversations with the Dalai Lama on brain science and Buddhism. Lion, New York
Damasio A (1999) Feeling of what happens – body and emotion in the making of consciousness. Harcourt, Brace, and Co., New York
Damasio A (2010) Self comes to mind—constructing the conscious brain. Pantheon, New York
Darwin C (1859) The origins of species. Clowes, London
Darwin C (1979) The origins of species. Gramercy, New York
Darwin C (2008) Le corail de la vie—carnet B (1837–1838) (trans: Rovere M). Payot, Paris
Deacon T (2012) Incomplete nature—how mind emerged from matter. Norton, New York
Donaldson J (2015) Missing link: the evolution of metaphor and the metaphor of evolution. McGill-Queen's University Press, Québec
Eco U (1976) A theory of semiotics. Indiana University Press, Bloomington
Fechner GT (1966) Elements of psychophysics. Vol. I, Translated by H.E. Adler. New York: Holt, Rinehart & Winston.
Emmet D (1932) Whitehead's philosophy of organism. Macmillan, London
Ford LS (1984) The emergence of Whitehead's metaphysics. State University of New York Press, Albany
Gethin R (2004) Cosmology. In: Buswell RE (ed) Encyclopedia of Buddhism. Macmillan, New York
Gibson J (1979) The ecological approach to visual perception. Psychology Press, New York
Hayward JW, Varela FJ (2001) Gentle bridges: conversations with the Dalai Lama on the sciences of mind. Shambala, Boston
Heft H (2001) Ecological psychology in context—James Gibson, Roger Baker and the legacy of William James's radical empiricism. Psychology Press, New Jersey
Heft H (2017) William James' psychology, radical empiricism, and field theory: recent developments. Philosophical Inquiries 2:111–130
Higuera CR, Kull K (2017) The biosemiotic glossary project: the semiotic threshold. Biosemiotics 10:109–126
Hoffmeyer J (2008) Biosemiotics: signs of life and life of signs. University of Scranton Press, Scranton
Hoffmeyer J (2011) Theories of signs and meaning: views from Copenhagen and Tartu. In: Emmeche C, Kull K (eds) A semiotic biology—life is the action of signs. Imperial College Press, London
Humphreys C (2005) A popular dictionary of Buddhism. Curzon, London
James W (1909) A pluralistic universe. Longmans, Green, and Co, New York
James W (1977) In: McDermott JJ (ed) The writings of William James – a comprehensive edition. Chicago University Press, Chicago
James W (1983) The principle of psychology. Harvard University Press, Harvard (UK)
James W (2010) Humana immortality. A Digireads.com Books, Boston
Jonas H (2001) The phenomenon of life. Northwestern, Evanston
Koutroufinis S (2014) Introduction. In: Koutroufinis S (ed) Life and process—towards a new biophilosophy. De Gruyter, Berlin
Kull K (2004) Uexküll and the post-modern evolutionism. Sign Syst Stud 32(1/2):99–114. https://doi.org/10.12697/SSS.2004.32.1-2.04
Leibniz G (1996) New essays on human understanding (trans: Remnant P, Bennett J). Cambridge University Press, Cambridge

Lewin R (2005) Human evolution: an illustrated introduction. Blackwell, Cambridge, MA
Logan RK (2007) The extended mind: the emergence of language, the human mind and culture. University of Toronto Press, Toronto
Manetti G (2002) Philodemus' 'De signis': an important ancient semiotic debate. Semiotica 138 (1/4):279–297
Marshall ME (1974) William James, Gustav Fechner, and the question of dogs and cats in the library. Journal of the History of the Behavioral Sciences 10(3):304–312
McDermott JJ (ed) (1977) The writings of William James—a comprehensive edition. Chicago University Press, Chicago
Merrell F (1997) Peirce, signs and meaning. Toronto University Press, Toronto
Mesle CR (2008) Process-relational philosophy—an introduction to Alfred North Whitehead. Templeton, West Conshohocken, PA
Nicholson DJ, Dupré J (eds) (2018) Everything flows: towards a processual philosophy of biology. Oxford University Press, Oxford
Pattee HH, Rączaszek-Leonardi J (2012) Laws, language and life—Howard Pattee's classic papers on the physics of symbols with contemporary commentary. Springer, Heidelberg
Peirce CS (1958) Peirce Charles Sanders: selected writings. Dover, New York
Peirce CS (1966) Charles S. Peirce: collected papers. Dover Publication, New York
Perry RB (1976) The thought and character of William James as revealed in un-published correspondence and notes, together with his published writings. In: Volume II Philosophy and Psychology. Oxford University Press, London
Pickering J (2012) Panpychism. In: Favareau D, Cobley P, Kull K (eds) A more developed sign—interpreting the work of Jesper Hoffmeyer. Tartu University Press, Tartu
Ricou J, Pollock JA (2012) The tree, the spiral and the web of life: a visual exploration of biological evolution for public murals. Leonardo 45(1):18–25
Sebeok T (1979) Signs – an introduction to semiotics. University of Toronto Press, Toronto
Sebeok T (2001) Signs: an introduction to semiotics. University of Toronto Press, Toronto
Sellars W (1981) Naturalism and process. Monist 64(1):66–90
Sharov AA (2013) Minimal mind. In: Swan L (ed) Origins of mind, biosemiotics 8. Springer, New York, London, pp 343–360
Sonesson G (2006) The meaning of meaning in biology and cognitive science: a semiotic reconstruction. Σημειωτκή-Sign Systems Studies 34(1):135–213
Tattersall I (1999) L'émergence de l'homme. Gallimard, Paris
Thompson E (2007) Mind in life—biology, phenomenology and the sciences of mind. Cambridge University Press, Cambridge, UK
Urbaniak R (2000). Web of life. https://www.naturalfrequency.net/Ray/webofLife.htm Accessed on 15 Jan 2021
von Uexküll J (1926) Theoretical biology. Harcourt, Brace & Company, New York
von Uexküll J (1982) The theory of meaning. Semiotica 42(1):25–82
Ward D, Silverman D, Villalobos M (2017) Introduction: the varieties of enactivism. Topoi 36:365–375
Weber M (2013) The polysemiality of the concept of "pure experience". In: Weber M (ed) Whitehead's pancreativism: Jamesian applications. De Gruyter, Berlin, pp 91–114. https://doi.org/10.1515/9783110329766.91
Whitehead AN (1919) The concept of nature. Cambridge University Press, Cambridge
Whitehead AN (1948) Science and the modern world. New American Library, New York
Whitehead, AN (1978) Process and Reality. The Free Press, New York

How Information Gets Its Meaning

Vinicius Romanini

> *All icons, from mirror-images to algebraic formulae, are much alike, committing themselves to nothing at all, yet the source of all information. They play in knowledge a part iconized by that played in evolution according to the Darwinian theory, by fortuitous variations in reproduction.*
>
> —Peirce, MS 599, 42

Abstract This chapter's primary purpose is to discuss how Peirce's semiotic and general philosophy can help us understand the biological, evolutionary, and even universal roots of our human and social communication skills and culture. Peirce's concept of information is central to this, since it allows us to understand how we learn from experience, interpret symbols, and share our beliefs through communication. I approach the subject by exposing Peirce's late theory of signs and how Peirce himself related semiosis and evolution to the concept of information. In fact, in some of his later manuscripts, Peirce defined a symbol as related to biological heredity and the icon as a sign that plays the same role in culture that random variation in reproduction plays in Darwin's theory. As Peirce claimed that information must be embodied in icons and expressed by indices to be communicated

---

Note: Throughout this chapter, I use the following coding system: "MS" followed by a number identifies the relevant Peirce manuscript number (following the Robin catalog); this is followed by a comma, a space, and the page number(s). "CP" means "Collected Papers" (see Peirce et al. (1932, 1933, 1935) and Peirce and Burks (1958)) and is followed by the volume number, a period symbol, and the paragraph number. "EP" refers to "The Essential Peirce" (see Houser and Kloesel (1992) and Peirce Edition Project (1998)) and is followed by the roman numeral "I" or "II" indicating the volume number, a colon, a space, and the page number(s). "W" refers to "Writings of Charles S. Peirce" (see Peirce et al. (1982–present)) and is followed by the volume number, a colon, a space, and the page number(s). In instances in which the date is known and is relevant to my argument, a comma, a space, and the year follow the page number(s).

V. Romanini (✉)
School of Communication and Art, University of São Paulo, São Paulo, Brazil
e-mail: vinicius.romanini@usp.br

© Springer Nature Switzerland AG 2021

E. Pagni, R. Theisen Simanke (eds.), *Biosemiotics and Evolution*, Interdisciplinary Evolution Research 6, https://doi.org/10.1007/978-3-030-85265-8_6

through symbols, its meaning can be understood only as a result of a universal semiosis that links our cognition to natural phenomena without a solution of continuity.

Keywords Information · Meaning · Semiotics · Peirce · Shannon · Fisher · Symbol · Icon · Communication · Universal evolution

1 Introduction

To our common sense, information is the occurrence of something unexpected. We all agree with this definition, although it lacks clarity. Unexpected to whom? On what grounds? And about what? On second thought, the use of "informative" as a predicate seems to be more appropriate than the hypostatized use of "information" as an abstract noun. After all, we find surprisingly informative the occurrences that sensitize our perceptual apparatus, and we do so through synthetic judgments. We do not intuit "information" as an intellectual object per se, without experience, as we can do with the aprioristic predicates of "space" and "time." The intrinsic semantic dynamism of information, always contextual, should exclude a purely syntactic and structural definition. Conversely, information seems to be naturally pragmatic and a consequence of semantic processes linked to psychical phenomena. However, the concept of information has lost its meaning and has even been reified as an elemental component of the universe. From a general predicate under the logic domain, information has come to be considered a measurable and countable quantity under the domain of physics and mathematics.

It is not my purpose here to elucidate how such a metamorphosis happened. Here I will only acknowledge that the result of this process is that for contemporary Western science, the concept of information has been reduced to that of entropy and algorithmic complexity. I guess that, to a great extent, this is a consequence of the hegemony of cybernetics in scientific culture after the union of the theories of Alan Turing and Claude Shannon. The infamous anecdote is that Shannon was advised by John von Neumann, the founding father of cybernetics, to equate his logarithmic–probabilistic concept of information with entropy to reduce the possibility of critical questionings of his seminal work on the mathematical theory of communication "since no one understands what entropy means." However, we must doubt if this clever stratagem, which seems to have worked so far, did not betray Boltzmann's noble intentions to seek a mathematical–scientific foundation for the empirical discoveries of thermodynamics.

I will proceed by going backward in the story by providing short accounts of Shannon Information (the currently dominant scientific paradigm) and Fisher Information (its close predecessor). I will then introduce Peirce's definition of information as a logical quantity of symbols and show how it relates to perception and experience. I will also point out some similarities with the other two previous descriptions through logarithmic calculations and conclude by introducing Peirce's extreme realism as the grounds for scientific metaphysics. Here final causation is the core idea to bring information and meaning under the same frame. Pragmatism might be defined as a method to extract real information and infuse it into our shared beliefs, as our representations evolve (albeit never achieving a definitive, ultimate state) toward a genuine final opinion.

Let us start with Shannon Information, which is commonly described as a measure of the number of binary distinctions that would be necessary to bring a disordered system into an orderly state or to describe it completely. Since it ultimately depends on the system's physical disorder, it can indeed be reduced to entropy. As the entropy of a system necessarily grows in time because of the second law of thermodynamics, the amount of information necessary to describe it increases proportionally. Moreover, since entropy is a statistical average quantity expressed by the logarithm of the sum of probable states of a system, so is Shannon Information (Shannon 1948). The larger the number of possible, probable states, the larger the ignorance about which state is the current actual one, and the larger the amount of information needed to describe the system by a sequence of binary distinctions. When entropy is maximal, Shannon Information is simply log n (the logarithm of n with base 2), where n is the number of probable states. Conversely, if a well-known system is in a state so well ordered that it does not change at all in time (totally redundant), then no Shannon Information is needed to describe it.

The contemporary hegemony of Shannon's definition tends to cloud other earlier and equally interesting mathematical formulations. For instance, in the 1920s, the British statistician Ronald Fisher (1922) had already proposed a logarithmic–mathematical definition of the concept of information, with the advantage that it preserved an observer's experience as ballast. Fisher developed his concept of information as a way to measure the degree of confidence in observational experience. Mathematically, Fisher Information is always about observing data concerning a parameter ***a*** (Frieden 2004; Frieden and Romanini 2008). The data ***x*** collected during any observation varies randomly because no real measurement is free of noise or imperfection in the measuring device. Fisher Information can be expressed as $I = <[(d/dx) \log p(x)]\ 2>$, where < > means average and p(x) means a probability function about a variable x and d/dx is its variation. The variable ***x*** is defined as data carrying information about ***a***. Fisher Information is then a measure of the "width" of the probability function p(x) or its density. That is why it can also be expressed as the curve of an amplitude of probability. A wide amplitude curve means less information, and a narrow amplitude curve means more information. Another important feature is that the variation of x expressed by d/dx can be interpreted as the expression of dynamics in time, a property that is lacking in Shannon Information.

Fisher Information is therefore local and measured from the logarithm of the registrations' density made on an observable parameter. Fisher offered a mathematical way to measure our experiments' effectiveness in the physical world, while

Shannon proved that everything solid melts into the air, so to speak. If Fisher had a ground wire connected to physical existence, Shannon offered us bits that navigate freely in the virtuality of the real but without any sense. Crushed by the uni-dimensionality of the binary digital code, framed by restricted Boolean logic (in which the principles of identity and the excluded third are essential), and cluttered with the numerical language of 0s and 1s that allows computers to perform algorithmic tasks, information has gained ontological accuracy but lost its meaning (Machado and Romanini 2011). A non-local and systemic amount has been transformed and wholly dissociated from the conditions of possibility of an observer's/interpreter's experience.

The story could be different if the semiotic concept of information developed by Charles Sanders Peirce had been more widely studied and understood before Fisher and Shannon started their studies. To our knowledge, Peirce was arguably the first scientist ever to deal with the concept of information scientifically and systematically.[1] As early as 1865, he defined information as a third logical quantity of symbols, in addition to denotation (extension, or also breadth) and connotation (comprehension, or also depth). During his half-century of philosophical production, Peirce's understanding of symbols evolved from the Kantian and nominalistic approach of his youth to an Aristotelian and extreme realist position in his late writings. Then the class of symbols covers everything capable of transmitting a form from an antecedent to a consequent in the process of triadic relations that Peirce calls semiosis, the action of signs—be it a communicative assertion, the work of bees, the action of a complex molecule in our immune system, or even the growth of crystals. In the limit, every conceivable universe where chance events are subsumed in new creative patterns is analogous to a developing symbol. The corollary is that information is the outcome of universal semiosis, which is pervasive in the cosmos.

2 Peirce's Definition of Information

From 1883 to 1891, as Peirce tried to make ends meet after a series of professional failures and personal setbacks, he worked as a freelance lexicographer for the *Century Dictionary*,[2] contributing as a writer or editor of more than 15,000 definitions. One of his entries was the definition of "Information" reproduced below.

[1] This was probably first noticed by David Hawkins (1967: 47): "It may be of some interest to notice that Charles Peirce employed the term 'information' long before the recent development of theory, in a way that finds expression within the theory."

[2] See http://www.global-language.com/century/.

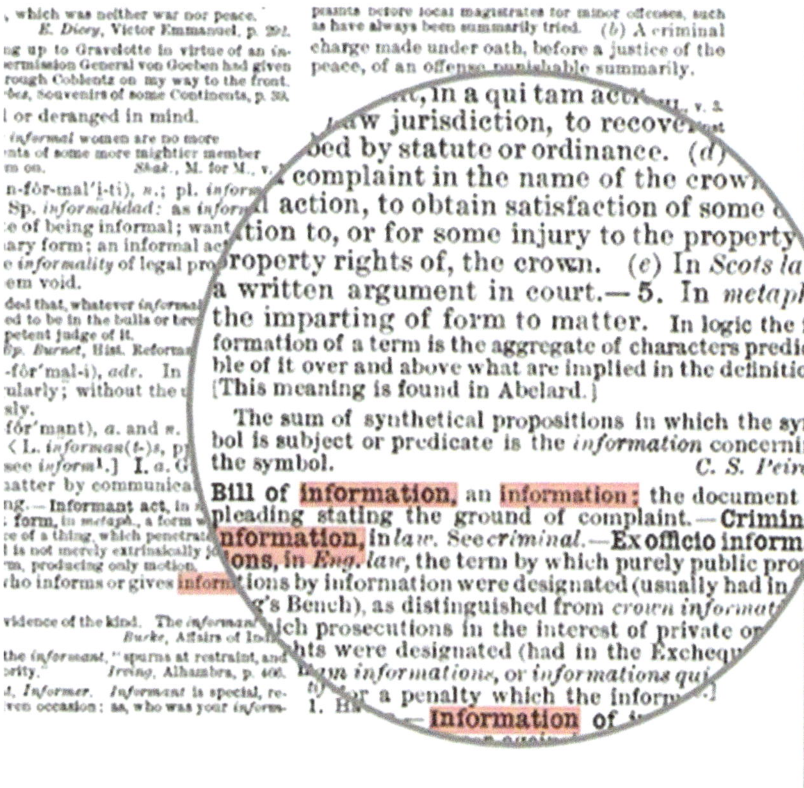

As the entry reveals, Peirce's concept of information is at the same time:

1. Logical in the traditional sense ("the aggregate of characters predicated of it over and above what is implied in its definition," as was already found in Abelard's nominalism).
2. Semiotic, understood as the expanded doctrine of logic ("the sum of synthetical propositions in which the symbol is either subject or predicate"). Here Peirce quotes himself, a clear indication that he recognizes the originality and uniqueness of his contribution to the concept.
3. Metaphysical ("the imparting of form to matter"), which resonates with Aristotle's hylomorphism.

Let us start with the first sense of information, the logical one. As mentioned above, Peirce introduced information as the third quantity of symbols as early as 1865 when he was 26 years old. Symbols can be concepts, the terms of propositions (such as the subject and predicate) but also any other general representation used to communicate, no matter how complex it is:

> Symbols or general representations ... connote attributes and so connote them as to determine what they denote. To this class belong all words and all conceptions. Most combinations of words are also symbols. A proposition, an argument, even a whole book may be, and should be, a single symbol. (W1: 467, 1866)

Peirce then explains his "Abelardian" definition of information:

> The information of a term is the measure of its superfluous comprehension. That is to say that the proper office of the comprehension is to determine the extension of the term. For instance, you and I are men because we possess those attributes—having two legs, being rational, &c.—which make up the comprehension of man. Every addition to the comprehension of a term lessens its extension up to a certain point, after that further additions increase the information instead. (W1: 467, 1866)

The quotation above exposes Peirce's originality. Although he adopts the Abelardian definition of information, Peirce fits it into the traditional scheme of inverse proportionality of comprehension and extension, which was formulated by the logicians of Port Royal. For them, comprehension and extension (also known as depth and breadth, and sometimes even connotation and denotation) were taken to be in a way that the growth of one of them would diminish the other. Instead of writing the usual "Comprehension × Extension" formula, Peirce creates a new one: "Comprehension × Extension = Information." He defines information as the quantity that can increase comprehension without diminishing extension or can increase extension without diminishing comprehension, or can even simultaneously increase both. So if I say the complex term "Caucasian man," I have lessened the extension while increasing the comprehension in comparison with saying the simple term "man" only. However, if I say "non-black Caucasian man," the addition of "non-black" as an increase in comprehension is superfluous and does not lessen any amount of extension, for "Caucasian" implies being "non-black."

So Peirce concludes, "(i)f we learn that S is P, then, as a general rule, the depth of S is increased without any decrease of breadth, and the breadth of P is increased without any decrease of depth" (CP 2.420). In sum, in the 1860s, information for Peirce is a correlated quantity of a triadic relation, which also includes an extension (all real things that can be predicated by a symbol) and comprehension (all general predicates involved in the definition of a symbol). Since this process of conceptual synthesis can be carried on indefinitely without ever coming to an end, Peirce's quantity of information implies a fallible state of knowledge, "which may range from total ignorance of everything except the meanings of words up to omniscience."

3 Information in Perception, Cognition, and Experience

Beginning in the 1870s (when he wrote his famous cognitive series of articles), Peirce's concept of information became an essential part of his theory of perception and cognition, which relates to the second aspect of his entry in the *Century Dictionary*. It was continuously generalized to cover the whole of experience, from perception to scientific inferences. From now on, we will see Peirce's understanding that not only symbols but also indices (quantifiers) and then icons (diagrams) are important for a complete theory of information. Information is grounded on habitual forms that are not available to reason, such as in the case of instinct. Peirce explains that a state of information is directly dependent on the knowledge we gather in experience, since perception, the parting point of experience, is the only door we have to let in new information. In CP 7.587, he explains that "perception is the possibility of acquiring information, of meaning more," pointing out that the reality of his three categories and correspondent types of signs (icons, indexes, and symbols) should be accounted for to produce a complete theory of information as a cognitive process of learning from experience—of meaning more.

Whenever we have an informative synthetic judgment such as the proposition "The orange in my hand is sweet," we see that the predicate "sweet" gains extension. In contrast, the subject "the orange in my hand" gains comprehension, for the class of known sweet things now has a new element. In turn, the object "orange" denoted by me now has a new and increased comprehension—i.e., besides being known to be a tropical, round, juicy, citric, orange-colored fruit, it is now known to be sweet. All scientific reasoning proceeds in the same manner: we note some interesting facts in our experience and then search for an explanatory hypothesis capable of subsuming these novelties under a larger and more comprehensive logical whole.

While keeping information as the ampliative result of a synthetic judgment, as an interpretant of a symbol produced by experience, Peirce is now advocating a much fuller view of information as connected to non-conscious feelings, volitive actions, and pragmatic consequences. The meaning of a concept, word, term, symbol, etc., is the sum of all general effects that would naturally precipitate from its adoption by a community of interpreters, and the meaning of symbols may grow as novel general effects are incorporated above its usually known (i.e., familiar) definition, which is always merely verbal:

> By information, I mean all that knowledge that we collect from the experience of ourselves and of others. Now I call any acquisition of Knowledge "information," which has logically required any other experience than the experience of the meanings of words. I do not call the knowledge that a person known to be a woman is an adult nor the knowledge that a corpse is not a woman, by the name of "Information," because the word "woman" means a living adult human being having, or having had, female sexuality. Knowledge that is not Informational may be termed "verbal." (MS 664, 19, 1910)

3.1 Collateral Information

At this point, we must introduce the concept of collateral experience, or collateral information, which covers precisely the amount of information that interpreting minds must possess as a repertoire before engaging in the process of interpretation.

> By collateral observation, I mean previous acquaintance with what the sign denotes. Thus, if the Sign be the sentence "Hamlet was mad," to understand what this means one must know that men are sometimes in that strange state; one must have seen madmen or read about them; and it will be all the better if one specifically knows (and need not be driven to presume) what Shakespeare's notion of insanity was. All that is collateral observation and is no part of the Interpretant. But that which the writer aimed to point out to you, presuming you to have all the requisite collateral information, that is to say just the quality of the sympathetic element of the situation, generally a very familiar one—a something you probably never did so clearly realize before—that is the Interpretant of the Sign—its "significance." (CP 8.179)

To sum up, collateral information is the previous knowledge about the object of representation that provides adequate comprehension of a symbol. As a familiarity, it depends on a "mental habit," or a belief that grounds the meaning of all predicates implied in a symbol's definition. In short, collateral information is what we already know about the object, while semiotic information is the new learnings about it.

3.2 *Is Collateral Information Logarithmic?*

The familiarity of collateral information relates to the redundancy of mental habits (our beliefs) as much as the novelty of information relates to the unexpectedness of perception. There can even be a logarithmic account of this relation in Peirce's concept of chance. Suppose the inquiry task is to break wrong old habits (usually grounded on unscientific methods such as blind reliance on authority, stubbornness, and aprioristic misconceptions[3]) and build new and more accurate ones. In that case, the research work seems to involve a minimax optimization equation quite similar to logarithmic relations. In Peirce's own words:

> [Knowledge has a "money value" that] increases with the fullness and precision of the information, but plainly it increases slower and slower as the knowledge becomes fuller and more precise. The cost of the information also increases with its fullness and accuracy, and increases faster and faster the more accurate and full it is. It therefore may be the case that it does not pay to get any information on a given subject; but, at any rate, it must be true that it does not pay (in any given state of science) to push the investigation beyond a certain point in fullness or precision. (CP 1.122)

In situations such as this, what we see is a game-like dispute between the inquirer and the object of inquiry, which we may consider as Nature in general. The best strategy for the inquirer is to formulate a hypothesis as close as possible to an even choice and eliminate (as quickly and as much as possible) wrong answers to a question:

> The game of twenty questions is instructive. In this game, one party thinks of some individual object, real or fictitious, which is well-known to all educated people. The other party is entitled to answers to any twenty interrogatories they propound which can be answered by Yes or No, and are then to guess what was thought of, if they can. If the questioning is skillful, the object will invariably be guessed; but if the questioners allow themselves to be led astray by the will-o-the-wisp of any prepossession, they will almost as infallibly come to grief. The uniform success of good questioners is based upon the circumstance that the entire collection of individual objects well-known to all the world does not amount to a million. If, therefore, each question could exactly bisect the possibilities, so that yes and no were equally probable, the right object would be identified among a collection numbering 2 to 20. Now the logarithm of 2 being 0.30103, that of its twentieth power is 6.0206, which is the logarithm of about 1,000,000 $(1 + .02 \times 2.3)$ $(1 + .0006 \times 2.3)$ or over one million and forty-seven thousand, or more than the entire number of objects from which the selection has been made. Thus, twenty skillful hypotheses will ascertain what two

[3] See Peirce (1877) for an account of how we fixate our beliefs.

> hundred thousand stupid ones might fail to do. The secret of the business lies in the caution which breaks a hypothesis up into its smallest logical components, and only risks one of them at a time. (CP 7.220)

For Peirce, the natural choice is to consider the logarithm of chance as the best measure of our attachment to a belief. He wants to ground his theory of information on experience, and chance is defined as the simple relation between the number of successes over the number of failures in a given experiment:

> Any quantity which varies with the chance might, therefore, it would seem, serve as a thermometer for the proper intensity of belief. Among all such quantities there is one which is peculiarly appropriate. When there is a very great chance, the feeling of belief ought to be very intense. Absolute certainty, or an infinite chance, can never be attained by mortals, and this may be represented appropriately by an infinite belief. As the chance diminishes the feeling of believing should diminish, until an even chance is reached, where it should completely vanish and not incline either toward or away from the proposition. When the chance becomes less, then a contrary belief should spring up and should increase in intensity as the chance diminishes, and as the chance almost vanishes (which it can never quite do) the contrary belief should tend toward an infinite intensity. Now, there is one quantity which, more simply than any other, fulfills these conditions; it is the logarithm of the chance. (CP 2.676)

Amazingly, Peirce concludes that an even chance (1/1) means a probability of (1/2), which he calls "randomness." It is not hard to imagine how chance can be translated in probabilistic statistics in such a way as to place the concept of total randomness as maximum entropy, where a maximum number of binary-choice experiments would have to be performed by a scientific mind. Both Fisher Information and Shannon Information would have a place in this translation, and this is work waiting to be accomplished by information theorists.

4 Sensations, Non-conscious Abductions, and Instinct

In our usual cognitions, the most fundamental level of information is given by qualities of feeling, which are purely iconic and yield only emotional effects. They are the cradle of all sorts of information and the origin of all sensations that, ultimately, ground our predicates. We will not deviate too much into Peirce's pragmatism, but we must recall that abductions or hypotheses are the starting point of all knowledge. During perception, abductions are responsible for synthesizing new predicates, which come first as hypothetical sensations and develop as accepted general terms. In perception, Peirce explains, non-conscious synthetic inferences reduce a multitude of complex feelings and outward irritations into one generalized sensation, which becomes the general primary state of information of a particular mind. This embodied iconic information turns out to be the primordial quantity of Peirce's logic system, which grew to become his general semiotics.

From our perceptual judgments to our most complex scientific inferences, the whole process can be described as a semiotic endeavor to capture the real qualitative forms and present them as assertions (synthetic judgments) composed of separated predicates and subjects. This means that concepts are not produced by any transcendental synthesis that welds together sense impressions, as Kant stated. On the contrary, they begin with a feeble qualitative hypothesis that springs from unconscious inferences as undifferentiated wholes—or "percepts." They only become assertions much later in the pipeline of experience, when subject and predicate are dissociated as different quantities (breadth and depth) and then reunited by reasoning.

> The judgement [sic], "This chair appears yellow," separates the color from the chair, making the one predicate and the other subject. The percept, on the other hand, presents the chair in its entirety and makes no analysis whatever. (CP 7.631)

This *reductio ad unum* is conjectural, non-conscious, and immediately connected to reality, which explains how "wild guesses" given in instinct are usually more reliable in practical matters than piecemeal intellectual reasoning:

> Instinct is capable of development and growth—though by a movement which is slow in the proportion in which it is vital; and this development takes place upon lines which are altogether parallel to those of reasoning. And just as reasoning springs from experience, so the development of sentiment arises from the soul's Inward and Outward Experiences. Not only is it of the same nature as the development of cognition; but it chiefly takes place through the instrumentality of cognition. (CP 1.637)

5 Information in Its Semiotic Minute Aspects

Peirce's theory of signs involves a complex network of sign types connected with an intricacy of philosophical and metaphysical positions regarding the nature of the real and our ability to know from experience, which could not be tackled unless we took a long detour into Peirce's method of pragmatism. However, that is not the aim of this chapter. For our limited purposes, it suffices to say that its most important feature is the triadic character of its relations, bearing directly on the problem of information we are dealing with. In Peirce's doctrine, the sign is a logical entity that could be broadly defined as a medium that conveys information from mind to mind (or quasi-mind, for the mind here is also defined as a logical entity).

The sign's task is to create in the mind of an interpreter a representation of something other than itself. "Object" is what the sign represents, the aspect of the sign relation that provides the sign with its "aboutness." The term "object" here is not necessarily an existent thing or fact, such as a chair or the falling of a specific rock, although it can certainly be any of these too. The sign's "object" is defined as a logical position and can be translated as the "subject" of representation without losing much logical meaning. More psychological approaches would prefer "subject" as the best term, but Peirce stresses that once the Copernican revolution of putting thought before the thinker is done (since we are in thoughts and not the other way around), "object" and "subject" become fundamentally identical in semiosis.

The receiver is the interpreter, or, if we get rid of the psychological component, we might call it the interpretant, which can be understood as the general tendency, or power, that any symbol possesses to determine specific kinds of effects, even if this tendency is never actualized, let alone developed to its full consequences. A message in a bottle is a symbol even if it is buried forever in the deep ocean. Nevertheless, Peirce's late extreme realism goes further and professes that a hidden fossil fish in a rock is also a potential symbol, albeit in a latent state until a proper reasoning mind eventually actualizes its possible general interpretant:

> If, for example, there be a certain fossil fish, certain observations upon which, made by a skilled paleontologist, and taken in connection with chemical analyses of the bones and of the rock in which they were embedded, will one day furnish that paleontologist with the keystone of an argumentative arch upon which he will securely erect a solid proof of a conclusion of great importance, then, in my view, in the true logical sense, that thought has already all the reality it ever will have, although as yet the quarries have not been opened that will enable human minds to perform that reasoning. For the fish is there, and the actual composition of the stone already in fact determines what the chemist and the paleontologists will one day read in them. ... It is, therefore, true, in the logician's sense of the words, although not in that of the psychologist's, that the thought is already expressed there. (EPII: 455, 1911)

Since the object must be logically absent to be represented in the sign, the efficient cause of this representation must be a "real form," or general potentiality, that in semiosis becomes an actual dynamic agent—or utterer of the communicated form. The third aspect of the sign relation is the interpretant, which is a consequence (possible, actual, or eventual) of the object's representation by the sign. As the effect of signification, the interpretant is taken to be the final cause of the process of semiosis. In a more direct phrase, the sign's function is to transmit the form of the object to the interpretant. In this sense, the sign is the medium of communication of a form par excellence:

> That which is communicated from the object through the sign to the interpretant is a Form. It is not a singular thing; for if a singular thing were first in the object and afterward in the interpretant outside the object, it must thereby cease to be in the object. The Form that is communicated does not necessarily cease to be in one thing when it comes to be in a different thing, because its being is a being of the predicate. The Being of a Form consists in the truth of a conditional proposition. Under given circumstances, something would be true. The Form is in the object, entitatively we may say, meaning that that conditional relation, or following of consequent upon reason, which constitutes the Form, is literally true of the object. In the sign the Form may or may not be embodied entitatively, but it must be embodied representatively, that is, in respect to the Form communicated, the sign produces upon the interpretant an effect similar to that which the object itself would under favorable circumstances. (MS 283; partially reprinted in EPII: 371–397, 1906)

The form present in the semiotic object is not so hidebound by strict habits as to be considered dead and then easily coded into bits, but is an active one that influences and guides the whole semiosis process. A dead form might be usual for the automatic regulation of machines. However, the form that runs in the artery of semiosis is closer to Leibniz's "vis viva," although with a significant difference: it is not conservative, nor does it strictly obey the principle of the least action; on the contrary, it is in continuous development and evolution due to the very process of semiosis in which it takes part—a property that could well be called continuous self-organization, or self-formation. In-formation, or the internalization of novel forms by the sign, in the process of continuous transformation of the sign.

This objective account of the form as an emanation from a dynamic object, which is ultimately the very essence of reality (its "species," as the scholastic philosophers would put it), is the basis for Peirce's so-called objective idealism. This objective idealism, in turn, is an interpretation of Plato's philosophy that accepts most of Aristotle's contribution as much as a particular reading of Darwinism, by which chance variations in the characteristics of a "species" are fortuitous (i.e., undetermined) as much as "finious," or directed to general end states, and this is the basis for the third aspect of his definition of "Information" for the *Century Dictionary*, as shown above.

Final causation, or the law of association to produce more generalized predicates, is taken by Peirce to be as much a critical logical rule in semiotic processes as the pure chance that Darwinists take to be the sole rule of natural selection. This radically new interpretation of Darwinism, introduced in the core of Platonism, can account for the growth of symbols, its continuous development, and self-organization toward more complex forms that, ultimately, grant the increase of concrete reasonableness in the real.

Contrary to the traditionally accepted account of the increase of entropy in the universe leading to a complete dissolution of all forms, Peirce advocates a balance between dissolution and evolution that, if in the short run it condemns every particular form of life to ultimately end, in the long run would favor intelligence by multiplication of the possible types of life. This means that semiosis and mind, as well as growth and development, are all-pervasive features of our universe. Without the influx of a symbol, Peirce states, the universe would be unintelligible and devoid of life. The evolutionary laws of nature favor life and intelligibility, and every thought, even if mistaken, must represent some aspect of the real and participate in this teleological process:

> A symbol is an embryonic reality endowed with the power of growth into the very truth, the very entelechy of reality. This appears mystical and mysterious simply because we insist on remaining blind to what is plain, that there can be no reality which has not the life of a symbol. (EPII: 323–324)

6 The Symbol as the Vehicle of Information

By symbol, I understand, first, a general sign. Symbols do not exist per se but depend on instantiations in replicas to gain physical embodiment. The typical example is a linguistic word, which does not depend on being written or spoken by some individual to be real, although only these existential instantiations can make it effective. This notwithstanding, a symbol must also represent some general aspects of the object or complex of objects represented, the latter being generally the case.

This general aspect, which is similar to the Platonic notion of the "idea," is also the general form that the symbol conveys as an interpreting power.

In a late manuscript, Peirce explains what he understands by the word "symbol," leaving no doubt that he is not classifying his semiotics as a branch of social psychology or even anthropology at large. In this undated but certainly late manuscript fragment, Peirce brings up the essential features of the nominalistic definition of a symbol (a human-made dyadic match-word—arbitrary, conventional) but then explains how he departs from and expands on it to embrace his realistic account of signification:

> Symbols: The word "symbol" has already many meanings; and I shall ask leave to add a new one. Among its early significations, perhaps the original one, is that of a match-word, a somewhat arbitrarily adopted word or phrase, by which persons of one party recognize one another. It is nearly in this sense that the church creeds are called symbols. So, a flag is a symbol, etc. I think, then, that I shall not wrench the word too much if I use it to mean a sign to which a general idea is attached by virtue of a habit, which may have been deliberately instituted, or may have grown up in a natural way, and perhaps have been acquired with one's mother milk, or even by heredity. (MS 797)

The central point is that every symbol must be a sign with "a general idea attached to it by virtue of a habit." As we have seen, this general idea grounding the symbol is its form or "species" in the mode of the conditional future—the real "would-be" not only capable of evolution but the very leitmotif of every evolutionary process. It is Peirce's habit.

A piece of colored cloth hanging on a pole cannot be a flag if it is not cognizable by an interpreter or community of interpreters. By spatial instantiation, replicas of the same type of flag can be found in many places throughout a nation. Another example is the sun, which is a singular thing but nevertheless a logical general, for its repetitive appearances in time, due to astronomical revolutions, grant the necessary cognitive generality for it to be considered what Peirce calls a "legisign." All physical laws of nature are legisigns because their final interpretants are necessarily habitual. Similarly, all natural classes fall into the same typology, such as chemical elements and biological taxonomies. They are intrinsically true and thus independent of any thinking about them.

However, not every thought, albeit habitual to some degree, has the same necessary nature. The nature of thought is that of being conditionally true. Its condition rests on the normativeness of logic: correct reasoning would in the long run produce habitual final interpretants, but at any stage of inquiry, the provisory dynamic interpretants of any thought are all imperfect to some degree. What is undeniable is their potentiality for being developed to a final true opinion, a power that is expressed by their habitual immediate interpretants. The distinctiveness of thought is that it has the power of information, the power of forming itself. It does so because the idea attached to it is an aspect of the object it professes to represent. Peirce calls this idea internal to the sign "immediate object."

These characteristics explain why the habitual "would-bes" of symbols are always modally vague and non-determined in a certain measure (not so much as to render them useless, but sufficiently to let iconic possibilities creep in and produce novelty), depending on the context to produce particular determinations. These determinations, as they rule out possibilities and define some range of possible types of outcomes, produce variety under invariance that can be mathematically expressed by differential equations. The symbol is then analogous to the double-faced Greek goddess Tyche (Luck). One of her faces captures information coming from the past, embodied by icons inhabiting the indices of experience. In fact, what determines the symbol is the index that it must involve, which materially connects it with the concrete existential context, the hic et nunc of reality. The other face envisions the future and creates conjectures trying to bring the perceived icons and indices into the unity of an evolving concept. A living symbol must be explained as a continuously evolving sign functioning as the vehicle or medium for a flow of information from the wholly determined past toward a vague and undetermined future—the core of semiosis.

This is why the growth of symbols takes place by the always defective but also self-corrective embodiment of the forms of the dynamic objects they represent. If and when this embodiment is completed, the symbol would reach its entelechy, or the perfect final interpretant. This entelechy would be its "ultimate final interpretant," which might be defined as a habit in perfect harmony with the super-order, or super-habit, that rules the laws of nature. Since even the physical laws and biological classes are in evolution, we must accept that the ultimate final interpretant of symbols in an evolutionary universe is to develop alongside them, breaking old habits and embracing new ones as reality itself unfolds toward complexity and concrete reasonableness.

We see then that information is a process intrinsic to the symbol, although the other types of representation also play important roles as symbols semiotically involve them. Icons are essential to embody the form or idea to be communicated by the symbol, while indices are needed to point out the objects to which this idea might be applied. Peirce defines the denoted object as the source of information, which occupies the position of the emitter (Peirce uses "utterer" for the source and "utterance" for the message). Since our main concern here is information in communication, we must consider what turns a symbol into a conveyor of information, such as an assertion (the expression of a particular belief in a definite context) or a proposition (the general form of an informative symbol, usually diagrammatic, which can be asserted in different syntaxes). Peirce calls these informative symbols "dicent" signs or "dicisigns," such as propositions. One of the examples most worked out by Peirce is the weather-vane (or weather-cock), a device capable of informing the wind's direction.

> The reference of a sign to its object is brought into special prominence in a kind of sign whose fitness to be a sign is due to its being in a real reactive relation—generally, a physical and dynamical relation—with the object. Such a sign I term an index. As an example, take a weathercock. This is a sign of the wind because the wind actively moves it. It faces in the very direction from which the wind blows. In so far as it does that, it involves an icon. The wind forces it to be an icon. A photograph which is compelled by optical laws to be an icon of its object which is before the camera is another example. It is in this way that these indices convey information. They are propositions. That is, they separately indicate their objects; the weathercock because it turns with the wind and is known by its interpretant to do so; the

> photograph for a like reason. If the weathercock sticks and fails to turn, or if the camera lens is bad, the one or the other will be false. But if this is known to be the case, they sink at once to mere icons, at best. It is not essential to an index that it should thus involve an icon. Only, if it does not, it will convey no information. (MS 7, 17–18)

The bunch of metal pieces or wood that compose the weather-cock convey no information if it is unassembled, rotten, or broken. The weather-cock can convey information about the wind only if it receives the influx of a symbol given by the interpreters' community that recognizes it as an informative device about wind conditions. A particular weather-cock becomes then the replica of an active symbol, and its pointing to a particular direction becomes an index—although not a pure blind one, but one more appropriately understood as selective or as a quantifier, capable of denoting the real direction of the wind. Its logical validity is grounded on the fact that it is a part of a greater whole, which is the real direction in which the wind is blowing.

> If the weather-cock is functioning correctly, this index also involves an icon that represents the real form of the blow of the wind. And the symbol would be true if the habitual "would-be" that accompanies it correctly represents this iconic information presented by the weather-cock. In logical terms, the symbol connotes truly what it truly denotes. Moreover, the wind that blows the weather-cock is in the past of any conceivable observer that would collect the information, while the pragmatic consequences of the assertion made by the apparatus are always in its future. There must be then a continuous schema, or syntax, linking the real possibilities of the icon at the perceptive level to the icon of the logical consequences. The former enters the knowledge through perceptual judgments, and the latter becomes conscious information by diagrammatic reasoning, where relations are represented in the form of thinking. This flow of information from the real form of the object to the general form of the interpretant in the symbol must then be continuous in time, and the logical schema of time must account for the being of a proposition. (MS 664, 10–13, 1910)

The result is that living symbols, capable of growth and development, must be in a dynamic process of embodying the form of their object in particular situations through experience. The information it generates then also becomes a continuous transformation of its intrinsic forms. This is the kernel of Peirce's synechism, the doctrine that continuity and generality are the bases of reality. Without the real continuous predicates that weld together every mind in a "commens," or commind, no communication would be possible, because there would be no common ground among the minds of the community of interpretants that is created by the very work of symbols, in a way that:

> No object can be denoted unless it be put into relation to the object of the commens. A man, tramping along a weary and solitary road, meets an individual of strange mien, who says, "There was a fire in Megara." If this should happen in the Middle United States, there might very likely be some village in the neighborhood called Megara. Or it may refer to one of the ancient cities of Megara, or to some romance. And the time is wholly indefinite. In short, nothing at all is conveyed, until the person addressed asks, "Where?"—"Oh about half a mile along there" pointing to whence he came. "And when?" "As I passed." Now an item of information has been conveyed, because it has been stated relatively to a well-understood common experience. Thus, the Form conveyed is always a determination of the dynamical object of the commind. (EPII: 478, 1906)

The assumption of a continuous flow of information that ultimately welds and fuses every intelligent mind that participates in and shares the same experience also opens what Peirce affirms to be the "lock" that Kant put in the door of philosophy when asked his most important question: How are a priori synthetic judgments possible? By "a priori," Peirce explains, Kant meant universal, and by synthetic judgments, Kant meant based on experience. Communication, or Rhetoric (as Peirce classified the third branch of Semiotics, after Speculative Grammar and Critical Logic), is to be understood not as a mere application of semiotics to real communicative situations but as a core discipline to understand how knowledge can be gathered from experience and shared by a community. This knowledge can be shared not only by a community of conscious rational minds but even by a community of uni-cellular living beings co-evolving through Darwinian reproduction and natural selection. Peirce says:

> Analogous to the increase of information in us, there is a phenomenon of nature—development—by which a multitude of things come to have a multitude of characters, which have been involved in few characters in few things. (CP 2.420)

However, we can further generalize to reach his cosmological view of information as a natural phenomenon connected as much as to matter as it is to mind, which is not surprising since his metaphysical account of reality does not separate mind and matter like we cut logs with an ax. Peirce's universal evolutionism is much more universal than contemporaneous proposals because Peirce sees some degree of life and iconic feelings whenever chance is involved in breaking old habits and establishing new ones, even when we talk about laws of nature and universal constants—all taken by him as a result of evolution:

> Consider the life of an individual animal or plant; or of a mind. Glance at the history of states, of institutions, of language, of ideas. Examine the successions of forms shown by paleontology, the history of the globe as set forth in geology, of what the astronomer is able to make out concerning the changes of stellar systems. Everywhere the main fact is growth and increasing complexity. (EPI: 307–308)

7 Summary and Conclusion

In natural languages, traditional logic, and human culture, semiotic information is a property of symbols. These symbols can be terms, propositions, or any complex arguments of any length (a book, or even a whole library) used to communicate the form of a cognizable object (taken as anything capable of being represented by generally known predicates) to an interpretant, which is the general consequence of such representation. However, in a more universal and (extreme) realistic approach to the nature of symbols, semiotic information can result from the communication of a general form from past to future, allowing for the growth of complexity. Information is generated as an interpretant of a symbol as it learns from experience—i.e., as it develops, evolves, and gains complexity in time. Ultimately, a portion of space-time can be seen as a medium of communication where information flows from past to future, and living organisms can be seen as very specialized portions of space-time where information flows at a very high rate, escalating quickly to create systems far from thermodynamic equilibrium (or pure randomness). Since every symbol also has both an iconic and an indexical part, so does information. In the iconic part, information is embodied as a novel predicate—unexpected and often fortuitous variations given by continuous perception or experience. The creative novelty in chemical reactions and the fortuitous variations in biological reproduction are good examples of iconic embodiments outside our human species' cultural realm. In the indexical part, information is expressed as subjects to which novelty is assigned, such as a unique crystal or living individual where novelty manifests. In the action of symbols as fully fledged propositions (a sentence, a codified gesture, a traffic sign, etc.), information is generalized and shared to all participants of a community in which it is adopted and used. In the action of symbols as living beings or developing universes, information is realized when new species appear in biological evolution and in the irreversibility of events in space-time as the future comes to the past. In all of these continuous symbolic processes, new general attributes become predicates of known general objects, allowing for permanence, survival, and being—and that is how information becomes meaningful.

References

Fisher RA (1922) On the mathematical foundations of theoretical statistics. Philos Trans R Soc A 222(594–604):309–368. https://doi.org/10.1098/rsta.1922.0009

Frieden BR (2004) Science from Fisher information: a unification. Cambridge University Press, Cambridge

Frieden BR, Romanini V (2008) Eluding the demon—how extreme physical information applies to semiosis and communication. Cognitio-Estudos (PUC-SP) 5:52–63

Hawkins D (1967) Taxonomy and information. In: Cohen RS, Wartofsky MW (eds) Proceedings of the Boston Colloquium for the philosophy of science 1964/1966. Boston Studies in the Philosophy of Science, vol 3. Springer, Dordrecht. https://doi.org/10.1007/978-94-010-3508-8_2

Houser N, Kloesel CJW (1992) The essential Peirce, volume 1: selected philosophical writings (1867–1893). Indiana University Press, Bloomington

Machado I, Romanini V (2011) Semeiotics of communication: from semiosis of nature to culture. Biosemiotics 5:47–60. https://doi.org/10.1007/s12304-011-9117-1

Peirce CS (1877) The fixation of belief. Popular Science Monthly 12(Nov):1–15

Peirce CS, Burks AW (1958) Collected papers of Charles Sanders Peirce, volumes VII and VIII: science and philosophy and reviews, correspondence and bibliography. Harvard University Press, Cambridge

Peirce CS, Fisch MH et al (1982–present) Writings of Charles S. Peirce: a chronological edition, Volume I 1857–1866, Volume II 1867–1871, Volume III 1872–1878, Volume IV 1879–1884, Volume V 1884–1886, Volume VI 1886–1890, Volume VIII 1890–1892. In: Edited by the Peirce Edition Project. Indiana University Press, Bloomington. (1982, 1984, 1986, 1989, 1993, 2000, 2010)

Peirce Edition Project (1998) The essential Peirce, volume 2: selected philosophical writings (1893–1913). Indiana University Press, Bloomington

Peirce CS, Hartshorne C, Weiss P (1932) Collected papers of Charles Sanders Peirce, volumes I and II: principles of philosophy and elements of logic. Harvard University Press, Cambridge

Peirce CS, Hartshorne C, Weiss P (1933) Collected papers of Charles Sanders Peirce, volumes III and IV: exact logic (published papers) and the simplest mathematics. Harvard University Press, Cambridge

Peirce CS, Hartshorne C, Weiss P (1935) Collected papers of Charles Sanders Peirce, volumes V and VI: pragmatism and pragmaticism and scientific metaphysics. Harvard University Press, Cambridge

Shannon C (1948) The mathematical theory of communication. Bell Syst Tech J 27:379–423, 623–656. http://cm.bell-labs.com/cm/ms/what/shannonday/shannon1948.pdf

Part II
Semiosis and Evolution

Inclusive Fitness Teleology and Darwinian Explanatory Pluralism: A Theoretical Sketch and Application to Current Controversies

Philippe Huneman

Abstract Formal Darwinism (FD) [Grafen (2002) J Theor Biol 217:75–91; (2007) J Evol Biol 20:1243–1254.] is a theoretical framework for articulating optimization models in behavioral ecology and allele dynamics modeling in population genetics. It yields a teleology centered on inclusive fitness maximization ("IF teleology"), which captures the many aspects of teleology in Darwinian thinking [Huneman (2019b) Stud Hist Philos Sci Part C 76:101188. 10.1016/j.shpsc.2019.101188] and supports an explanatory pluralism in evolutionary biology. Based on this framework, the present chapter intends to show how the major distinctions regarding kinds of explanation identified in evolutionary biology can be connected and systematized through such explanatory pluralism. Then I will show that it can be redescribed in terms of Aristotle's four causes, and finally, it makes sense of the use of two distinct notions of causation. The rest of the paper analyses two examples where this FD-based pluralism and the correlated use of IF teleology allow one to cast a light on current controversies regarding evolutionary theory: the disputed need to overcome the Modern Synthesis of evolution because of non-genetic inheritance, biased variation, or niche construction; and the opposition of kin selection and multilevel selection regarding the evolution of altruism.

Keywords Explanation · Natural selection · Inclusive fitness · Teleology · Causation · Pluralism · Aristotle

1 Introduction

In this chapter, I will argue that in the context of the "Formal Darwinism" elaborated by Alan Grafen (2002, 2007, 2015), a framework for thinking of explanatory and causal pluralism in evolutionary biology can be designed. Formal Darwinism (FD) is

P. Huneman (✉)
Institut d'Histoire et de Philosophie des Sciences et des Techniques (CNRS), Paris, France

Université Paris I Panthéon Sorbonne, Paris, France

© Springer Nature Switzerland AG 2021

E. Pagni, R. Theisen Simanke (eds.), *Biosemiotics and Evolution*, Interdisciplinary Evolution Research 6, https://doi.org/10.1007/978-3-030-85265-8_7

a view that establishes isomorphisms between this teleology, mathematically understood in terms of "optimization programs," and the population genetics, which models allele dynamics in a population at the genetic level (those isomorphisms hold between programs and the Price equation, see Grafen 2002, 2007, 2014; Huneman 2014a, 2015). The FD-based pluralism articulates teleology and mechanisms of efficient causes. In Huneman (2019b), I already argued that the teleological kind of explanation could be accounted for in terms of "inclusive fitness," where inclusive fitness (of an organism) measures the benefit of a trait or a strategy as the contribution to the offspring directly produced by the focal organism, plus the contribution to offspring produced by other organisms, weighted by the relatedness coefficient.[1] This teleological explanation based on inclusive fitness was there shown to capture the two dimensions of *design* and *contrivance* proper to living organisms—namely, organisms seem to be designed, and their parts are contrived toward an apparent purpose. This, as Okasha (2018) demonstrated, makes for a "'unity of purpose," which in most cases allows the biologist to describe organisms as agents that maximize some magnitude related to survival and reproduction and named fitness.

Besides, various distinctions of explanatory strategies, and causal concepts, have been proposed in evolutionary biology: Mayr's division of ultimate and proximate causes, functions vs. mechanisms, Tinbergen's "four questions" (Tinbergen 1963), and others. In this chapter, I will use Formal Darwinism as a tool for pluralism to systematically assess and articulate those divisions. I will start by sketching the main distinctions between explanatory types in evolutionary biology. Then in a second section, I will trace them back to an explanatory or causal pluralism first stated by Aristotle, who (according to most of the usual translations of his *Physics*) spoke of the "four causes," even though one may consider that he meant here "the four explanantia," *aitiai*. Given that Aristotle's distinction is highly concerned with what we call the difference between final causes or teleology and efficient causes or mechanisms, the IF teleology understood as the genuine Darwinian teleology, according to Huneman (2019b), will in the third section proves instrumental in making sense of this quadripartition in a Darwinian context. Explanatory pluralism is not an epistemic virtue by itself, and that it is not philosophically or scientifically productive unless a formal and conceptual articulation is provided for the different modalities of causation, which is done here.

In the last two sections, I wish to show that this pluralism may be a fruitful framework by considering two major controversies in evolutionary biology. These controversies concern the purported need to expand or extend the classical Modern Synthesis framework (Müller 2017) and the controversies over the proper account of the evolution of prosocial traits such as altruism of hymenopteran insects (who do

---

[1]On inclusive fitness, see Birch (2017b), which explores all dimensions of Hamilton's paper's legacy, who coined the main guidelines of the philosophy of social evolution, including this notion of inclusive fitness and the parent notion of kin selection (see below Sect. 5.2.). The coefficient of relatedness is notably difficult to evaluate and even define, but it is mostly thought to measure the statistical association between individuals at a specific locus of their genome.

not reproduce but work for the queen) or monkeys emitting alarm calls for the tribe at the cost of their lives.

Those are huge controversies, and this chapter will not solve them. However, it will focus on some implicit assumptions therein regarding causal concepts and explanatory perspectives and argue that the IF teleology, brought in the debates, can contribute to making sense of some theoretical divides and overcome disagreements due to the fact that authors talk past each other.

2 Darwinian Teleology and the Pluralism of Causes

By providing an isomorphism between the Price equation in population genetics (an equation analytically describing the change of phenotypic value or allele frequencies between two generations in functions of their frequency and fitness; see Gardner 2008) and "optimization programs" in behavioral ecology (namely, a mathematical description of the hypothetical choice of the best fitness-enhancing strategy by the organism), Formal Darwinism (Grafen 2002, 2007) allows an explanatory pluralism in evolutionary biology. Phenomena can equivalently be understood through optimization schemes or in terms of allele dynamics, and explanations can—depending upon available information and the nature of the explananda—be run at the gene or the organism level (Huneman 2014a, 2014b). This pluralism allows for making sense of a specific kind of teleological explanation, in terms of maximization of inclusive fitness, called IF Teleology (Huneman 2019b). The genuine Darwinian teleology is indeed based on this notion of maximizing inclusive fitness. Such account of teleology allows one to make sense of what I called "intrinsic teleology" (Huneman 2019b) approaches, such as self-organization approaches, or organizational views of functions (e.g., Mossio et al. 2009). It also allows the understanding of plasticity as leading evolution by biologists such as West-Eberhard (2003) or philosophers such as Walsh (2015) and of the etiological theory of functions (e.g., Wright 1973; Millikan 1984; Neander 1991; Griffiths 1993). It makes sense of traits that are apparently not benefitting their bearers, such as the peacock's tail or the worker bee's sterility, and indifferently understands teleological statements regarding organisms or traits (or alleles) (Huneman 2019b). A conceptual analysis of teleology in terms of IF maximization, embedded within the explanatory pluralism based on Formal Darwinism, provides us with a framework to revisit the various notions of causation used by biologists. I will do this in this section.

Since Aristotle, philosophical reflection on causation and explanation has been accompanied by the awareness that causation occurs in many senses. Aristotle famously distinguished four kinds of causes (*aitiai*[2]). Modern science is said to have restricted the "four causes" to the latter, namely the antecedent efficient cause,

[2]Even though it can also be translated by "explanation." See above.

mostly understood as a mechanical impetus. Nevertheless, since then, biologists struggled with causal claims that seem to exceed this restriction, especially by allowing for final terms—goals of *embryological processes, adaptations* that seem to require explanation through design, as Paley famously argued (Huneman 2015, on this argument), *functions*, which explain X by invoking some effect of X, therefore exemplifying the logics of final cause. Analytic philosophers put a great weight of attention on functional statements—especially since the paraphrases by Nagel (1961) or Hempel (1959) in mechanistic terms seemed inconclusive (McLaughlin 2001). However, Kant's major attempt to account for biological judgment represented a first significant treatment of the same problem (Kant 1790; Ginsborg 2014; Huneman 2006).

In this section, I will show that FD pluralism, essentially tying IF teleology with dynamics, can be integrated within a general scheme of explanatory pluralism in evolutionary biology and pinpoint the specific role of teleology. The main idea in the two next sections will be that acknowledging Formal Darwinism and the foundations it provides for Darwinian teleology allows one to build a framework encompassing the main explanatory differences in evolutionary theory.

In the current terms of the philosophy of science, explanatory pluralism means that several types of explanation for aspects of a single phenomenon are different but together legitimate. In this chapter, following Aristotle and the usual translation of "*aitia,*" I consider that explanations search for causes (Salmon 1984; Woodward 2003) so that explanatory pluralism here goes with causal pluralism. In other contexts, I have challenged the exclusivity of causal explanations, understood in the sense of causation entirely restricted to production or to a sense that is based on production,[3] for instance, in Huneman (2010a, 2018). Here I equate causation in a very general sense with explanation. This chapter contributes to an updating of Aristotle's four causes, which justifies my take here. Moreover, given that his quadripartition of causes is very general, I will start by considering biology in general, even though most of my argument will concern evolutionary biology. If one agrees that, following the usual credo, "Nothing makes sense in biology except under the light of evolution," then the proper Darwinian teleology provides us with the genuine meaning of biological teleology, and I can legitimately proceed from evolutionary biology to biology.

To this extent, I start by reviewing five explanatory/causal pluralisms seen in the literature about biology (Sect. 1), and then I will consider the role of Formal Darwinism and IF teleology in their context (Sect. 2).

A. A major variety of pluralism goes with the difference stated by Ernst Mayr (1961) between *"proximate" and "ultimate" causes*. Proximate causes occur in an individual organism's lifetime, whereas ultimate causes pertain to past generations of populations of the organism's species (see Beatty 1994; Ariew 2003). Evolutionary investigations search for ultimate causes, while physiological or

[3] See Glennan (2017) for an account of how production will always be the fundamental meaning of causation.

molecular genetic investigations search for proximate causes. It has been argued that advances in Evo-Devo and molecular genetics challenge this distinction because, for example, if development (embryogenesis, or, more ecologically, niche construction) is relevant to evolution, then the two kinds of causes overlap (Laland et al. 2014; see also Pigliucci and Scholl (2015) for a more nuanced reading of Mayr). However, I think that this difference is still to be considered as a reference point; it is indeed still used to organize textbooks, lectures, or other presentations.

B. A second difference between kinds of causes holds between a *function* and the *mechanism* that realizes the function. The mechanisms of the cells, glands, and vessels in the kidney explain why the elimination of toxins occurs. The fact that the function of kidneys is eliminating toxins explains why kidneys are here. Therefore, functions and mechanisms are two kinds of causes, so indissociably related that they look like two faces of the same medal.

A caveat here: philosophers argued over the concept of function for decades; some favor the etiological concept of function, according to which "F is the function of X" means "F has been selected for doing X" (Wright 1973; Neander 1991). This has several shortcomings (e.g., Walsh 2002; Enç 2002), and many refined versions of the etiological view have been proposed, including Griffiths (1993), Kitcher (1993), Garson (2017), Huneman (2013a). The other prominent family of accounts of functions says that "F is the function of X" means that (a) there is a system (X) in which an R-ing activity is carried on, (b) F is part of S, and X is the contribution of F to R-ing, and (c) X contributes to explain R-ing in a causal way (Cummins 1977, 2002). Therefore, this second account of functions ascribes to functional concepts another explanatory role than the one I considered here (namely, explaining the presence of X), which is the explanatory role considered in etiological views of function. Thus, the current division (B) of causes concerns only the function in the sense of etiological theories.

C. A third difference is precisely the one on which Formal Darwinism focuses: *optimization vs. dynamics*; both would explain why kidneys eliminate toxins, but the selective stories at stake are told in different perspectives, dynamics, and optimization, which ultimately resort to, respectively, a physical and an economical language. It is unclear whether optimization approaches are causal explanations—it has been argued, by Rice (2012) and Huneman (2018), that they are indeed not mechanisms and possibly not causes. In any case, this divide concerns a central explanatory dualism in evolutionary biology.

Lastly, dynamics and optimization seem to be more related to the "ultimate causes," *sensu* Mayr, because they are about evolution. Proximate causes—for example, mammal maternal care behavior—can be understood in terms of a decision-making process oriented towards maximization of a proxy for fitness. However, at the same time, they can be understood in terms of dynamics (such as a process of behavioral conditioning).

D. This difference between types of causes also cuts across another important distinction, which is sometimes believed to make sense of the difference between disciplines like population genetics and disciplines like behavioral ecology. The former considers *how a trait evolves*, namely, by selection or by drift, or by mutation or migration; therefore, they do not consider *why the trait is here for*. The latter asks about the reasons *why those traits or these genes have the fitness they have*; they investigate the "causes of selection" (Wade and Kalicz 1990). In fitness terms, this means that some approaches enquire about the causes of fitness (behavioral ecology), and therefore about adaptation; others, like population genetics, investigate the dynamics of evolution and thereby take fitness values as given, notwithstanding their causes. Thus, their models can be applied to very different ecological situations, provided that the distribution of fitness values is always similar, as well as the population structure, even though they do not consider the nature of adaptation (namely, what adaptations are for, ecologically speaking), which will vary according to ecological settings.

E. Finally, biological explanations can be either more *focused on traits* or more *focused on organisms*. This difference pervades all the three other elements of the above distinctions. Such difference when it comes to teleology is refracted into the difference between the "intrinsic teleology" account (which includes the organizational account of function, *sensu* [Mossio et al. 2009]), and the selected effects (or etiological) account of functions, because the former is focused on organisms as teleologically oriented, while the second is focused on traits (as targets of selection), as explained in Huneman (2019b).

To sum up, any causal explanations can be either ultimate or proximate (A); and ultimate explanations can either focus on traits or organisms (E), and they can be couched either in terms of optimality considerations or in terms of gene dynamics (C). Optimality generally focuses on the reasons for fitness values, while gene dynamics is concerned with the processes through which evolution occurs (D). Finally, something can be, in general, explained by a function, or the explanation can appeal to a mechanism that explains the function (B). Any genuine explanation can be classified according to each of these distinctions (Fig. 1). The next section will detail the relations between these dimensions of distinctions.

3 Formal Darwinism and Explanatory Pluralism

In this section, I argue that the Formal Darwinism (hereafter FD), a theoretical framework first designed by Grafen (2002, 2007, 2014), shows how all these distinctions can be interrelated, and finally, how an explanatory pluralism proper to evolutionary biology could be articulated.

Regarding distinction **D** (*how a trait evolves/why is it for*), the first thing is that the dynamics of gene frequencies in evolutionary theory mostly corresponds to the *working of selection*. In contrast, the optimization approach corresponds to the *reasons for selection*: it wonders why the fitness (inclusive) is higher with this trait.

However, FD shows how these two questions are related: the workings of selection as a dynamics of allele frequency change exactly underpin the realization

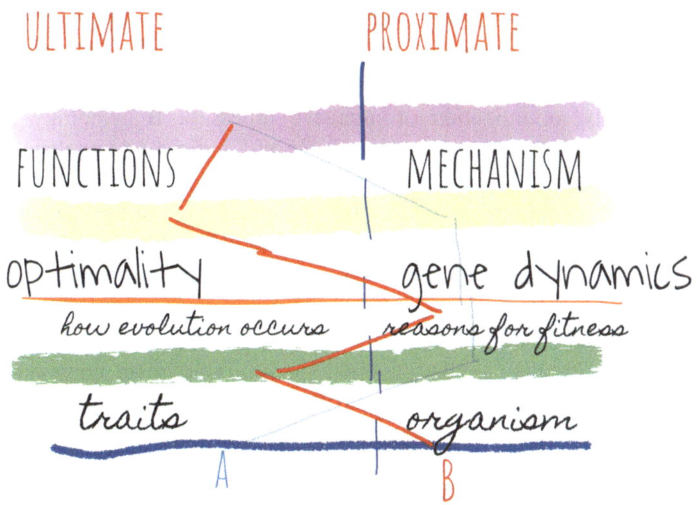

Fig. 1 The five dimensions of explanatory pluralism (and how two given explanations A and B behave regarding those distinctions)

of an optimum (under constraints) by an organism seen as a strategy-choser (see Okasha (2018) for a systematic investigation of this model of the strategy-choser). Hence it can account for the perspective difference **D** within a unified setting—"account for," here, meaning that it shows how the two members of the distinction define legitimate explanations likely to be articulated together in a non-reductive way.

Regarding **E** (*trait focused—organism focused*), FD establishes a link between organism-focused and trait-focused approaches because it explains why the emergence of traits occurs in a designed way, i.e., by contributing to the overall design of the organism (as argued in Huneman 2019b). It yields a genuine teleology—IF teleology—that embraces both trait-oriented perspectives and organism-oriented perspectives. Hence it accounts for **E** within a unified framework, which is the same framework articulating the two sides of **D**.

As to **B** (mechanism-function), teleological explanation pertains to the function of traits; the function/mechanisms distinction often seems to correspond to the distinction **A** between ultimate and proximate causes.[4] In general, however, FD shows that, within the same functional attribution, a more fine-grained distinction stands between a particular mechanism or dynamics of allele frequencies and the teleological explanations that uncover the function's target. In other words, the etiological

[4] As is apparent in the example of the kidneys: their function is eliminating toxins; the mechanism is a complex dynamic of filtering that implies the osmotic properties of cell membranes, and that can be studied at various levels of integration—tissues, cells, metabolic pathways—within the lifetime of the organism.

function F—as when one says "X has the function F"—is carried on by a molecular, cellular, and physiological mechanism. Nevertheless, this function itself results from another kind of mechanism, namely the population-level dynamics of gene frequencies across generations. In turn, such a mechanism can also be explained in terms of teleology, namely, optimization of inclusive fitness or IF teleology (Huneman 2019b). Therefore, the pair "mechanisms/functions" indicates an explanatory difference more complex than a mere correspondence with the pair "ultimate/proximate" explanation (e.g., of the kidney), since the "ultimate" part of the explanation—the "function" in an etiological sense—can be understood simultaneously in terms of a population-level mechanism (namely, as an allele frequency change) and as an optimization.

Such optimization is the maximization of inclusive fitness, which has been shown to be the most systematic sense of teleology in a Darwinian context (Huneman 2019b). Thus, finally, when it comes to the difference **A** between ultimate and proximate in general, dynamics and optimization as understood by FD *both* pertain to ultimate causation. Thereby the distinction **C** further divides each member of distinction **A** into two sub-categories: *dynamical* and *optimal*. The category "Ultimate cause" therefore has to be divided into a dynamic of alleles, which is, so to say, a "*proximate ultimate*" cause, and IF teleology, which is an "*ultimate ultimate*" cause, and both are connected through the isomorphisms stated by Formal Darwinism.

Thus, there are two lessons from this examination of the relationship between differences in explanatory types. First, the five explanatory pluralisms **A-E** that pervade evolutionary biology, and even biology in general (whence one considers distinction **A**), can be *articulated through FD*. Secondly, the same reasoning shows that there are *grades of teleologically ultimate explanations* and that IF provides the "ultimate ultimate" cause in biology.

4 The Four Causes, Revisited

Aristotle famously divided causes into four kinds: *efficient* (material events which produce the consequent); *formal* (the essence of the event, i.e., the usual answer to the question "what is it?"); *final* (the goal of the event); *material* (the substance of the event).[5] Those categories have been mainly given up with the scientific revolution

[5] Here is the text where Aristotle first set this distinction. "In one way, then, that out of which a thing comes to be and which persists is called a cause, e.g., the bronze of the statue, the silver of the bowl, and the genera of which the bronze and the silver are species. In another way, the form or the archetype, i.e., the definition of the essence, and its genera, are called causes (e.g., of the octave the relation of 2:1, and generally number), and the parts in the definition. Again, the primary source of the change or rest; e.g., the man who deliberated is a cause, the father is cause of the child, and generally what makes of what is made and what changes of what is changed. Again, in the sense of end or that for the sake of which a thing is done, e.g., health is the cause of walking about. ("Why is

and the change in the ideas of explanations, causation, and law brought about by modern physics; however, the Aristotelian inspiration may sometimes help draw the conceptual space of a scientific field. Especially, Tinbergen explicitly thought about it by distinguishing four causes, even though they do not correspond one-to-one to Aristotle's causes (Tinbergen 1963[6]). According to him, a biological behavior calls for explanations in terms of its development, its mechanisms, its evolutionary history, and finally, its adaptive meaning or function. Here, "function" would correspond to the final cause, "development" to the efficient cause, "mechanism" to the material cause ("formal cause" is harder to get into the picture).

In the case of Formal Darwinism, gene frequencies' dynamics would correspond to the "efficient cause" since it is the temporal process that brings about the trait along evolutionary time. The "formal cause" would be the function of the trait, namely *what it does* that explains *why it is here*. Furthermore, the "final cause" would be the maximization of inclusive fitness, which is even more general than the function itself. In this case, according to our typology, *population genetics* would consider both the material and the efficient causes (respectively, the alleles and their dynamic); the *behavioral ecology* considers the formal cause, namely the function, and the final cause—namely, to leave as many offspring as possible through direct and indirect ways, i.e., maximizing inclusive fitness.

Thus, within evolutionary biology, FD allows us to recover a quadruple partition of causes that updates an Aristotelian-style metaphysical division. Moreover, as in Aristotle's physics, all those explanations are compatible: taken together, they constitute a complete explanation of the phenomenon under focus.

However, the pluralism we just considered until now was about *types of causes*, like Aristotle's partition, which could map onto *types of explanations*. Nevertheless, there is another philosophical pluralism, which concerns the *concept of causation* (or the nature of causation itself, if one wants) (e.g., Godfrey-Smith 1994). Here, the main distinction is about two extensive families of theories of causation, which corresponds to two intuitions: "A causes B" can mean (1) that "if A had not occurred, B would not have occurred," or it can mean (2) "there is a physical process through which B stems from A." The first families can be called "difference-making causation" (Menzies 2004). They comprehend the so-called counterfactual theories of causation, first elaborated by Lewis (1973) (e.g., Collins et al. 2004), the "manipulationnist" theories of causation—namely, an intervention on variable A changes, *mutatis mutandis*, variable B—, and the probabilistic theories of causation—i.e., the difference A makes on B is probability raising. The second family of accounts is the "process causation," first elaborated by Salmon's theory of transmitted mark (Salmon 1984), then refined (Dowe, 2000). Hall (2004) argued that the two families, corresponding to the two very general concepts of causation, are

he walking about?" We say: "To be healthy," and, having said that, we think we have assigned the cause.)" (Physics, 194b24-195a3, tr. Barnes.)

[6] See also Hladký and Havlíček (2013) on the relation between this quadripartition and Aristotle's four causes.

heterogeneous—especially, features of the latter such as locality are not possessed by the former and reciprocally. Glennan (2017) argued that Salmon's focus on fundamental physics is unnecessary and that the process account may hold even though not all causal processes stand at the level of particle physics. In any case, if Hall is right regarding concept dualism, this entails a general pluralism in scientific explanations, when causal statements can sometimes be uttered according to one concept but not to the other.[7]

This difference directly impinges upon evolutionary biology. Actually, Formal Darwinism states an equivalence between these two kinds of causation: *dynamics* of gene frequencies concerns causation in the sense of a *process*, while *optimization* clearly concerns causation in the sense of *difference-making*, since the traits that make the highest difference upon inclusive fitness will evolve. Therefore, the *explanatory pluralism* stated above in Sect. 1, which systematizes the distinctions **A–E**, is supplemented along another dimension by a *causal pluralism* in the form of equivalence between difference-making and process causation in evolutionary biology, to the extent that equivalences do hold between behavioral ecology and population genetics (instantiating those two distinct concepts of causation) established by FD.

Until then, I proposed that the pluralism proper to FD, especially the kind of teleology defined on this basis (IF Maximization), makes room for a reassessment of Aristotle's causal pluralism. The major distinctions regarding explanations in biology (labeled **A–E**) were translated into a unity of teleology and mechanisms based on the ultimate teleology provided by FD, namely IF teleology. IF also allows one to make sense of the uses of two concepts of causation, the difference-making concept and the process concept, and provides possible connections between those two causal schemes.

To sum up this examination of the relation between Aristotle's classical theory of causes and the current practical distinctions one can make between various explanatory types in evolutionary biology, I will say that various kinds of explanations, distinguished by considering the pluralism involved by FD, could be rethought along the lines of Aristotle's division of causes. The last section of this paper will now apply this view in order to show that it could contribute to solving current controversies in the field because many debates are either triggered by confusion between kinds of causes or by a disagreement about which one should be in priority handled, for methodological or pragmatic reasons.

[7]In Huneman (2012, 2013b), I argued that natural selection could be understood as a causal explanation when causation is taken in the sense of difference-making, and I defended an account of this causation in terms of counterfactuals—but not in the sense of production processes.

5 Pluralism and the Current Controversies

5.1 Pluralism and the Current Controversies: (a)—The So-Called Extended Synthesis

There is an ongoing controversy about whether one should extend or expand evolutionary theory (Pigliucci and Müller 2011, Pigliucci 2007, Jablonka and Lamb 2005, Laland et al. 2014; Müller 2017) in the light of findings by Evo-Devo or behavioral ecology or ecology; or not (e.g., Lynch 2005; Wray et al. 2014).[8] The alternative theories are very varied. However, they all insist, in different ways and with distinct emphases, that the Modern Synthesis cannot integrate new findings of developmental and molecular biology and genomics. These findings include the non-genetic forms of inheritance such as parental effects (Bonduriansky and Day 2009) or epigenetics (Jablonka and Raz 2009; Danchin et al. 2011; Danchin and Pocheville 2014); the role of organisms in shaping their environment (niche construction, Odling-Smee et al. 2003); the complexities of genomic systems (Griffiths and Stotz 2013); or the prevalence of phenotypic plasticity (West-Eberhard 2003) and developmental biases (Raff 1996; Brakefield 2006; Uller et al. 2018). Many of these claims challenge the gene-centered view of the Modern Synthesis, which conferred to population genetics the crucial role of modeling the process of evolution by natural selection.[9] On the contrary, they call for a "return of the organism" (Bateson 2005; Huneman 2010b). I already argued that FD pluralism might conciliate the organism-centered and the gene-centered views because it holds together gene dynamics and organismal teleology (see Huneman 2014b). Here, I consider another avenue of conciliating claims regarding the pluralism of explanation and causation, which is the focus of this chapter.

As summed up by Huxley to Mayr in 1951, the position that the alternative to MS intends to challenge is the following: "Natural selection, acting on the heritable variation provided by the mutations and recombination of a Mendelian genetic constitution, is the main agency of biological evolution."[10] Among many biologically disputed issues, one philosophical argument coined by Modern Synthesis's challengers is about causes of adaptation: the claim is that cumulative selection on small variation is not the actual cause of adaptation. Based on the interpretation of natural selection called "the statisticalist interpretation" (Walsh et al. 2017), which is out of the scope of this chapter, Denis Walsh developed a very sophisticated

[8]Various approaches to the alternative theories are collected in Huneman and Walsh (2017), in which. I tried to show which empirical data would be required to trigger a real revolution of the explanatory scheme proper to the Modern Synthesis, rather than a piecemeal rearrangement. The perspective chosen in this chapter does not contradict this more extended argument.

[9]For example: "The core of the synthetic theory is pretty much just the theory of population genetics" (Beatty 1986, p. 125).

[10]Julian Huxley to Ernst Mayr, 3 September 1951. Papers of Ernst Mayr. HUGFP 14.15 Box 1. Harvard University Archives, Cambridge, MA.

argument according to which natural selection causes the spreading of adaptive traits but does not cause their emergence (see also Walsh 1998). Instead, the latter is due to some individuals' developmental process, especially adaptive plasticity (Walsh 2010) or possibly self-organization (Walsh 2003). In the introduction of their extended synthesis book, Pigliucci and Müller (2011) try to give a general account of the rationale for extending the evolutionary synthesis. The general idea is that Darwin and the Modern Synthesis thinkers had a *statistical* view of evolution by natural selection, mainly counting representations of genes generation after generation because they did not know the mechanisms of variation and production of new traits. However, now that we can access these mechanisms, the overall picture of evolution and adaptation changes: hence we switch from a "statistical" conception of causation in evolution to a "mechanical" conception of causation.[11] As they say, "the shift of emphasis from statistical correlation to mechanistic causation arguably represents the most critical change in evolutionary theory today." Interestingly, this distinction between views is grounded on the distinction stated above between the concepts of causation: according to Pigliucci (2007), the "extended synthesis" would move us towards a process view of causation and explanation, instead of the difference-making view, implicit in any probabilistic account. Thus, for the same reason, the FD pluralism sketched here allows one to conciliate them and solve the dispute, as I will indicate now.

Even if they are quite different, both reformist claims—Walsh and the 'statisticalists', Pigliucci, Müller and the Extended synthesis—mentioned here say that evolution by natural selection, according to MS, is a *statistical* explanation. However, the underlying mechanisms are the genuine *causes* of adaptation (for Walsh) or evolution (for Pigliucci and Müller). This view overlooks the complex picture of evolution by natural selection provided by FD. There is, in fact, a dynamic of gene frequencies, which corresponds to a statistical explanation and especially brackets the variation mechanisms. Even so, the inference that new science of mechanisms would give a better or more accurate account of adaptation or evolution is not justified, since we just saw that this statistical explanation is equivalent to another one couched in non-statistical terms: the teleological one, or *optimization*. Therefore, the need to extend the MS cannot be attributed to the fact that MS was only a dynamic of genes, statistically modeled, because this is only one side of the FD equivalence. On the other side, IF teleology explains adaptations as such, as it unravels the final cause that accounts for why a given trait is there—namely, its maximizing inclusive fitness.

To be more precise, the emphasis put by Walsh (2010, 2015), as by many biologists and philosophers who support a radical reform of the Modern Synthesis, on mechanisms of variation versus natural selection, envelops the assumption that natural selection only explains the *spreading* of the traits, but not their *origin*. There is a longstanding debate on this issue (Neander 1995; Walsh 1998), opposing what

[11] I gave a direct extended critique of this argument in Huneman (2019a), based on analysis of some explanatory practices in postgenomic evolutionary biology, but here I focus on the metaphysics of causation.

Neander has coined the "Negative view" (i.e., the above-mentioned assumption about what selection explains) and a "Creative view" of selection. I will not survey it here; let us just notice that the Darwinians of the Modern Synthesis were concerned by this issue and that, as Beatty (2016, 2019) has argued, one of the hallmarks of their view is the commitment to the idea of a "creativity of natural selection." This commitment means that for them—and against, first, the Mendelians and then, some the opponents of Mayr or Dobzhansky, such as Goldschmidt—natural selection was not only a sieve that prevents the less fit from spreading and therefore lets the fittest spread. It also shapes the adapted traits because it constructs the gene pool from which new variants are built across generations.

The FD pluralism advocated here includes a teleological-explanation-based account for this "creativity" view. Why? The whole population genetics modeling of the action of selection can be seen as a *statistical* explanation of what happens. Therefore, it is mostly concerned by the *spreading* of the adapted traits since those traits are originally brought about by mutation and recombination in some genotypes. However, if we switch to a teleological view—and this perspective is always available, because of the nature of IF teleology, as rooted in the equivalences between allele dynamics and optimization (Grafen 2002, 2007)—, then, the adapted traits themselves (and not their spreading) are here because they maximize inclusive fitness. The teleological argument indeed is formulated as: "since trait X maximizes inclusive fitness, it will be the strategy adopted by organisms of the kind considered, in the environment and population under focus"; and under this view, no population-level process statistically described is among the explanantia. Hence, against the "negative" view, which is presupposed by the arguments of those who think Modern Synthesis should be overcome by a mechanistic and non-statistical understanding of adaptation, natural selection really explains the fact that the adaptive traits are there.[12]

To sum up, here, the IF teleology allows one to downplay one of the epistemological arguments put forth by defenders of an alternative to the Modern Synthesis—namely, the opposition between a supposedly statistical knowledge in MS and a mechanistic knowledge brought for by its alternatives—, because not all aspects of the overall explanatory picture of MS have been considered by such critical account.

5.2 Pluralism and the Current Controversies: (b)—Explaining Altruism: The Multilevel/Kin Selection Controversy, Sketching an Answer

For three decades in evolutionary biology, a huge debate has been going on about what is the cause—or, at least, what best explains—some prosocial traits, i.e., the

[12] In fact, to be successful, the argument should consider cumulative selection underlying complex adaptation as the trait maximizing inclusive fitness. However, this issue is not central here. The importance of cumulative selection to justify teleology is highlighted in Huneman (2019b)

traits that are costly for their bearer and beneficial for others (often called *altruistic*, see West et al. 2007). The apparent evolutionary problem here is that these altruist traits should be counter-selected, but in fact, they are everywhere: sterile workers in hymenopterans insects, alarm calls in antelope or monkeys, helping behaviors in many species, etc. The emergence and maintenance of these traits raise, therefore, a profound problem for classical Darwinism.

Two main theories have been advanced to understand this: the first one (suggested by Hamilton 1963) is the kin selection/inclusive fitness theory, according to which a trait evolves if its benefit is higher than the cost mitigated by the "relatedness." A proxy for relatedness is kinship. However, relatedness is, in general, more complicated than this (Taylor and Frank 1996; Frank 2006). The famous "Hamilton's rule" summarizes this account of altruism: $b > rc$—where b is the benefit for the receiver, c the cost paid by the focal actor, and r their relatedness (even though one can model kin selection by considering the payoff undergone by the focal actor due to the actions of the others modulo relatedness (Taylor et al. 2007)). The rule can compare the two components of inclusive fitness, namely the direct ($-c$) and indirect (br) fitness payoffs.

The other view is called Multilevel Selection (MLS) theory (Sober and Wilson 1998; Wilson 2001; Okasha 2006; Damuth and Heisler 1988). It is grounded in the following idea: if one supposes that evolution occurs in a population divided into groups, altruists compared to selfish individuals or behavioral strategies score worse. However, groups including many altruists fare better than groups with a lower proportion of altruists since altruists, per definition, invest resources, energy, and time in the group's welfare. Therefore, evolution can be considered as resulting from the addition of intragroup competition (which favors selfish elements, see e.g. Burt and Trivers 2006) and intergroup competition (which favors altruists): when the second term is higher, altruism can evolve.

Some authors have claimed that one of the following views is just a particular case of the other. Sober and Wilson (1998) held that multilevel selection is the most general theory and includes kin selection as a case where groups are kin groups; Nowak (2006) claimed that kin selection is only one among other explanations of cooperation and that it is overstated (Nowak et al. 2010). West et al. (2007), Lehmann and Keller (2006), Lehmann et al. (2007), and others argued that the evolutionary mechanism is always a kin selection. In turn, Abbot et al. (2011), West et al. (2010), Ferrière and Michod (2011) argued that the kin selection/inclusive fitness theory is the most powerful and accurate one.[13]

The main argument of MLS tenants is that it captures the real causal structure of evolution, while kin selection may correctly represent what happens to genes, but does not capture anything except a shadow of the causal processes (Sober and Wilson 1998). Interestingly they invoke the same distinction between causal and pseudo-causal distinction (mentioned above Sect. 5.1) as Walsh (2003) did

[13]Nevertheless, see Birch (2017a) for a comprehensive account of what Nowak et al. (2010) really meant about the kin selection and how the controversy mostly focuses on something else.

regarding adaptations. In effect, they say that what causes the evolution of altruism is, in real life, competition between groups, while, by overlooking real interactions between organisms, the kin selection view cannot see this. This view justifies their "averaging fallacy" argument, which in essence means that computing the fitness of organisms by averaging their reproductive success in different contexts (i.e., groups) may give a correct estimation of final gene frequencies. However, it neglects the causes of this final frequency, i.e., their belonging to specific groups, since, by definition, the averaging neglects this fact. This argument connects to Sober and Lewontin (1982) critique of genic selectionism, where allelic selectionist is said to be blind to the level of real causes, which stand at the level of the genotypes; the classic example here is the case of the superiority of heterozygotes, as happens with recessive alleles for malaria resistance in sickle cell anemia.[14]

On the contrary, supporters of inclusive fitness would say that the causes of evolution are occurring at the level of alleles increasing or decreasing in frequency in accordance to their contribution to the fitness of the relatives so that in all cases presented by supporters of MLS (e.g., Traulsen and Nowak 2006), what really happens is a process of kin selection (Lehmann et al. 2007). They add that multilevel selection is most of the time mathematically intractable (West et al. 2007), which provides a substantial methodological advantage to kin selection models. Notwithstanding this argument, the two camps clearly do not focus on the same causal aspects: the MLS supporters argue that the causal story is the competition between groups, so their causally relevant facts are intergroup competition and intragroup competition; the kin selection theorists argue that the causally relevant facts are direct fitness benefits and indirect fitness benefits. Therefore, in the first approach, what is crucial to determine is the population's partition into groups. In the second one, what is crucial is relatedness because it allows one to define and measure indirect payoffs. Furthermore, it seems that relatedness is tractable in an easier way than group partitioning, which could explain that most accounts of social evolution are couched in terms of kin selection (West et al. 2010).

However, there are equivalences between these stories (for an in-depth analysis of those equivalences, see Birch 2017a). As Kerr and Godfrey-Smith (2002) have shown, a multilevel approach of evolution, in general, is mathematically equivalent to an approach where the fitness of individuals is contextualized over groups and then computed across generations. More simply, the more the intergroup competition increases relative to the intragroup competition, the more Multilevel Selection one has for the prosocial (altruists) traits: but this means that the between-group variance relatively overcomes the intragroup variance. This fact, in turn, entails that relatedness increases in each set of individuals interacting with a focal individual and then in general, which in the end means that kin selection increases; and reciprocally (Frank 2006; Foster et al. 2006; West et al. 2007). Let us call (E) this equivalence.

[14]Because being heterozygote is a property of the genotype, not the allele; hence the relevant causal property stands at the genotypic, not the allelic, level.

6 Conclusions

In this paper, I considered the uses of a view of evolutionary theory based on Formal Darwinism, which confers an overarching role to a specific teleological explanation understood in terms of the maximization of inclusive fitness. On this basis, I presented a general scheme for making sense of explanatory pluralism in evolutionary biology, integrating five classical views of explanatory differences, and then, both the four types of causes and the two notions of causation, and sketched the position occupied by the Darwinian IF teleology into this scheme.

This is not only a general investigation of the scientific image of biology aimed at philosophers; more importantly, it is intended to provide a framework in which some controversies can be solved, and some current challenges to classical evolutionary thought can be addressed and assessed. This is why I indicated how an awareness of such explanatory and causal pluralism—updating (for the former) the Aristotelian quadripartition—may help biologists in dealing with two massive controversies in evolutionary theory. These controversies concern the call for an alternative explanatory scheme integrating non-genetic inheritance, facilitated variation or niche construction, and the debates about the proper model and theory for accounting for the evolution of prosocial traits. Because Aristotle's distinctions have been fruitful in understanding science across the ages, I think that an attempt to update these distinctions by considering various explanatory practices in evolutionary biology could be useful.

Acknowledgements This paper stems from conversations and work with Andy Gardner, without whom nothing would have been possible. I am hugely grateful to him, and thank him for his reading of the first version of the text. I also thank Elena Pagni, as well as two anonymous reviewers for their reviews. I am thankful to Andrew McFarland for language checking.

Editorial Note: The Evolution of Meaning—From Neo-Darwinism to Biosemiotics

Richard Theisen Simanke

Huneman's chapter presents an original, thoughtful, and compelling argument for explanatory pluralism in the biological sciences. Assuming Formal Darwinism as a theoretical framework and focusing on the crucial concept of inclusive fitness maximization, the author argues that this perspective provides a consistent strategy for articulating the different causal pluralisms proposed within the field of evolutionary theory. Aristotle's theory of causation—arguably the first systematic model of explanatory pluralism in the history of philosophy of science—is called upon as a historical point of comparison, and the author presents his argument as reconstruction and updating of the Aristotelian views, especially on the complementary role played by teleology and mechanism in causation. Underlying the author's views is the conviction that explanatory pluralism is not an epistemic virtue by itself and that it is not philosophically or scientifically productive unless one provides a formal and conceptual articulation for the different modalities of causation. The chapter still contains a consistent proposal not only to articulate the different kinds of causes within each pluralistic model but also to relate these models to one another.

Since this collection's general subject matter addresses both evolutionary biology and biosemiotics, it seemed adequate to add here some remarks concerning how an argument for explanatory pluralism and the integration of different theoretical models relates to the biosemiotic view of living beings as meaning-producing entities and natural interpretive systems.

First of all, one must note that Neo-Darwinian evolutionism, however prevalent it may be in the context of contemporary life sciences, is but one of the great paradigms or theoretical models that one can distinguish in this field. Biosemiotics is another model of comparable scope and complexity, with all the doctrinal and methodological implications entailed by such condition. These models foster the emergence of research programs on specific questions in the field of biological knowledge and the formation of research communities endorsing these views of life and the corresponding conceptions of science. Other such models can be mentioned, like those defined by the core concepts of autopoiesis or artificial life.

All these theories organize biological research, with repercussions extending from the more concrete and applied issues to the more general and abstract ones. These more abstract questions touch the borders between biological sciences and the philosophy of biology, including the problem of defining *life* as such. Many authors (El-Hani 2008; Emmeche 1997) have argued that these paradigms provide the possibility for a situated and circumscribed definition of the very concept of life, relatively to the theoretical model informing the investigation of its phenomena. This strategy could overcome the supposed intractability of defining life, often rejected as a metaphysical and unscientific problem. Thus, from the viewpoint of artificial life, being alive is defined as a property of systems capable of reacting automatically and adaptively, in an open-ended way, to unpredictable changes in their environments (Bedau 1996). For autopoiesis, living beings are organizationally closed but structurally open (both materially and energetically) networks, whose components produce the network itself and its boundaries and are, in turn, recursively produced by them (Maturana and Varela 1980). From a Neo-Darwinian perspective, if one wishes to define life in accordance to the major theoretical principles (which is not something mandatory to do in this framework), life would consist in the property of being self-replicating entities likely to evolve through random variation of inheritable traits and a posteriori natural selection of those traits favoring survival and differential reproduction (Dawkins 1976, 1983). Finally, biosemiotics sees life as meaning-production through interpretation of natural sign-systems or, in Claus Emmeche's (1998, p. 11) synthetic formulation, as the "*functional interpretation of signs in self-organised material code-systems making their own Umwelts*" (author's emphases).

One of these models' common denominators is that they all recognize the evolution of species, even if departing from the orthodox Neo-Darwinian view of evolution, at least in some of its aspects. Biosemiotics also acknowledges the species transformation over time but changes the focus from the evolution of structures, functions, and behavior to the evolution of systems of signs and the organisms' interpretation capacities. At least since Terrence Deacon's works, the integration of evolutionary and semiotic perspectives has come to the foreground of the debate concerning human evolution and its distinctive features (Deacon 1997; Schilhab et al. 2012). In this context, Huneman's equally integrative approach, even if formulated from within the Darwinian evolutionary framework, can contribute to bringing closer two of the main theoretical models in contemporary biological sciences with their respective conceptions about the fundamental nature of living beings, as seen above.

Particularly significant is Huneman's recovery and updating of Aristotelian biological philosophy. The rediscovery of Aristotle's philosophy of nature is a striking feature in contemporary science, especially biology, although not exclusively (Feser 2019; Simpson et al. 2018). Besides providing a representative historical illustration for the kind of explanatory pluralism claimed by Huneman, Aristotelian naturalism makes it possible to consider the teleological dimension of biological explanation in terms compatible with a scientific attitude and thus reconcile two types of explanatory strategies—mechanism and finalism—often regarded as incompatible.

As intentional acts, the symbolic communication and sign interpretation privileged by biosemiotics contain a dimension of intentionality in the phenomenological sense of the term, even if considered natural phenomena (Hoffmeyer 2012). Thus, they require some teleology modality, even if it is not the transcendent finality presupposed by vitalist and metaphysical views of life, evolutionary or otherwise. Again, reference to classical conceptions of life and nature can provide a model to conceive of the immanence of meaning in the lifeworld without simply anthropomorphising it. As Merleau-Ponty remarks in the opening of his courses on nature at the Collège de France, referring to classical Greek thought:

> There is nature wherever there is a life that has a meaning but where, however, there is not thought (. . .). Nature is what has a meaning, without this meaning being posited by thought; it is the autoproduction of a meaning. Nature is thus different from a simple thing. It has an interior, is determined from within (. . .). Yet nature is different from man; it is not instituted by him (. . .). (Merleau-Ponty 2003, p. 3)

There are elements in this passage that allow for adding a biosemiotic and autopoietic perspective to the scientific knowledge of nature—and of life, in particular. The evolutionary element—generally absent from classical thought—complements and makes this multidimensional theoretical model more comprehensive. However, if Neo-Darwinian evolutionism must participate in this process, one must trim some of its edges and challenge the necessity of some of its doctrinal commitments. It is to this task that Huneman's work presented here represents an invaluable contribution.

References

Abbot P, Abe J, Alcock J, Alizon S, Alpedrinha JAC et al (2011) Inclusive fitness theory and eusociality. Nature, Nature Publishing Group 471(7339):E1–E4
Ariew A (2003) Ernst Mayr's 'ultimate/proximate' distinction reconsidered and reconstructed. Biol Philos 18(4):553–565
Bateson P (2005) The Return of the whole organism. J Biosci 30(1):31–39
Beatty J (1986) The synthesis and the synthetic theory. In: Bechtel W (ed) Integrating scientific disciplines. Nijhoff, The Hague, pp 125–136
Beatty J (1994) The proximate/ultimate distinction in the multiple careers of Ernst Mayr. Biol Philos 9:333–356
Beatty J (2016) The creativity of natural selection? Part I: Darwin, Darwinism, and the mutationists. J Hist Biol 49:659
Beatty J (2019) The creativity of natural selection? Part II: the synthesis and since. J Hist Biol 52:705–731. https://doi.org/10.1007/s10739-019-09583-4
Bedau M (1996) The nature of life. In: Boden M (ed) The philosophy of artificial life. Oxford University Press, Oxford, pp 332–357
Birch J (2017a) The inclusive fitness controversy: finding a way forward. R Soc Open Sci 4:170335
Birch J (2017b) Philosophy of social evolution. Oxford University Press, New York
Bonduriansky R, Day T (2009) Nongenetic inheritance and its evolutionary implications. Annu Rev Ecol Evol Syst 40:103–125
Brakefield PM (2006) Evo-devo and constraints on selection. Trends Ecol Evol 21:362–368
Burt A, Trivers R (2006) Genes in conflict: the biology of selfish genetic elements. Belknap Press of Harvard University Press, Cambridge, MA
Collins J, Hall N, Paul L (eds) (2004) Causation and counterfactuals. MIT Press
Cummins R (1977) Functional analysis. J Philos 72:741–765
Cummins R (2002) Neo-teleology. In: Ariew et al. [2002], pp 157–172
Damuth J, Heisler IL (1988) Alternative formulations of multilevel selection. Biol Philos 3:407–430
Danchin E, Charmantier A, Champagne FA, Mesoudi A, Pujol P, Blanchet S (2011) Beyond DNA: integrating inclusive inheritance into an extended theory of evolution. Nat Rev Genet 12:475–486
Danchin E, Pocheville A (2014) Inheritance is where physiology meets evolution. J Physiol 592 (11):2307–2317
Dawkins R (1976) The selfish gene. Oxford University Press, Oxford
Dawkins R (1983) Universal Darwinism. In: Bendall DS (ed) Evolution from molecules to man. Cambridge University Press, Cambridge, pp 403–428
Deacon T (1997) The symbolic species: the co-evolution of language and the brain. W. W. Norton & Co, New York and London
Dowe P (2000) Physical causation. Cambridge University Press, Cambridge
El-Hani CN (2008) Theory-based approaches to the concept of life. J Biol Educ 42(4):147–149
Emmeche C (1997) Autopoietic systems, replicators, and the search for a meaningful biologic definition of life. Ultimate Reality Meaning 20(4):244–264
Emmeche C (1998) Defining life as a semiotic phenomenon. Cybern Human Knowing 5(1):3–17
Enç B (2002) Indeterminacy of function attributions. In: Ariew A, Cummins R, Perlman M (eds) Functions: new essays in the philosophy of psychology and biology. Oxford University Press, New York, p 291
Ferrière R, Michod R (2011) Inclusive fitness in evolution. Nature 471(7339):E6-8
Feser E (2019) Aristotle's revenge: the metaphysical foundation of physical and biological science. Editiones Scholasticae, Neunkirchen
Foster KR, Wenseleers T, Ratnieks FLW (2006) Kin selection is the key to altruism. Trends Ecol Evol 21:57–60

Frank SA (2006) Social selection. In: Fox CW, Wolf JB (eds) Evolutionary genetics: concepts and case studies. Oxford University Press, Oxford, pp 350–336
Gardner A (2008) The price equation. Curr Biol 18(5):R198–R202
Garson J (2017) A generalized selected effects theory of function. Philos Sci 84(3):523–543. https://doi.org/10.1086/692146
Ginsborg H (2014) The normativity of nature: essays on Kant's critique of judgment. Oxford University Press
Glennan G (2017) The new mechanical philosophy. Oxford University Press, New York
Godfrey-Smith P (1994) A modern history theory of functions. Noûs 28:344–362
Grafen A (1984) Natural selection, kin selection and group selection. In: Krebs J, Davies N (eds) Behavioural ecology: an evolutionary approach. Blackwell, Oxford, pp 62–84
Grafen A (2002) A first formal link between the Price equation and an optimisation program. J Theor Biol 217:75–91
Grafen A (2006) Optimisation of inclusive fitness. J Theor Biol 238:541–563
Grafen A (2007) The formal Darwinism project: a mid-term report. J Evol Biol 20:1243–1254
Grafen A (2014) The formal darwinism project in outline. Biol Philos 29(2):155–174
Grafen A (2015) Biological fitness and the fundamental theorem of natural selection. *Am Nat* 186(1):1–14
Griffiths P (1993) Functional analysis and proper functions. Br J Philos Sci 44(3):409–422. https://doi.org/10.1093/bjps/44.3.409
Griffiths P, Stotz K (2013) Genetics and philosophy: an introduction. Cambridge University Press, Cambridge
Hall N (2004) Two concepts of causation. In: Collins J, Hall N, Paul LA (eds) Causation and counterfactuals. The MIT Press, Cambridge, MA, pp 181–204
Hamilton W (1963) The genetic evolution of social behavior. J Theor Biol 7:1–16
Hempel CG (1959) The logic of functional analysis. In: Gross L (ed) Symposium on sociological theory. Harper and Row, New York, pp 271–287
Hladký V, Havlíček J (2013) Was Tinbergen an Aristotelian? Comparison of Tinbergen's four whys and Aristotle's four causes. Hum Ethol Bull 28(4):3–11
Hoffmeyer J (2012) The natural history of intentionality: a biosemiotic approach. In: Schilhab T, Stjernfelt F, Deacon T (eds) The symbolic species evolved. Springer, Dordrecht, pp 97–116
Huneman P (2006) From comparative anatomy to the 'adventures of reason'. Stud Hist Phil Biol Biomed Sci 37(4):649–674
Huneman P (2010a) Topological explanations and robustness in biological sciences. Synthese 2010 (177):213–245
Huneman P (2010b) Assessing the prospects for a return of organisms in evolutionary biology. Hist Philos Life Sci 32(2/3):341–372
Huneman P (2012) Natural selection: a case for the counterfactual approach. Erkenntnis 76 (2):171–194
Huneman P (2013a) Weak realism in the etiological theory of functions. In: Huneman P (ed) Functions: selection and mechanisms, vol 13. Springer, Dordrecht, pp 105–113
Huneman P (2013b) Assessing statistical views of natural selection: is there a room for non local causation? Stud Hist Phil Biol Biomed Sci 44:604–612
Huneman P (2014a) Formal Darwinism and organisms in evolutionary biology: answering some challenges. Biol Philos 29:271–279
Huneman P (2014b) A pluralist framework to address challenges to the modern synthesis in evolutionary theory. Biol Theory 9(2):163–177
Huneman P (2015) Redesigning the argument from design. Paradigmi 33(2):105–132
Huneman P (2018) Diversifying the picture of explanations in biological sciences: ways of combining topology with mechanisms. Synthese 195(1):115–146
Huneman P (2019a) The multifaceted legacy of the human genome program for evolutionary biology: an epistemological perspective. Perspect Sci 27(1):117–152

Huneman P (2019b) Revisiting Darwinian teleology: a case for inclusive fitness as design explanation. Stud Hist Philos Sci Part C 76:101188. https://doi.org/10.1016/j.shpsc.2019.101188
Huneman P, Walsh D (2017) Challenging the modern synthesis: adaptation, development, and inheritance. Oxford University Press, New York
Jablonka E, Lamb M (2005) Evolution in four dimensions. MIT Press, Cambridge
Jablonka E, Raz G (2009) Transgenerational epigenetic inheritance: prevalence, mechanisms, and implications for the study of heredity and evolution. Q Rev Biol 84:131–176
Kant I (1790) Critique of judgment (1987) (trans: Pluhar WS). Hackett, Indianapolis. (often translation revised) (noted CJ)
Kelly R (1985) The Nuer conquest. The structure and development of an expansionist system. The University of Michigan Press, Ann Arbor
Kerr B, Godfrey-Smith P (2002) Individualist and multi-level perspectives on selection in structured populations. Biol Philos 17:477–517
Kitcher P (1993) Function and design. Midwest Stud Philos 18(1):379–397
Laland K, Tobias U, Feldman M, Sterelny K, Müller GB, Moczek JO-S (2014) Does evolutionary theory need a rethink? Yes, urgently. Nature 514:161–164
Lehmann L, Keller L (2006) The evolution of cooperation and altruism—a general framework and a classification of models. J Evol Biol 19(5):1365–1376
Lehmann L, Keller L, West S, Roze D (2007) Group selection and kin selection: two concepts but one process. Proc Natl Acad Sci USA 104:6736–6739
Lewis D (1973) Causation. J Philos 70(17):556–567
Lynch M (2005) The origins of genome architecture. Sinauer Assocs., Inc, Sunderland, MA
Maturana HR, Varela F (1980) Autopoiesis and cognition. Reidel, Dordrecht
Mayr E (1961) Cause and effect in biology. Science 134:1501–1506
McLaughlin P (2001) What functions explain. Functional explanation and self-reproducing systems. Cambridge University Press, Cambridge
Menzies P (2004) Difference-making in context. In: Collins J, Hall N, Paul LA (eds) Causation and counterfactuals. The MIT Press, Cambridge, MA, pp 139–180
Merleau-Ponty M (2003) Nature: course notes from the Collège de France. Northwestern University Press, Evanston, IL
Millikan RG (1984) Language, thought, and other biological categories. MIT Press, Cambridge, MA
Mossio M, Saborido C, Moreno A (2009) An organizational account of biological functions. Br J Philos Sci 60:813–841
Müller GB (2017) Why an extended evolutionary synthesis is necessary. Interface Focus 7:20170015
Nagel E (1961) The structure of science. Routledge & Kegan Paul, London
Neander K (1991) Function as selected effects: the conceptual analyst's defense. Philos Sci 58:168–184
Neander K (1995) Pruning the tree of life. Br J Philos Sci 46:59–80
Nowak MA (2006) Five rules for the evolution of cooperation. Science 314(5805):1560–1563. https://doi.org/10.1126/science.113375
Nowak MA, Tarnita CE, Wilson EO (2010) Evolution of eusociality. Nature 466:1057–1062
Odling-Smee J, Laland K, Feldman M (2003) Niche construction. The neglected process in evolution. Princeton University Press, Princeton
Okasha S (2006) The levels of selection in evolution. Oxford University Press, Oxford
Okasha S (2018) Goals and agents in evolution. Oxford University Press, Oxford
Pigliucci M (2007) Do we need an extended evolutionary synthesis? Evolution 61(12):2743–2749
Pigliucci M, Müller G (2011) Evolution: the extended synthesis. MIT Press, Cambridge
Pigliucci M, Scholl R (2015) The proximate–ultimate distinction and evolutionary developmental biology: causal irrelevance versus explanatory abstraction. Biol Philos 30(5):653–670
Raff R (1996) The shape of life: genes, development, and the evolution of animal form. University of Chicago Press, Chicago

Rice CC (2012) Optimality explanations: a plea for an alternative approach. Biol Philos 27(5):685–703
Salmon WC (1984) Scientific explanation and the causal structure of the world. Princeton University Press
Schilhab T, Stjernfelt F, Deacon T (eds) (2012) The symbolic species evolved. Springer, Dordrecht
Simpson WMR, Koons RC, Teh NJ (eds) (2018) Neo-Aristotelian perspective on contemporary science. Routledge, New York and London
Sober E, Lewontin RC (1982) Artifact, cause and genic selection. Philos Sci 49(2):157–180
Sober E, Wilson DS (1998) Unto others. Harvard University Press, New Haven
Taylor PD, Frank SA (1996) How to make a kin selection model. J Theor Biol 180:27–37
Taylor P, Wild S, Gardner A (2007) Direct fitness or inclusive fitness: how shall we model kin selection? J Evol Biol 20:301–309
Tinbergen N (1963) On aims and methods of ethology. Z Tierpsychol 20:410–433
Traulsen A, Nowak MA (2006) Evolution of cooperation by multilevel selection. Proc Natl Acad Sci USA 103:10952–10955
Uller T, Moczek AP, Watson RA, Brakefield PM, Laland KN (2018) Developmental bias and evolution: a regulatory network perspective. Genetics 209(4):949–966
Wade MJ, Kalisz S (1990) The causes of natural selection. Evolution 44:1947–1955
Walsh D (1998) The scope of selection: Sober and Neander on what natural selection explains. Australas J Philos 76(2):250–264
Walsh D (2002) Brentano's chestnuts. In: Ariew A, Cummins R, Perlman M (eds) Functions: new essays in the philosophy of psychology and biology. Oxford University Press, New York, p 314
Walsh D (2003) Fit and diversity: explaining adaptive evolution. Philos Sci 70:280–301
Walsh D (2010) Two neo-Darwinisms. Hist Philos Life Sci 32(2-3):317–339
Walsh D (2015) Organisms, agency and evolution. Oxford University Press, New York
Walsh D, Ariew A, Mattehn M (2017) Four pillars of statisticalism. Philos Theory Pract Biol 9:1. https://doi.org/10.3998/ptb.6959004.0009.001
West SA, Griffin AS, Gardner A (2007) Social semantics: altruism, cooperation, mutualism, strong reciprocity and group selection. J Evol Biol 20:415–432
West S, Wild J, Gardner A (2010) The theory of genetic kin selection. J Evol Biol 24:1020–1043
West-Eberhard MJ (2003) Developmental plasticity and evolution. Oxford University Press, Oxford
Wilson DS (2001) Evolutionary biology: struggling to escape exclusively individual selection. Q Rev Biol 76:199–206
Woodward J (2003) Making things happen: a theory of causal explanation. Oxford University Press
Wray GA, Hoekstra H, Futuyma D, Lenski R, Mackay TFC, Schluter D, Strassmann J (2014) Does evolutionary theory need a rethink? No, all is well. Nature 514:161–164
Wright L (1973) Functions. Philos Rev 82:139–168

The Origins and Evolution of Design: A Stage-Based Model

Juan Mendoza-Collazos, Jordan Zlatev, and Göran Sonesson

Abstract Within a broader definition of design, as the conception and planning of everything that is artificial, the place of tools is central. With the help of cognitive semiotics in general and Donald's theory of the evolution of the human mind in particular, we propose a stage model for the evolution of design, consisting of four stages: Proto-design, Simple design, Complex design, and Advanced design, some of which are further divided into substages. We argue that nonhuman animals are capable of Proto-design, but it is only with the advent of stone tool technology in hominins that we witness Simple design. The stages are individuated based on evidence from archaeology and cognitive science, but also on the particular semiotic resources that are novel for each stage: mimesis-based gestures, speech, drawings, and polysemiotic communication. The model suggests both continuities and relative discontinuities in the evolution of design, and more generally, in human bio-cultural evolution.

Keywords Bio-cultural evolution · Tool-making · Cognitive semiotics · Design theory · Design semiotics · Derived agency

J. Mendoza-Collazos (✉)
Cognitive Semiotics, Center for Language and Literature, Lund University, Lund, Sweden

Universidad Nacional de Colombia, Bogotá, Colombia
e-mail: juan.mendoza@semiotik.lu.se

J. Zlatev · G. Sonesson
Cognitive Semiotics, Center for Language and Literature, Lund University, Lund, Sweden
e-mail: jordan.zlatev@ling.lu.se; goran.sonesson@semiotik.lu.se

© Springer Nature Switzerland AG 2021

E. Pagni, R. Theisen Simanke (eds.), *Biosemiotics and Evolution*, Interdisciplinary Evolution Research 6, https://doi.org/10.1007/978-3-030-85265-8_8

1 Introduction

What is design, and how has it evolved? There are at least two different ways to answer the first question. In a narrow sense, design is a particular kind of practice within industrialized societies serving to produce complex artifacts (Bayasid 2004; Acha 2009). According to a broader definition, design involves "the conception and planning of the artificial" (Buchanan 1992: 14). We may understand "the artificial" to mean material resources that go beyond those offered by the living body; in this context the place of tools is central. Taking the broader view in order to answer the second question, we here present an evolutionary stage-based model: going from the Proto-design of nonhuman apes to the Advanced design of human beings through the strategic use of signs in polysemiotic communication, understood as "the combination of a number of different semiotic systems [e.g. Language, Gesture, Depiction] within an integrated communicative system" (Zlatev et al. 2020: 157). In particular, we explore the relationship between cognitive skills and semiotic resources involved in tool-making, from the dawn of humanity to the present, drawing some implications for current design practices.

The arguments that we present are based on studies of cultural evolution from fields including archaeology (Bar-Yosef 2017; Dal Sasso et al. 2014; Stout et al. 2011; Wynn 2012), comparative psychology (Stout and Chaminade 2007), evolutionary theory (Richerson and Boyd 2005; Henrich 2017; Heyes 2018), cognitive science (Casakin and Kreitler 2011; Donald 2012; Itkonen 2005; Schön 1983; Vaesen 2012), and cognitive semiotics (Zlatev 2008, 2016, 2019; Dunér and Sonesson 2016; Sonesson 2019). Notably, such studies show *differences* in the cognitive capabilities and semiotic resources involved in tool-making at different stages of human bio-cultural evolution. Rudiments of the abilities required for tool use and tool-making are found in other primates, notably those capable of social learning (Tomasello and Call 1997), as well as some other animals (Hunt et al. 2006; McGrew 2013). However, these abilities are only involved in forms of Proto-design, serving as the basis for the emergence of proper design. The first steps into Simple design in hominins (Semaw et al. 1997; Wynn 2012) also do not show noticeable qualitative changes compared to Proto-design, as we show in section "A Model for the Evolution of Design." Nevertheless, with the mimetic cultures of *Homo ergaster/erectus*, something genuinely new emerged: bodily mimesis (Donald 1991, 2013; Zlatev 2019). From then on, the incremental bio-cultural evolution of design has led to levels of fantastic complexity. This, in short, is the basis for the stage-based model of the evolution of design presented in section "A Model for the Evolution of Design." Before that, we briefly characterize the science of cognitive semiotics and explain why it is appropriate as a framework for studying the evolution of design.

2 The Relevance of Cognitive Semiotics to the Study of Design

As the transdisciplinary science of meaning-making, cognitive semiotics aims to integrate facts and findings, as well as concepts and methods, developed within semiotics, linguistics, cognitive science, and other human sciences. The home site of the International Association for Cognitive Semiotics (IACS) specifies that:

> /c/ognitive semiotics investigates the properties of our meaningful interactions with our surroundings in all domains. We integrate perspectives, methods and insights from cognitive science, cognitive linguistics and semiotics ... using experimental methods, as well as classical text analysis and theory. (iacs.dk)

As Zlatev (2015: 1043) observes, while cognitive semiotics "may indeed focus on what is specific about human forms of semiosis, there is widespread agreement that this can only be properly understood in a comparative and evolutionary perspective." As pointed out by Sonesson (2009a) and Tønnessen (2016), this clearly implies that cognitive semiotics overlaps with biosemiotics, as understood by either Hoffmeyer (2008) or Barbieri (2008). For example, both cognitive semiotics and biosemiotics agree on the claim that life is a precondition for any kind of meaning-making (Zlatev 2003; Sheets-Johnstone 2011; Niño 2015). However, there are also differences, as cognitive semiotics emphasizes the triangulation of first-, second-, and third-person methods (Zlatev 2015) and is more indebted to Husserlian phenomenology than to Peircean semiotics (Sonesson 2009b). A specific difference that is especially relevant in the present context is that while meaning may be coextensive with life, cognitive semiotics researchers usually hold that there are *different layers of meaning-making*. These come in a particular sequence, both in evolution and in child development. This difference does not mean that there must be total breaches between these layers of meaning. Instead, it means that, at their core, they are differentially structured and contain features that are irreducible to lower levels. This basic idea is reflected in the various versions of the *Semiotic Hierarchy* (Zlatev 2009, 2018; Sonesson 2015a; Konderak 2018).

Starting from the "bottom," biological meaning in the *Umwelt* is distinct from the physical world as such, but also forms the stratum on which the *natural Lebenswelt*, characterized by perceptual intentionality, emerges. In turn, this layer serves as a basis for the *cultural Lebenswelt*, which presupposes a community of individuals with shared habits and conventions. A particular kind of *cultural Lebenswelt* is built up from *signs*, which requires the ability to handle meanings differentiated into expression and content, or in other words: constituted of two items, one of which is more directly experienced, while the other is in more focus (Sonesson 2009a). These are eventually organized into different sign systems, including Language, Gesture, and Depiction systems,[1] which become increasingly complex with cultural evolution (Zlatev 2016, 2019).

---

[1] Capital letters are used to distinguish the sign systems themselves from particular manifestations of them, such as individual languages, culturally shaped gestures, or practices such as painting (see Zlatev et al. 2020).

Tools and other artifacts are clearly physical objects. In this sense, properly speaking, they are outside of the Semiotic Hierarchy, that is, meaningless in themselves. However, given that they provide different *affordances for action*, they are clearly something more than mere objects. They may even seem imbued with a kind of shadowy life, which has led some scholars to treat them as "agents" or "actants" (Latour 1996, 2005; Malafouris 2013). However, they can only be such because human beings or other living creatures with intrinsic intentionality and agency infuse them with meaning by using them to deal with the world. Thus, tools function through "derived agency" (Niño 2015; Mendoza-Collazos 2016) or "remote intentions" (Sonesson 1999), i.e., a kind of secondary agency assigned to them by actual living agents. It is the evolution of the design of such artifacts that we explore in the following section.

3 A Model for the Evolution of Design

An influential evolutionary theory of human cognitive evolution, also for cognitive semiotics (Dunér and Sonesson 2016; Zlatev 2016), has been developed by Merlin Donald in several publications (e.g., Donald 1991, 2012, 2013). This model claims that the earliest hominins shared essentially the same kind of *episodic* cognition as nonhuman apes. What made a critical difference with the genus *Homo* was the capacity for *bodily mimesis*[2] understood as a "unified neuro-cognitive adaptation that formed the early foundation of a mind-sharing culture" (Donald 2012: 181). This adaptation allowed significant modifications in procedural learning and memory capacities and the ability to perform deliberate and repeated rehearsal of previous actions toward an imagined end state. Along with this, it allowed for the first time to communicate with *signs* (see Section "Introduction"), rather than only *signals*, which are predominantly innate and thus limited, dyadic associations between vocal and bodily expressions, on the one hand, and certain environmental stimuli, on the other (Zlatev et al. 2020). The next stage was established with the ability to use more or less complex narratives, and this implies the presence of languages with relatively complex grammars, at least around 200,000 years ago, with the evolution of *H. sapiens*. The final stage was brought by a purely cultural transition, based on external representations such as notations, and hence called "theoretical culture" by Donald. This stage is the one for which we have a clear historical record based on writing.[3]

Our stage-based model is clearly dependent on this theory, but it is more specific by focusing on the evolution of design. As shown in Table 1, it also distinguishes

[2] As pointed out by Zlatev (2007: 320): "The adjective 'bodily' distinguishes bodily mimesis from the broader concept of mimesis with Aristotelean roots."

[3] There are reasons for contesting several aspects of Donald's model, but this is not our topic at present. See papers in Dunér and Sonesson eds. (2016).

Table 1 A stage-based model of the evolution of design

#	Design stage	Technology	Conception and planning	Examples	Key semiotic resource	Time	Species
1	Proto-design	Bending, shaping, combining materials for tools	Intention in action	Short-term use of *objet trouvé* tools	Signals	~20 MYA	Nonhuman apes
2				Basic tool-making			
3	Simple design	Mode 1		Basic knapping		~3 MYA	Hominin
4		Mode 2	Prior intentions	Bifaces	Mimesis	~1.5 MYA	*Homo*
5		Mode 3	Planned conception	Symmetrical hand axes	Language (speech)	~200 KYA	*H. sapiens*
6	Complex design	Early craftsmen		Decoration, identity	Drawing	~10 KYA	
7		Craft guilds		Brand identity, styles	Polysemiotic, simple	Up to 300 YA	
8	Advanced design	Industrial revolutions		Branding, product identity	Polysemiotic, complex	Current	

between four major stages, though these do not entirely overlap with Donald's stages, especially the difference between Simple and Complex designs. The first three stages can be further subdivided, and we present such finer divisions when there is evidence to support them. We delineate (sub)stages based on evidence from archaeology (the column "Technology") as well as evidence for the evolution of new semiotic resources, such as the distinction between signals and signs given above. The stages should be viewed as approximate thresholds, which add semiotic features while at the same time accumulating features from previous stages.

The rightmost columns provide approximate timestamps to make this an empirical model that can be at least in principle validated or falsified based on data and the likely biological species with which the particular design stage is associated. The column for "Key Semiotic Resource" indicates the most prominent semiotic system for each stage, which does not necessarily correspond to the time of its emergence. For instance, drawing certainly occurs long before its systematic use for decoration of Neolithic tools (stage 6).

In the remainder of this section, the four main stages are outlined with evidence in their support.

3.1 Proto-Design

By Proto-design, we mean a stage that is a precursor to design proper (which we define below), present in at least two hominid species (*Pongo* and *Pan*), and possibly in some other animals (e.g., Hunt et al. 2006). Tool use in primates depends on extractive foraging and manipulative skills that enable the rapid acquisition of novel capabilities. Some species of primates like *Macaca* and *Papio* use feeding tools in the wild, and two species (*Pongo* and *Pan*) are known to make—and not only use—feeding tools routinely (van Shaik et al. 1999). Some nonhuman primates clearly display some capacity for stone tool manufacture, with a varied repertoire of actions that include direct pounding, stabbing, levering, puncturing, and the use of hammers and anvils. The motor patterns necessary for cracking nuts using natural hammers with control and moderated force are also impressive. Another pre-hominin feature is familiarity with causal and functional aspects of applying tools in striking objects (Whiten et al. 2009).

Proto-design is thus characterized, at an early substage, by the ability to use so-called *objets trouvés* (found objects) or *manuports* (Leakey 1979: 60), and in a later substage, by a primary way of making tools in the wild for short-term use: employing shaping, bending, or combining materials (McGrew 2013). In either case, it requires a degree of *innovativeness* (Whiten et al. 2009) as the meaningfulness of an object needs to be perceived. To perform this, the animals in question would require at the very least adaptations in perceptual-motor control (oriented to objects) and understanding of causality, though these may differ from those of human beings (Vaesen 2012). Nevertheless, nonhuman animals are incapable of communicating with signs in the wild, while many have successfully learned simple sign systems through human enculturation (e.g., Pepperberg 2009). Their dominant semiotic resource is that of signals, which are not about external events (i.e., they are dyadic rather than triadic) and lack the clear conscious differentiation between expression and object that allows for denotational reference (Hurford 2007; Zlatev et al. 2020). This stage thus corresponds quite nearly to the episodic cognition stage in Donald's model.

3.2 *Simple Design*

Donald (1991) considered the mimetic stage to begin with *Homo ergaster/erectus* ca. 1.5 MYA, but new evidence has led him to push its onset back to the dawn of the *Homo* genus, or perhaps even to some australopithecines (Donald 2013). The Oldowan complex is considered the first clear evidence of hominin tool-making from 2.5 million years ago (Semaw et al. 1997). Its products include sharp-edged stone flakes produced by striking one cobble (the core) with another (the hammer-stone). These products testify to a more systematic and long-term production process than with the improvised proto-design from the previous stage. The lifestyle in Oldowan culture, involving communal processing of large food items and extensive food sharing and tolerated scrounging (scarcely present in other species) formed the necessary social conditions for the emergence of design proper, complete with invention and transmission (van Shaik et al. 1999: 738).

Simultaneously, there were hardly any qualitative cognitive or semiotic differences compared to the stage of Proto-design (see the dashed line in Table 1). Stout and Chaminade (2007) used functional brain imaging with modern people having no specific skills for making stone tools to investigate the processes involved in Oldowan-like tool-making, and found that:

> Both behavioral and brain activation data indicate that the initial stages of Oldowan tool-making skill acquisition are primarily concerned with perceptual-motor adaptation to task constraints and especially the discovery and exploitation of *object affordances, rather than with executive planning and problem solving* (...) the acquisition of such sensorimotor capabilities clearly depends upon a combination of neural preconditions with motivated and effortful practice. (Stout and Chaminade 2007: 1098, our italics)

This stage corresponds to the "Mode 1" technology in Table 1. Wynn (2012) pinpoints three modes of tool-making in the Paleolithic period, thus within the Simple design stage. In Mode 1, the conception of what is to be produced fuses with the situated action of tool-making. Schön's (1983) notion of *reflection-in-action, as understood in the design field,* is useful to describe such a fusion between action and assessment: processes of reflection (e.g., thinking for solutions, ongoing assessments) are embedded in the very act of construction and need not precede them. This concept echoes the notion of *intention in action*, introduced by Searle (1983), as described in *material engagement theory*, characterized by the binding of materials and makers in a creative action that demands few if any "internal

to visualize the problem and to obtain a global insight into the variables at play. The construction of 3D models can involve virtual 3D computer-generated or physical models. Designers, up to our own days, use easy-to-shape materials to find solutions by creating an instance of the artifact called a *prototype*. Mental conceptualizing and such external models form a coupled system used to achieve cognitive offloading and better deal with the increased complexity of design variables (Liu and Stasko 2010).

As before, the stage of Advanced design preserves and enhances the cognitive skills and semiotic resources from previous stages: material engagement, bodily mimesis, speech, and drawing. However, what is arguably new is the use of complex polysemiotic combinations of language, gesture, and depiction into integrated communicative systems. Such strategic use of polysemiotic communication creates a brand style that applies everywhere, from advertising to corporate buildings, uniforms, vehicles, and products. Simultaneously, the design practice becomes divided and specialized into multiple professions, such as graphic design, industrial design, and interior design.

4 Conclusions

Our proposed model of the evolution of design is not comprehensive but it demonstrates several important aspects of human cognitive-semiotic evolution. First, it implies both continuity and discontinuity between nonhuman animals and human beings. A stage-based model is necessary to help showing that while it may be hard to draw the boundary between two stages, the difference appears to be qualitative when new properties accumulate. Nevertheless, there has never been a "saltation," a jump from one stage to another.

Second, it suggests that the evolution of design took place in conjunction with a myriad of other conditions, summarized in the concept of *biocultural coevolution* (Richerson and Boyd 2005; Dunér 2016). Within this intricate network of conditions, bodily mimesis apparently played an important role in the early stages, or in what we called Simple design. In later stages, designers conceptualize and communicate with sophisticated polysemiotic systems, but these are arguably grounded in bodily mimesis itself (Zlatev 2019). The result is a complex artificiality, which clearly distinguishes human beings from other species (e.g., Tallis 2011).

Third, and consequently, we have argued that design proper is an exclusively human activity. It is one (even though not the only) of the differences in making us human. This claim is also supported by reviews of the literature on cognitive bases of tool-making, which show that taken together, the set of cognitive skills in human beings needed for tool-making is unique (Vaesen 2012). Design can thus be seen as the integrative capacity underlying the wide range of cognitive skills for tool-making as one uniquely human trait.

The gradual improvement of tools attested by the archaeological record shows how the design evolved from humble origins into an enhanced artifact creation capacity. It permitted our engagement with long-term use tools, leading us to technological complexity and cumulative culture. While what we have called Advanced design is a complex manifestation of polysemiotic communication with strategic manipulation of signs, even Simple design involved prior intentions in a situated, enactive meaning-making in our human ancestors, which are still part and parcel of the practice of modern designers. Likewise, the cognitive skill for intention in action may have evolved into ("external") representations but continue to be used in everyday life. With the help of cognitive semiotics, we have shown how these different kinds of meaning-making practices can both coexist at given evolutionary periods and yet need to be differentiated from one another, as they did not all appear at once.

Acknowledgments The first author would like to thank David Dunér, Josie Dixon, and Vaughan Phillips for feedback on an early version of this chapter; and the History Museum at Lund University for an inspiring environment of a variety of tools from the Stone Age. In addition, we would like to thank Aaron Stutz and Przemyslaw Żywiczyński, as well as two anonymous reviewers, for their fruitful comments.

References

Acha J (2009) Introducción a la Teoría de los Diseños. Editorial Trillas, Ciudad de México

Barbieri M (2008) Life is semiosis. Cosm Hist J Nat Soc Philos 4(1–2):29–52

Bar-Yosef O (2017) Can archaeology tell us about the evolution of cognition and language? J Neurolinguistics 43:222–227

Bayasid N (2004) Investigating design: a review of forty years of design research. Des Issues 6 (1):16–29

Buchanan R (1992) Wicked problems in design thinking. Des Issues 8(2):5–21

Casakin H, Kreitler S (2011) The cognitive profile of creativity in design. Think Skills Creat 6:159–168

Dal Sasso G, Maritana L, Salvatori S, Mazzolia C, Artioli G (2014) Discriminating pottery production by image analysis: a case study of Mesolithic and Neolithic pottery from Al Khiday (Khartoum, Sudan). J Archaeol Sci 46:125–143

Donald M (1991) Origins of the modern mind: three stages in the evolution of culture and cognition. Harvard University Press, Cambridge, MA

Donald M (2012) The mimetic origins of language. In: Tallerman M, Gibson K (eds) The Oxford handbook of language evolution. OUP, Oxford, pp 180–183

Donald M (2013) Mimesis theory re-examined, twenty years after the fact. In: Hatfield G, Pittman H (eds) Evolution of mind, brain and culture. University of Pennsylvania, Philadelphia, pp 169–192

Dunér D (2016) Science: the structure of scientific evolutions. In: Dunér D, Sonesson G (eds) Human Lifeworlds: the cognitive semiotics of cultural evolution. Peter Lang Edition, Frankfurt am Main, pp 229–266

Dunér D, Sonesson G (2016) Encounters: the discovery of the unknown. In: Dunér D, Sonesson G (eds) Human Lifeworlds: the cognitive semiotics of cultural evolution. Peter Lang Edition, Frankfurt am Main, pp 267–300

Henrich JP (2017) The secret of our success: how culture is driving human evolution, domesticating our species, and making us smarter. Princeton University Press, Princeton

Heyes CM (2018) Cognitive gadgets: the cultural evolution of thinking. The Belknap Press of Harvard University Press, Cambridge, MA
Hoffmeyer J (2008) Semiotics: an examination into the signs of life and the life of signs. University of Chicago Press, Chicago
Hunt G, Rutledge B, Gray R (2006) The right tool for the job: what strategies do wild new Caledonian crows use? Anim Cogn 9:307–316
Hurford JR (2007) The origins of meaning: language in the light of evolution. OUP, Oxford
Itkonen E (2005) Analogy as structure and process: approaches in linguistics, cognitive psychology and philosophy of science. Benjamins, Amsterdam
Konderak P (2018) Mind, cognition, semiosis: ways to cognitive semiotics. UMCS Press, Lublin
Latour B (1996) On Interobjectivity. Mind Cult Act 3(4):228–245
Latour B (2005) Reassembling the social: an introduction to actor-network theory. Oxford University Press, Oxford
Leakey M (1979) Olduvai Gorge. Book Club Associates, London
Lenninger S (2016) Pictures: perceptions of realism in the service of communication. In: Dunér D, Sonesson G (eds) Human Lifeworlds: the cognitive semiotics of cultural evolution. Peter Lang Edition, Frankfurt am Main, pp 97–121
Liu Z, Stasko J (2010) Mental models, visual reasoning and interaction in information visualization: a top-down perspective. IEEE Trans Vis Comput Graph 16(6):999–1008
Malafouris L (2013) How things shape the mind: a theory of material engagement. MIT Press
McGrew W (2013) Is primate tool use special? Chimpanzee and new Caledonian crow compared. Phil Trans R Soc Biol Sci 368:20120422
Mendoza-Collazos J (2016) Design semiotics with an agentive approach: an alternative to current semiotic analysis of artefacts. In: Zlatev J, Sonesson G, Konderak P (eds) Meaning, mind and communication—explorations in cognitive semiotics. Peter Lang, Frankfurt am Main, pp 83–99
Mendoza-Collazos J, Sonesson G (2020) Revisiting the life of things in Amazonia: a cognitive semiotic approach to the Agency of Artefacts. Pub J Cogn Semiot 9(2):30–52. https://doi.org/10.37693/pjos.2020.9.22012
Niño D (2015) Elementos de Semiótica Agentiva. Utadeo, Bogotá
Pepperberg I (2009) The Alex studies: cognitive and communicative abilities of grey parrots. Harvard University Press, Cambridge, MA
Richerson PJ, Boyd R (2005) Not by genes alone: how culture transformed human evolution. University of Chicago Press, Chicago
Schön D (1983) The reflective practitioner: how professionals think in action. Basic Books, New York
Searle J (1983) Intentionality: an essay in the philosophy of mind. Cambridge University Press, Cambridge
Semaw S, Renne P, Harris J, Feibel C, Bernor R, Fesseha N, Mowbray K (1997) 2.5-million-year-old stone tools from Gona, Ethiopia. Nature 385:333–336
van Shaik C, Deaner R, Merril M (1999) The conditions for tool use in primates: implications for the evolution of material culture. J Hum Evol 36:719–741
Sheets-Johnstone M (2011) The primacy of movement. Benjamins, Amsterdam
Sonesson G (1999) Postphotography and beyond. From mechanical reproduction to digital production. In: Sonesson G (ed) *VISIO*, 4, 1: *Postphotography*, p 11–36
Sonesson G (2009a) New considerations on the proper study of man—and, marginally, some other animals. Cogn Semiot 4:133–168
Sonesson G (2009b) View from Husserl's lectern. Cybern Hum Know 16(3–4):107–148
Sonesson G (2015a) Le jeu de la valeur et du sens. In: Biglari A (ed) Valeur. Aux fondements de la sémiotique. L'Harmattan, Paris, pp 85–100
Sonesson G (2015b) The mirror in-between picture and mind. A phenomenologically inspired approach to cognitive semiotics. Chin Semiot Stud 11(2):159–180

Sonesson G (2019) The evolution of thinking. Cognitive semiotics in between deep history and the history of mentalities. In: Dunér D, Ahlberger C (eds) Cognitive history. de Gruyter, Berlin, pp 35–74

Stout D, Chaminade T (2007) The evolutionary neuroscience of tool making. Neuropsychologia 45:1091–1100

Stout D, Passingham R, Frith C, Apel J, Chaminade T (2011) Technology, expertise and social cognition in human evolution. Eur J Neurosci 33:1328–1338

Tallis R (2011) Aping mankind: Neuromania, Darwinitis and the misrepresentation of humanity. Routledge, New York

Thompson E (2007) Mind in life: biology, phenomenology, and the sciences of mind. Belknap Press of Harvard University Press, Cambridge

Tomasello M, Call J (1997) Primate Cognition. Oxford University Press, Oxford

Tønnessen M (2016) Agency in biosemiotics and enactivism. In: Zlatev J, Sonesson G, Konderak P (eds) Meaning, mind and communication—explorations in cognitive semiotics. Peter Lang, Frankfurt am Main, pp 69–82

Vaesen K (2012) The cognitive bases of human tool use. Behav Brain Sci 35:203–262

Whiten A, Schick K, Toth N (2009) The evolution and cultural transmission of percussive technology: integrating evidence from palaeoanthropology and primatology. J Hum Evol 57:420–435

Wynn T (2012) The Palaeolithic record. In: Gibson K, Tallerman M (eds) The Oxford handbook of language evolution. Oxford University Press, Oxford

Zlatev J (2003) Meaning = life (+ culture). Evol Commun 4(2):255–299

Zlatev J (2007) Embodiment, language and mimesis. In: Ziemke T, Zlatev J, Frank R (eds) Body, language and mind, volume 1: embodiment. Mouton de Gruyter, Berlin

Zlatev J (2008) From proto-mimesis to language: evidence from primatology and social neuroscience. J Physiol PAR 102:137–152

Zlatev J (2009) The semiotic hierarchy: life, consciousness, signs and language. Cogn Semiot 4:170–201

Zlatev J (2015) Cognitive semiotics. In: Trifonas P (ed) International handbook of semiotics. Springer, Dordrecht, pp 1043–1067

Zlatev J (2016) Preconditions in human embodiment for the evolution of symbolic communication. In: Etzelmüller G, Tewes C (eds) Embodiment in evolution and culture. Mohr Siebeck, Tübingen, pp 151–174

Zlatev J (2018) Meaning making from life to language: the semiotic hierarchy and phenomenology. Cogn Semiot 11(1)

Zlatev J (2019) Mimesis theory, learning and polysemiotic communication. In: Peters M (ed) Encclopedia of educational philosophy and theory. Springer, Dordrecht

Zlatev J, Żywiczyński P, Wacewicz S (2020) Pantomime as the original human-specific communicative system. J Lang Evol 5(2):156–174. https://doi.org/10.1093/jole/lzaa006

Biosemiotics and Applied Evolutionary Epistemology: A Comparison

Marta Facoetti and Nathalie Gontier

Abstract Both biosemiotics and evolutionary epistemology are concerned with how knowledge evolves. (Applied) Evolutionary Epistemology thereby focuses on identifying the units, levels, and mechanisms or processes that underlie the evolutionary development of knowing and knowledge, while biosemiotics places emphasis on the study of how signs underlie the development of meaning. We compare the two schools of thought and analyze how in delineating their research program, biosemiotics runs into several problems that are overcome by evolutionary epistemologists. For one, by emphasizing signs, biosemiotics needs to delineate a semiotic threshold, which is a problem not encountered by evolutionary epistemologists. Instead, the latter recognizes that all organisms are knowers that evolve knowledge, which they recognize to extend toward phenomena produced by organisms such as behavior, cognition, language, culture, science, and technology. Secondly, biosemiotics attempts at continuing adaptationist notions on how organisms relate to their environment, while especially Applied Evolutionary Epistemology comes to redefine the nature of the organism–environment relationship in such a way that it recognizes the spatiotemporal boundedness of existence, which in turn makes adaptationist accounts obsolete.

Keywords Biosemiotics · Evolutionary epistemology · Applied evolutionary epistemology · Naturalized epistemology · Organism–environment relationship · Ontology-epistemology divide

---

M. Facoetti · N. Gontier (✉)
Applied Evolutionary Epistemology Lab, Centro de Filosofia das Ciências, Departamento de História e Filosofia das Ciências, Faculdade de Ciências, Universidade de Lisboa, Lisboa, Portugal
e-mail: nlgontier@fc.ul.pt

© Springer Nature Switzerland AG 2021

E. Pagni, R. Theisen Simanke (eds.), *Biosemiotics and Evolution*, Interdisciplinary Evolution Research 6, https://doi.org/10.1007/978-3-030-85265-8_9

1 Introduction

Evolutionary epistemology (EE) and biosemiotics are contemporary research fields similarly concerned with the evolutionary study of knowledge and knowing. In this chapter, we offer a comparative analysis of these schools of thought. We begin by briefly summarizing the basic tenets of EE (Sect. 2.1) and biosemiotics (Sect. 2.2), whereafter we highlight the contact points as well as the points of divergence that exist between these two schools (Sect. 3). We demonstrate that similarities are found in how both disciplines endorse evolutionary approaches to knowledge, but the two disciplines hold diverging views on how to understand evolutionary continuity in knowledge and semiosis (Sect. 3.1). Moreover, while both study knowledge from an evolutionary perspective beyond classic Neo-Darwinian theories of evolution, the disciplines emphasize different evolutionary mechanisms and processes (Sect. 3.2). Finally, we show that while both schools endorse constructivist and relativist approaches to knowledge (Sect. 4.1), especially Applied Evolutionary Epistemology (AEE) comes to redefine the organismal–environmental relationship in such a way that it overcomes problems of adaptation and semiotic threshold (Sect. 4.2).

2 Two Naturalized Approaches to Knowing and Knowledge

Both EE and biosemiotics share an interdisciplinary and evolutionary outlook on the study of knowledge. Biosemiotics thereby focusses on the *semantics* or *meaning-making aspects of knowledge*, which is understood as *sign* evolution (Favareau 2007; Hoffmeyer and Emmeche 1991; Kull 2015; Sharov 2016). This tradition's intellectual roots reach back to pragmatism (James 1907; Peirce 1992, 1999) and semiotics (Eco 1984; Sebeok 1994; von Uexküll 1940), which are schools that developed in the philosophy of mind, language, and linguistics.

These just mentioned traditions also influence evolutionary Epistemology (EE) but EE is more an outgrowth of naturalized epistemology (Quine 1969; Popper 1972; Rorty 1980). EE developed further in association with the rise of ethology (Lorenz 1977) and comparative psychology (Campbell 1974a). Instead of focusing on the semantics of meaning-making, EE focuses on *formally* identifying the *structure that grounds the evolution of epistemology*, which is understood in terms of *units, levels, mechanisms, or processes of knowledge evolution* (Bradie 1986; Campbell 1974a; Hull 1988; Gontier 2006b, 2012; Plotkin 1994). This tradition associates more with the philosophy of biology, evolutionary biology, behavioral and cognitive sciences.

Both biosemiotics and EE have, throughout their intellectual development, also actively integrated important insights coming from cybernetics (Wiener 1948), information theory (Shannon 1948), general systems theory (von Bertalanffy 1968), and hierarchy theory (Pattee 1973; Salthe 1985; Simon 1962). In what follows, we offer introductory overviews of both disciplines.

2.1 Evolutionary Epistemology

Evolutionary Epistemology (EE) is a research area of contemporary philosophy of science that studies the evolution of epistemology, i.e. the natural history of knowledge (Popper 1972), its evolutionary development or growth over time (Toulmin 1972; Hull 1988, Gontier 2006b).

Epistemology is understood as both *the evolved act of knowing* and as the outcome of knowing, which is *knowledge* that includes scientific theories but also behavior or cognition (Campbell 1974a; Hahlweg and Hooker 1989; Lorenz 1977; Munz 1993; Riedl 1984; Vollmer 1984, 1987; Wuketits 1992; 1998). In this regard, Bradie (1986) has distinguished EE into two research programs, EEM, that studies the Evolution of Epistemological Mechanisms responsible for knowing, and EET, that examines knowledge or the Evolution of Epistemological Theories.

The evolutionary study of *knowing* finds its intellectual roots in Naturalized Epistemology (Quine 1969) that understands knowing as a property of a cognitive organism (a knower) and, therefore, as something that can be studied from within the field of psychology. EE extends the latter's scope by recognizing that researching knowing is not merely a matter of psychological or cognitive research but also of *evolutionary* research. Not only humans or cognitive organisms, but *all biological organisms are knowers* that evolved *knowledge*. EE thus also includes studies of organisms such as bacteria or plants that do not have a brain but that nonetheless depict intelligible behavior. In this regard, EE also associates with the schools of ethology (Lorenz 1941; Tinbergen 1963) and comparative psychology (Campbell 1974a) that played crucial roles in "evolutionizing" the study of cognition and behavior, which are themes later also adopted by sociobiology and evolutionary psychology (reviewed in Gontier 2012).

For Evolutionary Epistemologists, *knowledge* is understood, on the one hand, to be *embodied* in evolved organisms and, on the other hand, to also *extend* biological organisms in phenomena preceded by them in time and phenomena produced by them throughout evolution (Bradie and Harms 2001; Campbell 1974a; Clark and Chalmers 1998; Facoetti 2020; Gontier 2006a; Hull 1988: Munz 1993; Plotkin 1994; Popper 1972; Toulmin 1972; Wuketits 1992, 2006). Examples of knowledge phenomena that precede life are molecular and biochemical processes that lie at the origin of life. Examples of extended knowledge are cognition, behavior, language, culture, science, and technologies portrayed and produced by living beings. These phenomena are understood to evolve over time, just like biological organisms do, and to embody knowledge.

Both embodied and extended knowledge also give new meaning to the concept of *theory* and its relation to the world (the epistemology-ontology divide). In this aspect, evolutionary epistemology is dividable into different schools of thought (Table 1).

Traditional EE (Popper 1963, 1972; Munz 1993) argues that knowing organisms provide unfalsified theories of an external world (Kant's *Welt an sich* or world-as-it-is-in-itself). Non-Adaptationist EE (Wuketits 2006; Diettrich 2006; Riegler 2006)

Table 1 Differences between traditional, non-adaptationist, and applied EE

Aspect of difference	Traditional EE	Non-adaptationist EE	Applied EE
Evolutionary framework	Strict-Neodarwinian	Systems theory (Eco-Evo-Devo)	Pluralistic
Explanation	Reductionist	Holistic	Integrative
Evolutionary hierarchy	Gene-focused	Organism-focused	Units, levels, mechanisms-focused
Causation	Upward	Up- and downward	Reticulate
Organism–environment relationship	Dualistic, adaptationist	Dialectic, non-adaptationist	Cognitive, ecological, and socio-cultural constructivism of bio-realities
Worldview	Hypothetical realism	Coherence theory and cognitive constructivism	Spatiotemporally-bounded realism

instead claims that organismal theories are cognitive constructs coherent and functional for the organism, but we cannot know how it relates to an outer world-in-itself or whether the organism is adapted to it. Applied EE understands organisms and their extensions to continually alter and construct bio-realities, making the notion of a world-in-itself invalid. We briefly discuss the different schools.

2.1.1 Traditional EE

Original research within EE (e.g., Lorenz 1941, 1977; Popper 1963) operated within strict Neo-Darwinian evolutionary schools of thought that we criticize today for overemphasizing gene-reductionism and adaptationism and for endorsing dualistic views on how organisms relate to the environment.

Following Neo-Darwinian schools of thought, early supporters of EE assumed that organismal behavior was genetically determined and that, unable to change their genetic makeup, organisms passively undergo selection from an active environment. The environment was assumed to weed out maladaptive organisms, so only adaptive organisms survive. The organismal–environmental relationship was thereby explained in *dualistic* terms: organisms and the environment were considered distinct and homogeneous entities that interact only through natural selection (reviewed in Gontier 2006b).

Such an evolutionary view enabled early evolutionary epistemologists to endorse *hypothetical realism*, whereby, biology-wise, organisms were understood as adapted to their environment, and, epistemology-wise, their evolved knowledge on the environment was assumed to somehow corroborate to it (Popper 1963).

2.1.2 Non-Adaptationist EE

Gene-reductionist and adaptationist perspectives became heavily criticized in biology. By building upon the works of Jakob von Uexküll (1921, 1928) as well as early developmental systems theory (Gould 1977; Maturana and Varela 1980) and ecology (Gould and Lewontin 1979), scholars reinstated an organismal-focused biology. This introduced a more *dialectic* understanding of how organisms interact with the environment. And it became recognized that organisms can construct *ecological niches* that are functional for the organism but these niches are not necessarily adapted to the outer world.

The systems-theoretical and constructivist approaches to evolution, in turn, inspired a new wave of evolutionary epistemologists who came to reject hypothetical realist accounts and to endorse instead *non-adaptationist* perspectives on knowledge (Diettrich 1998, 2001, 2004, 2006; Hahlweg and Hooker, 1989; Riegler 2001, 2006; Wuketits 1992, 2006; reviewed in Facoetti 2020, in press). For them, organismal knowledge, in so far as it is the outcome of evolution, is also constrained by its evolving cognitive apparatus and how it processes data from an outer and internal environment, what Lewontin (1983) called the "internalization of selection". That means that how organisms develop knowledge is determined by their biological makeup, and any act of cognition is, therefore, also an act of *construction*.

By integrating views from ecology, developmental biology, and systems theory, non-adaptationist EE has helped enlarge the scope of the Modern Synthesis toward what we now know as Eco-Evo-Devo (Ecological and Evolutionary Developmental theory). On their account, reductionism and adaptationism have to be replaced with more *holistic* explanations that allow for both up- and downward causation in development (Campbell 1974b) and a more *dialectic* perspective on the organism–environmental relationship (Wuketits 1989, 17).

Contrary to adaptationist accounts, *niche construction*, be it ecological or cognitive, demonstrates a system's ability to self-maintain and self-organize (Maturana and Varela 1980) sometimes *despite* inhospitable environments (Lewontin 1983, 2000; Gontier 2018). The recognition that organisms construct ecological (Lewontin 1982, 1983), cultural (Odling-Smee et al. 1996), and cognitive[1] (Magnani 2017) niches has made non-adaptationist evolutionary epistemologists assume that organisms preside over the construction of an experiential world. This world is not perforce connected with an outer reality (e.g., the Constructivist EE defended by Diettrich (2006) and the radical constructivism of Riegler (2006)). Instead, at an epistemological level, cognitive or behavioral traits need only be *functionally or internally coherent* (Wuketits 2006).

Nonetheless, non-adaptationist EE-ers continue to assume that there is a world-in-itself, and they are therefore faced with the question of how the constructed niches relate to such an "outer" world, ontologically speaking.

[1] Note that the concept cognitive niche is also used by Tooby and DeVore (1987) in an adaptationist context.

2.1.3 Applied EE

More recently, Gontier (2018) has demonstrated that non-adaptationist approaches to EE are right in rejecting hypothetical realism and in recognizing the importance of ecological, cognitive, and sociocultural niche construction in knowledge formation instead. However, they are flawed in understanding constructed niches as distinct from a world as it is in itself. Like Kant, they thereby remain conflicted on how and if the noumenal world (ontology) relates to the phenomenal world (epistemology).

Such perspectives continue to separate organisms from environments, and biotic niches from an abiotic world, thereby assuming that the latter somehow forms a more stable and even essential ontological entity, void of or unaffected by the evolving and living world. Such a view falls short of recognizing the temporal and evolutionary aspects of the world. Instead, on Gontier's account, accepting evolution means that there no longer exists a world-in-itself. Rather, organisms constantly recycle existing matter into a living and ever-changing world. In such a world, epistemology *defines* ontology.

Epistemology refers to evolved knowledge that comes in the form of organisms and what extends them, and this defines ontology as a living and spatiotemporally changing earth. That means that earth is not a single stable entity but a result of constructed environments or *biorealities*. According to Gontier (2018, 30):

> The living earth evolves in congruence with these expanding (generating and/or speciating) and contracting (degenerating and perishing) bio-realities that are dependent upon organismal and species survival, reproduction and extinction as well as the ecological materializations they bring forth in time and space (or spacetime). *Epistemology, understood not as theories but as the evolution of embodied knowledge in organisms and their extended niches that underlie bio-reality formation, therefore equals ontology, the currently living world.* (Gontier 2018, 30)

The goal of AEE (Gontier 2017, 7–15) therefore is to investigate how different, species-specific bio-realities evolve and underlie the current living and evolving earth by adopting a pluralistic stance on units (parts) and levels (wholes) that together make up complex hierarchies (bio-realities) that underlie the ontological layeredness and variety of the evolving world.

Gontier characterizes this approach as "Applied Evolutionary Epistemology" (AEE) because it thinks through the consequences of non-adaptationist EE, and it applies EE not only to living organisms but to all aspects of the living earth.

2.2 *Biosemiotics*

Biosemiotics, as the name suggests, combines *semiotics* (a field in linguistics and philosophy that studies semiosis or sign processes) with *biology* (the study of living beings) (cf. Favareau 2007, 33; Brier 2006, 13) to study how biological organisms develop meaning through signs.

Table 2 Semiosis

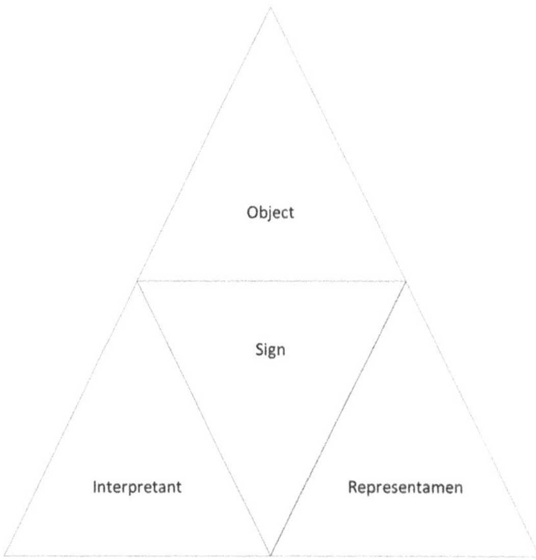

More specifically, biosemiotics can be defined as the "study of semiosis in living nature" and "biology as a sign system study" (Hoffmeyer & Kull 2011; Kalevi 1998; Kull 2009a, b; Kull et al. 2011b, 15).

The goal of biosemiotics is to develop "a scientific understanding of how the subjective experience of organisms—as enabled differently by each species' particular biological constitution—comes to play a genuinely causal role in the ongoing co-organization of nature" (Favareau 2007, 33). Within such a perspective, semiosis underlies communication, and knowing is understood as a communicative and meaning-making process between organisms and the world. For Kull (2014, 48), for example, "knowing is possible only due to semiosis, through semiosis," and just like EE, biosemiotics can therefore be defined as the study of knowledge and knowing in all living systems (Brier 2006, 1), but knowing and knowledge are understood exclusively through the study of signs.

The *semiotic* part of "biosemiotics" goes back to a particular branch of semiotics introduced by Charles S. Peirce in the late nineteenth century. Peircean semiotics (Table 2) rests upon a triadic notion of the sign, which is understood as a relation between a representamen, an object, and an interpretant (Peirce 1931–1935). It furthermore adheres to a threefold classification of the sign relation between object and representamen, which can take on the form of icons, indexes, and symbols. Biosemioticians draw upon such classification in various ways, by relating different kinds of signs to different modes of semiosis in a variety of organisms (cf. Sharov and Vehkavaara 2015; Kull 2009c, 2018).

The *biological* part of "biosemiotics" developed along the same lines EE and general evolutionary theory developed, from adaptationist-focused Neo-Darwinian schools of animal behavior and cognition to more extended views thereof. In addition, biosemiotics integrates the pre-constructivist view of Jakob von Uexküll (1921, 1928, 1937), which pivots on the *Umwelt* concept, the idea that the "*perceptual world*" ("all that a subject perceives") and the "*effector world*" ("all that he does") "together form a closed unit, the *Umwelt*" (von Uexküll 1934, 320). In this view, an *Umwelt* is "the set of features of the environment as distinguished by the organism, or the self-centered world that relates an organism with everything else" (Kull et al. 2011a, 38). Stated still differently, the *Umwelt* is the "species-specific model" of the world "incorporate[d]" by an organism (Sebeok 1996, 102; quoted by Kull et al. 2011a, 70).

With the *Umwelt* concept, von Uexküll put forward "a theory of meaning which considered animals as interpreters of their environment" (Sharov, n.d.) and signs as *adaptations* (Sharov 1999). Such a view is a form of pre-constructivism because it understands meaning as a form of "co-organization" (Favareau 2007, 33) between organism and environment. The concept also directly inspired Lewontin (1982), who introduced the niche construction perspective.

Gensini (2002, 123–127), for example, reports that semioticians agree that human semiosis has continuity with other species, firstly, by sharing the capacity for *Umwelt* formation and, secondly, because every living being is, like humans, capable of applying the "criterium of pertinence." This criterium refers to the capacity of distinguishing what is life-supportable from what is not (Gensini 2002, 124; translation ours). Finally, semioticians agree that different species-specific perceptual apparatuses and their associated categorical systems determine different species-specific relationships to species-specific worlds (Gensini 2002, 124).

In this regard, current biosemioticians share with supporters of non-adaptationist and applied EE an adherence to anti-reductionist views on evolution, where notions such as "function," "information," or "signal" are irreducible to chemical compounds or genes (Kull et al. 2011c, 31, 2011, 7; Sharov 1999). Especially biosemioticians have thereby also actively tried to incorporate teleological and vitalist notions (e.g., *entelechy* or *goal-directedness*) in biology by understanding organisms as living systems with selves, i.e. as self-organizing autonomous agents whose activities are "goal-directed" and "sign-dependent" (Sharov 2018, 197) (cf. Kull et al. 2011c, 27–32).

Furthermore, as theories within biosemiotics have developed and attempted to synthesize and integrate new schools of thought, von Uexküll's *environmental* Umwelt concept has been taken as synonymous with Maturana's notion of *a cognitive* domain or Hoffmeyer's *semiotic* niche concept (Brier 2015, 578) in attempts to bridge the gap between the kinds of organisms that exist and the kinds of environments they inhabit.

Beyond the Peircean and Constructive/Umwelt approaches to biosemiotics, a third group of biosemioticians have attempted to reduce biosemiotics to "code biology" (Barbieri 2014) by trying "to refashion the primary biosemiotic articulation from one of 'sign' to 'codes'".[2] The latter is comparable to the search for universal formulas and heuristics of natural selection in evolutionary epistemology, such as the "blind variation and selective retention" schemes introduced by Campbell (1974a), or the "generate-test-regenerate" scheme of Plotkin (reviewed in Gontier and Bradie 2017).

3 Digging Deeper for Similarities and Divergences

In this section, we dig deeper into the similarities and divergences between EEs and biosemiotics.

3.1 (Dis)Continuities in Knowledge and Semiosis

Traditional, non-adaptationist, and applied approaches to EE overcome traditional empiricist, rationalist, and naturalist perspectives on epistemology that understood knowledge as "human-bounded," "cognitive," and "language-like" (Gontier 2006b).

EE evolved by recognizing the presence of knowledge in other organisms, and knowledge is extended to both the abiotic world and phenomena that evolved with the rise of living beings, such as behavior, cognition, languages, cultures, and technocomplexes.

Like evolutionary epistemologists, biosemioticians accept the idea of a certain continuity between humans and the rest of the world, but they also continue to look for a semiotic threshold. However, this threshold has proven to be the Achilles heel of contemporary biosemiotics. It currently functions as a source of debate, critique,

[2] According to scholars of the International Society for Code Biology (http://www.codebiology.org/) (i.e., Barbieri together with Stefan Artmann, Joachim De Beule, Peter Dittrich, Almo Farina, Dennis Görlich, Hendrik Hofmeyr, Stefan Kühn, Chris Ottolenghi, Liz Swan, Morten Tønnessen, and Jan-Peter Wills), Peircean biosemiotics is incapable of providing "a scientific approach to the semiosis of Nature" (Barbieri 2014). Hence, by distancing themselves from Peircean biosemiotics and by working from within a new theoretical framework called "code biology," these scholars aspire to provide such "a scientific approach" through "a study of all codes of life with the standard methods of science" (Barbieri 2014). For a review of other critical perspectives on Peircean biosemiotics and biosemiotics in general, the reader may refer to (Favareau 2007), who points to the existence of both "informed" and "uninformed criticism" of such view. According to Favareau (2007, 45), whereas insiders of biosemiotics generally put forward the former (he cites Vehkavaara (2002, 2003) and Artmann (2005), among others), the latter is mainly proposed by "those critics from the outside" to whom he attributes a misunderstanding of the theory. As we will see, the Constructive approach to biosemiotics discussed in this paper (and of which Vehkavaara is one of the leading advocates) is critical of Peircean biosemiotics and, more specifically, of its understanding of the "sign" notion.

non-biotic or from-life-evolved phenomena such as technocomplexes as knowledge and information.

3.2 The Scope of Evolution and the Importance of the Organism

Both EE and biosemiotics understand knowledge/semiosis as an evolved phenomenon. Both agree that the knowledge that organisms gain or produce is fundamentally constrained by evolution (Gontier 2018, 541), and it bears with itself the marks of its evolutionary history (cf. Wheeler in Favareau et al. 2017, 10). While both adhered initially to adaptationist views, today, both fields expand narrow Neo-Darwinian theories of evolution and integrate ideas associated with extensions of the Modern Synthesis and non-Darwinian evolutionary theories.

Neo-Darwinism is a school of thought that understands natural selection as a two-way process. Genes underlie the formation of anatomical, functional, and behavioral traits, and these enable organisms to establish a perfect fit with their environment (Dawkins 1983). In such a view, organisms and environment interact only through the process of natural selection and ontogeny, the development of an organism from conception until death, becomes separated from phylogeny, the evolution of species. Such a view is dualistic because organisms are understood as passive entities that become selected only when adapted and discarded when not by an active environment (reviewed in Gontier 2006b).

Today, evolutionary epistemologists and biosemioticians distance themselves from such a Neo-Darwinian understanding of evolution. As a forerunner of evo-devo, von Baer's evolutionary developmental theory already inspired Jacob von Uexküll at the beginning of the twentieth century to understand organisms as active agents interacting with their environment rather than being passively selected by it (Kull 1999, 391). In turn, inspired by von Uexküll, today, evo-devo has also expanded toward ecological theory into the new field of eco-evo-devo where biosemioticians such as Sharov et al. (2016) and Kull et al. (2011c) find inspiration (cf. Kull 1999, 407). In such theories, these biosemioticians find "the agential properties of life and emphasiz[e] the autonomy and goal-directedness in the activity and communication of organisms" (Sharov 2018, 199). Because ultimately, biosemioticians aim to integrate evolutionary theory with a "theory of reference and interpretation" in order to explain "the teleological nature of living systems" (Hoffmeyer 2010, 368), that is "in order to make explicit. . .such unanalysed teleological concepts as function, adaptation, information, code, signal, cue, etc. . .," which for them do essentially characterize living processes (Hoffmeyer 2011, 64).

Like biosemioticians, evolutionary epistemologists have come to favor non-adaptationist theories of evolution over adaptationist or Neo-Darwinian ones. However, contrary to biosemioticians, most evolutionary epistemologists have refrained from incorporating teleology and purpose or natural finality into biology.

Non-adaptationist evolutionary epistemologists have thereby been fighting at the front line to embrace the Modern Synthesis's extensions that build upon the achievements of eco-evo-devo theories that include recognizing phenotypic plasticity, niche construction, and a more significant role for drift theory. These theories recognize organisms as active entities that can condition their chances of survival. Also reinstating an organismal point of view rather than a gene or environmental point of view, along these lines, constructivist approaches have come to replace a dualistic perspective on the organism–environment relation with a dialectic one in which organisms preside over the construction of a species-specific experiential world which is not necessarily connected with an external reality (reviewed in Gontier 2018, Facoetti, in press).

AEE, in particular, has also thought through the consequences of reticulate and constructivist evolutionary theories for evolutionary epistemology. Within this perspective, the relationship between organisms and their environment is "exclusive because there is no outer abiotic earth anymore" (Gontier 2018, 545).

In this context, it is essential to emphasize that both biosemiotic and non-adaptationist evolutionary epistemological schools of thought have been deeply influenced by the 1960s–1980s research in systems theory (von Bertalanffy 1968) and (bio-)cybernetics (Maturana 1970; von Glasersfeld 1981, 1984). This new wave of theoretical biological studies had a substantial impact on research in evolutionary biology, which, as Kull (1999, 405) reports, in the last decades of the twentieth century started to put an unprecedented emphasis on the "form and activity of organism[s]." Such theories distinguished themselves by proposing a holistic view, an organism-centered perspective that highlights the autonomous nature of organisms. Based on these new perspectives, the study of organisms as wholes ceased to be *reduced* to the study of their physicochemical parts. Organisms came to be viewed in all their complexity, as "integrated wholes," that is, as "hierarchically organized, stratified, multi-levelled systems with dynamic interaction between all levels" (Löwenhard 1989, 90). Moreover, organisms came to be understood as autonomous systems capable of self-organizing and self-regulating (cf. Maturana and Varela's [1980] concept of autopoiesis), sometimes even despite the environment. As seen, whereas biosemioticians have come to emphasize the teleological character (i.e., the goal-directedness) of such self-organizing and self-regulating activities, non-adaptationist scholars and applied evolutionary epistemologists have mostly put the accent on the independent character of such activities from an outer world.

4 Defining the Organism–Environment Relationship: Old and New Cosmologies

Both EE and biosemiotics reject reductionism, the idea that higher biological phenomena can be reduced to underlying physicochemical processes (Brigandt and Love 2017). In line with this and influenced by Bertalanffy's system theory, both non-adaptationist and applied EE, and biosemiotics, in its Peircean and Constructive interpretations, do furthermore maintain that there is autonomy at the organismal level as organisms are autonomous (autocatalytic, autopoietic, i.e. self-organizing) systems which cannot be explained merely by their parts.

Within such perspectives, the denial of a reductionist stance goes hand in hand with the rejection of Cartesian dualism, i.e. the idea of there being a rigid division between mind and body and between organisms and the outer world.

In taking on a non-Darwinian perspective on evolution and a biological line of thinking informed by research in biocybernetics and systems theory, both EE and biosemiotics come to reject *structural realism*—the idea that there exists an independent outer world of which organisms can know its structures. They also reject the notion of *truth as correspondence*—the idea that organismal theories about the world can be true if they correspond to or accurately describe its structures (cf. Diettrich). Today, both Non-Adaptationist and Constructivist evolutionary epistemologists (Wuketits, Diettrich, Riegler) and biosemioticians (both Peircean and Constructive ones) suggest similar alternatives to the traditional interpretation of the organism–environment relationship. In so doing, they maintain a distinction between ontology and epistemology. Applied EE, on the other hand, goes beyond the traditional ontology/epistemology divide.

4.1 Non-Adaptationist EE and Biosemiotics Offer Similar Epistemologies and Ontologies

The epistemological and ontological accounts that non-adaptationist EE and biosemiotics put forward to explain the organism–environment relationship are similarly characterized by the appeal to constructivism, relativism, idealism, and functional realism.

Within non-adaptationist EE, dualistic pictures of the organism–environment relation are rejected through the appeal to functional realist accounts (Wuketits) or complete[4] constructivist perspectives (Diettrich and Riegler). By embracing such positions, non-adaptationist evolutionary epistemologists maintain that organismal knowledge of the world is always *relative* to the organism's species-specific cognition and anatomy. Such knowledge, they hold, should not be evaluated on the grounds of its correspondence to the structures of an outer world, but in light of its coherence with the whole of an organism's experiences, whether it is life-supportable and functional to the survival and fitness of organisms (Wuketits

[4]We use the term "complete" to refer to Diettrich's Constructivist EE (CEE) and Riegler's Radical Constructivism. Both authors resort to such an adjective (Diettrich 1998; Riegler 2001) to mark the distance between their positions and von Glasersfeld's Radical Constructivism. Contrary to the latter, Diettrich and Riegler endorse a kind of constructivism "on all levels" (Riegler 2001, 7) (reviewed in Facoetti 2017, 3).

2006, 40), viable (Riegler 2001, 6), reproducible, and consistent (Diettrich 2001) (discussed in Facoetti 2020, in press).

On the one hand, Wuketits embraces relativism and moderate constructivism. For him, organisms and their environment are engaged in a mutual relationship, whereby the environment defines organisms' adaptation, and organisms define adaptability (i.e., by niche construction) (Wuketits 2006, 38) in light of their inner world or species-specific needs and experiences (Wuketits 2006, 43). In such a view, the organismal view of the world does not need to be in 1 to 1 correspondence with an outer world, but following the ideas of von Glasersfeld (1995), it needs to be functionally coherent for the organism, and it needs to enable survival in the environment. For Wuketits (2006, 44), according to a functional realist point of view that reality is what is functional to the survival of an organism, "there is no need for the belief in the *unknowable*—and thus no need to assume the existence of an unknowable *world-in-itself*." Wuketits (2006, 43) considers the notion of a world-in-itself "obsolete or at least redundant," as "what counts for any organism is that it copes with its *own world* properly." Nonetheless, "what an organism constructs must, one way or another, correspond to some aspects of reality" (Wuketits 1992, 158) because "any organism that would totally neglect the outer world and rely exclusively on its own constructions would not survive" (Wuketits 2006, 43).

On the other hand, relativism and complete constructivism characterize the views of Diettrich (1998, 2001, 2004, 2006) and Riegler (2001, 2006), both of whom distance themselves from von Glasersfeld's radical constructivism and hence from the idea that the experiential world that our mind "has priority in constructing (. . .) necessarily relate to an outer world" (Gontier 2018, 545). On von Glasersfeld's account, as framed by Diettrich (2004, 61), knowledge still has to meet external reality requirements, although "not necessarily by means of delineating environmental structures but rather functionally." Opposing this view, Diettrich and Riegler espouse an agnostic perspective, according to which we are neither in the condition of negating the existence of such reality nor claiming "its non-existence" (Riegler 2001, 3). By drawing upon Piaget (1970), Diettrich (1998) pictures the organism–environmental relation in terms of a reconstruction–construction process, whereby the act of assimilation (construction) alternates that of accommodation (adaptation/reconstruction) by giving rise to a continuous feedback circuit (reviewed in Facoetti 2020, in press). Along these lines, as Diettrich (2004, 61) reports, within such perspectives, knowledge is considered to be reliable if it "derive[s] from perceptions and their appropriate interpretation, but neither perceptions nor their (viable) interpretations need the evaluation by an external world." For his part, Riegler (2006, 51) proposes a radical "subject-centered" interpretation of the organism–environment relation according to which organisms actively construct species-specific experiential worlds which do not perforce relate to an external world.

Like new evolutionary epistemologists, biosemioticians reject dualistic views on the organism–environment relation by subscribing to a pragmatist/functionalist position (Sharov, Wheeler) or an idealistic approach to semiotics (Sebeok). In particular, Sharov (2016, 15) sees in constructivism "a valuable addition to biosemiotics because it emphasizes the activity of agents in self-construction, self-reproduction, and development of sign relations." Sharov (2016, 2018) draws upon Vygotsky's constructivist position ("historical-dialectical-monism" or "functional monism" [Liu and Matthews 2005, 397]) as well as on von Glasersfeld's radical constructivism (Sharov 2018, 203). Vygotsky's dialectical constructivist position allows Sharov "to go beyond the boundaries set by dualism, and to see how man and world, mind and reality can become *the source of growth and change for each other*" (Liu and Matthews 2005, 397; italics ours). Von Glasersfeld's radical constructivism, in turn, enables him to picture organisms as "self-constructed unities" whose "knowledge fits reality in the Darwinian sense [i.e. by adaptation], but does not necessarily include a representational relation to it" (Sharov 2018, 203).

Just as new evolutionary epistemologists, biosemioticians also embrace a relativist view, according to which knowledge is always relative to an organism's *Umwelt* and mediated by species-specific cognitive and perceptual faculties, apt to satisfy species-specific needs. Besides, similarly to new non-adaptationist scholars, biosemioticians replace structural realist perspectives on the knowledge-world relation with idealist or pragmatist/functionalist standpoints.

Sebeok (1994) espouses a semiotic idealist view that comes close to Riegler and Diettrich's stance. On Sebeok's account, the mind presides over the creation of a structured and meaningful world (Umwelt) "out of a vast and diverse crush of sense impressions" (Sebeok 1994, 37). To Sebeok (1994, 34):

> Whether there is a reality behind signs-perhaps what Heraclitus called logos, the repeatable structure that secures for any object its ideal unity and stability (...)—humanity can never be sure.

Within this perspective, as Deely (2013, 35–38) explains, the idea of an independent external world comes to be understood as "a species specifically human representation": "[w]e awaken not to a physical environment of pure *ens reale* but to an objective world which, like that of every animal, is a mixture of *ens rationis* and *ens reale*." The notion *ens reale* "is neither identical with 'the external world' nor the starting point as such of species-specifically human knowledge, but merely a recognizable dimension experienced within objectivity," our human-specific experiential world (Deely 2013, 38).

Sharov's Constructive biosemiotic view, on the other hand, comes close to that of Wuketits (1992, 2006), as he adopts a pragmatist or functionalist perspective which "does not view existence as objective (i.e., observer-independent) reality." According to Sharov (1999), "existence" should be "evaluated subjectively according to its expected effects on the existence of other things in the future": "food exists for an animal because it increases chances of [i.e. *it is functional to*] survival and reproduction (i.e., existence) of this animal." In Sharov's view, knowledge still has to fit (some parts of) reality—organisms must adapt to reality. Otherwise, they could not survive, although such knowledge does not need to represent reality accurately (Sharov 2018, 203).

Like Sharov, Wheeler (in Favareau et al. 2017, 10) espouses a functionalist point of view, according to which organisms construct a subjective world out of what is functional to their survival:

> The world of our senses does not reveal the whole world to us. Even at the level of simple objects, we sense only what we need to sense to get on with caring about being. This means that news of the world arrives in the form of signs that "stand in for" the world's inaccessible plenitude: signs that shine and compel us. (Wheeler in Favareau et al. 2017, 10)

Wheeler's view is part of the Peircean approach to biosemiotics, which traditionally maintains an outer world's existence, a purely physical and singular realm (Kull 2015, 523). According to Peircean biosemioticians (Kull, Pattee, Emmeche, Hoffmeyer), such an outer world is dominated by physical laws, which, by definition, are "universal," "singular," and "inexorable" (Pattee and Kull 2011, 220; Kull 2015, 522–523). Organisms appear to "locally escape the global behavior of physical laws, yet without ever disobeying them" (Pattee and Kull 2011, 222). In other words, organisms are "active systems of sign production, sign mediation and sign interpretation, that harness the physical laws in order to live and sometimes to make a more complex living" (Kull et al. 2011c, 1). The realm of living creatures, that is, the realm of semiosis, is dominated by semiotic rules or regularities (Pattee and Kull 2011; Kull 2018). The latter, in contrast to the universal and singular character of physical laws, are non-universal but "local" and "plural" (Pattee and Kull 2011; Kull 2015, 2018).

A supporter of the Peircean paradigm, Kull replaces the Cartesian division between mind and body and organism and the world with the distinction between "signness" and "non-signness" (cf. Kull 2015; Kull et al. 2011c; Pattee and Kull 2011; Kull 2018). For Kull (2018, 145), this division lies between the domain of "semiotic regularities" that can be learned and that of physical laws that are given. For Kull (2018, 142), the possibility of constructing (or representing) space, for example, is connected to the learning of particular signs relations. The construction of space is needed for "the modelling and construction of the *Umwelt*," and this process, in turn, depends upon learning sign-to-sign connections (Kull 2018, 142). The vegetative *Umwelt* is "an umwelt in which space cannot be constructed" as "there is no sign-to-sign connectedness due to the absence of associative learning" (Kull 2018, 142). However, in the animal *Umwelt*, "[t]he availability of indexes in addition to icons (as a result of associative learning) makes a huge difference in the representation of the world" as it becomes possible to construct space (Kull 2018, 143).

4.2 Applied Evolutionary Epistemology (AEE) Beyond the Ontology/Epistemology Distinction

As mentioned, new evolutionary epistemological and biosemiotic schools of thought rely on an old cosmology, which, as just seen, rereads the organism–environment relationship in light of the traditional ontology/epistemology distinction. Within such cosmological perspective, organisms either try to gain knowledge about an outer independent world ("the universe, earth, an abiotic environment, or a 'more fundamental' physicalist level") or, at the opposite pole, they appear to construct a world of their own in their minds (Gontier 2018, 550). Such cosmology and the underlying ontology/epistemology distinction, as Gontier (2018, 539) notices, were developed before "the recognition that we live in an evolving world that forms part of an expanding universe and possibly a multiverse." Today, due to advances in physics, socio-anthropological sciences, and evolutionary epistemological schools of thought, this understanding of the organism–environment relations appears no longer tenable (Gontier 2018, 546).

The newly developing cosmology shows that, contrary to what functionalist/dialectic/pragmatist approaches to EE (Wuketits) and biosemiotics (Sharov) maintain, "there is no single static cosmos 'out there' that organisms acquire knowledge on or adapt to" (Gontier 2018, 549). Such schools of thought draw upon current niche construction theories. These theories "understand organisms to primarily conform or adapt to a given and somewhat stable biological or socio-cultural environment which are the niches constructed and only in a later phase can individuals modify it" (Gontier 2018, 546). Gontier points out, however, that functionalist/dialectic/pragmatist schools of thought underestimate "the creative power organisms have in continuously bringing forth new niches, new bionts, and new holobionts." According to Gontier (2018, 546–547):

> Niche construction theory can fare much better by abandoning both its notions of adaptation and adaptability. These are non-evolutionary because they accept an outer, somewhat stable world. Adaptation or superorganic realms are concepts belonging to older cosmologies, they are not part of the new worldview that is developing. For the same reason, we shall also surpass Levin & Lewontin's (1985) Hegelian and Marxist dialectic position.

For the same reason, new evolutionary epistemological and biosemiotic positions that draw upon functional, dialectic, and pragmatist ideas should be surpassed. Also, in opposition to complete constructivist (Diettrich and Riegler) and semiotic idealistic (Sebeok) perspectives, the new cosmology demonstrates that organisms' constructive activities are not confined to their minds: "organisms reconstruct the earth, not just in their minds, they *embody* that knowledge in their anatomy and cognition, and they *extend* it onto their progeny and into the niches they construct" (Gontier 2018, 545). Moreover, contrary to those biosemioticians (Kull, Emmeche, Hoffmeyer, Pattee) who support the idea of an outer physical world, the "newly evolving cosmology" shows that there is no longer room for the idea of an abiotic world (Gontier 2018, 551):

> Ever since life evolved, life has rebuilt earth inside out, recycling existing matter, energy and space made in previous moments in time, into a living earth, up to the point that earth no longer exists as a purely physical "outside" entity. If that abiotic entity once existed, it now exists no more. Rather, it evolved into a living planet through the organisms that reconstruct it from its subatomic particles onward by reproducing and constructing new material life forms as well as extended and equally material niches. (Gontier 2018, 545)

In light of this, those evolutionary epistemological and biosemiotic standpoints that rely upon complete constructivist and semiotic idealist perspectives or that embrace the idea of an outer physical world should also be surpassed. Ultimately, the traditional epistemology/ontology should be abandoned:

> Organisms and the environments they build (epistemology understood as evolved knowledge) are what is real (ontologically), and the relation is exclusive because there is no outer abiotic earth anymore. Our living planet is not just hypothetically real, it is spatiotemporally real, or stated otherwise variant in time and space. (Gontier 2018, 545)

On the grounds of this new cosmology, Gontier's AEE proposes to go further than new EEs and biosemiotic schools of thought by framing the organism–environment relationship within an innovative, radical spatiotemporal realist perspective, in which constructivist and relativist notions come to acquire brand-new meanings. Within AEE, organisms are niches constructors that are also habitable zones of life for other organisms, engaged in the continual activity of creating new bio-realities and molding their extended niches (Gontier 2018, 550).

Within this view, the construction of niches no longer implies the notions of adaptation and adaptability, as the idea of an outer world comes to be supplanted by that of multiple evolving bio-realities which modify, recycle, and ultimately replace physical earth (Gontier 2018, 550). The latter, once "a purely physical or physico-chemical object," no longer exists as it has left the place to "expanding and contracting biologically-informed realities or bio-realities" (Gontier 2018, 551). On the other hand, since knowledge is identified with "*an evolving phenomenon that materializes into organisms and the overlapping biological realities they construe*" (Gontier 2018, 550), within AEE, knowledge comes to equal reality, or, in other words, epistemology comes to equal ontology, understood as "the current living world" (Gontier 2018, 545). In light of this, within AEE, the evaluation of knowledge does not longer imply a comparison with an outer physical world. Along these lines, AEE comes to subscribe to a relativistic or pluralistic account, according to which truth evolves over time and space and varies from organism to organism, from niche to niche (Gontier 2018, 554–557). Eventually, within AEE:

> The question is not who is right or wrong, but how distinct insights from different human cosmologies and other organisms together provide a deeper understanding of the complex and multiple realities that life has evolved up until today, and how we can move forward from there. (Gontier 2018, 557)

5 Conclusion

The comparison of EE and biosemiotics has provided a viewpoint to appreciate the respective extensions of these two disciplines, observe their specific conformation, and locate the points of mutual intersection and sharp separation.

Although EE and biosemiotics aim at different targets, in Sect. 3, we have demonstrated that it is possible to draw some parallels between these schools of thought, especially when considering Non-Adaptationist and Applied versions of EE in comparison with the Peircean and Constructive interpretations approaches to biosemiotics. A systematic analysis has furthermore revealed differences in how the disciplines understand and make use of the same ideas.

For one, both EE and biosemiotics recognize evolutionary continuity between humans and other beings. However, whereas biosemioticians preserve knowledge for inhabitants of the living realm, where it becomes identified with the domain of semiosis, Adaptationist, Non-Adaptationist, and Applied EE extend knowledge to phenomena produced by living organisms (behavior, cognition, languages, cultures, science, and technocomplexes), and especially with AEE, such a continuum is even further extended toward the non-living world. As such, all EEs avoid running into the conceptual problem faced by biosemioticians in having to demarcate a biosemiotic threshold in order to define the minimal semiotic entity. Biosemioticians' continued emphasis on semiosis or interpretation and meaning-making by individual agents of the world explains their lack of references toward technological, sociocultural, and linguistic phenomena, as well as to abiotic phenomena that lack selves. Evolutionary epistemologists, however, do not require the identification of agency or a semiotic threshold to identify information and knowledge. They can instead move about freely in identifying biotic and from-life-evolved phenomena as well as abiotic and non-biotic phenomena as embodying information and knowledge.

Secondly, in studying knowledge, both traditions have been similarly influenced by research in cybernetics and systems theory, and more recent approaches in EE and biosemiotics also resort to non-Darwinian theories of evolution. In so doing, both schools understand organisms as autonomous systems capable of self-organizing and self-regulating. Biosemioticians, however, have emphasized the teleological character (i.e., the goal-directedness) of self-organizing and self-regulating activities while Non-Adaptationist and Applied Evolutionary Epistemologists have highlighted the independent character of such activities from an outer world.

Thirdly, we have shown that the different schools all share an interest in the organism–environment relationship. Nevertheless, whereas non-adaptationist EE, as well as biosemiotics, resort to an old cosmology when defining the organism–environment relation, a cosmology that relies upon a sharp distinction between ontology and epistemology, AEE draws upon a new worldview, up-to-date with the latest research in biological, socio-anthropological, and physical sciences, a worldview in which epistemology and ontology ultimately come to coincide.

By analyzing how (A)EE and biosemiotics compare to one another, we have laid bare problems faced by biosemioticians and overcome by scholars active in non-adaptationist EE and AEE. On such grounds, we contend that biosemiotics necessitates a revision of the way it accounts for the relation between biological organisms and otherworldly phenomena. By drawing upon the recent history of AEE, biosemiotics might find the means to operate such change and continue to build up its path, alongside that of (A)EE.

Acknowledgements Parts of this paper were presented at the 19th Annual Gatherings in Biosemiotics that was held in Moscow, Russia, in 2019. We thank the audience for useful comments. We are furthermore grateful to the editors of this inspiring volume for inviting us to contribute. Gontier in addition acknowledges the financial support of FCT, the Portuguese Foundation for Science and Technology, Grant ID DL57/2016/CP1479/CT0066 and Project IDs: UID/FIL/00678/2019 & UIDB/00678/2020.

References

Artmann S (2005) Biosemiotics as a structural science: between the forms of life and the life of forms. J Biosemiotics 1:229–265

Barbieri M (2014) From biosemiotics to code biology. Biol Theory 9(2):239–249

von Bertalanffy L (1968) General system theory: foundations, development, applications. George Braziller, New York

Bradie M (1986) Assessing evolutionary epistemology. Biol Philos 1:401–459

Bradie M, Harms W (2001) (substantive revision: 2016). Evolutionary epistemology. Stanford encyclopedia of philosophy. http://plato.stanford.edu/entries/epistemology-evolutionary/. Accessed Jan 2020

Brier S (2006) Biosemiotics. Int Encycl Lang Linguist 2:31–40

Brier S (2015) Can biosemiotics be a 'science' if its purpose is to be a bridge between the natural, social and human sciences? Prog Biophys Mol Biol 119(3):576–587

Brigandt I, Love A (2017) Reductionism in biology. In Zalta EN (ed) The Stanford encyclopedia of philosophy. Spring. https://plato.stanford.edu/archives/spr2017/entries/reduction-biology/. Accessed 4 June 2019

Bruni LE (2008) Cellular semiotics and signal transduction. In: Barbieri M (ed) Introduction to biosemiotics. The new biological synthesis. Springer, Dordrecht, pp 365–407

Campbell DT (1974a) Evolutionary epistemology. In: Schlipp PA (ed) The philosophy of Karl Popper, vol I. La Salle, Chicago, IL, pp 413–459

Campbell DT (1974b) Downward causation in hierarchically organized biological systems. In: Ayala FJ, Dobzhansky T (eds) Studies in the philosophy of biology: reduction and related problems. Palgrave, London, pp 179–186

Clark A, Chalmers DJ (1998) The extended mind. Analysis 58(1):7–19

Dawkins R (1983) Universal Darwinism. In: Hull DL, Ruse M (eds) The philosophy of biology. Oxford University Press, Oxford, 15–35 [First published in Bendall DS (ed) (1998) Evolution from molecules to man. Cambridge University Press, Cambridge, pp 403–425]

Deely J (2013) The quasi-error of the external world: an essay for Thomas A. Sebeok, in memoriam. Cybern Hum Knowing 10(1):25–46

Diettrich O (1998) On some relations between cognitive and biological evolution. In: Van de Vijver G, Salthe S, Delpos M (eds) Evolutionary systems. Kluwer, Dordrecht, pp 319–340

Diettrich O (2001) A physical approach to the construction of cognition and to cognitive evolution. Found Sci 6(4):273–341

Diettrich O (2004) Cognitive evolution. In: Wuketits FM, Antweiler C (eds) Handbook of evolution, Vol. 1: the evolution of human societies and cultures. Wiley-VCH Verlag GmbH & Co. KgaA, Weinheim, pp 25–75

Diettrich O (2006) The biological boundary conditions for our classical physical world view. In: Gontier N et al (eds) Evolutionary epistemology, language and culture. Springer, Dordrecht, pp 67–93

Eco U (1984) Semiotics and the philosophy of language. Indiana University Press, Bloomington and Indianapolis

Facoetti M (2017) Evolution, knowledge, and reality: a defence of non-adaptationist evolutionary epistemology. Master thesis. https://dspace.library.uu.nl/handle/1874/353294. Accessed 22 May 2020

Facoetti M (2020) Donald Davidson's critiques of conceptual relativism applied to non-adaptationist evolutionary epistemology and refuted. Found Sci 25:357–374

Facoetti M (2021) United in diversity: an organic overview of non-adaptationist evolutionary epistemology. J Gen Philos Sci 52:211–225

Favareau D (2007) The evolutionary history of biosemiotics. In: Barbieri M (ed) Introduction to biosemiotics: the new biological synthesis. Springer, Dordrecht, pp 1–67

Favareau D et al (2017) How can the study of the humanities inform the study of biosemiotics? Biosemiotics 10:9–31

Gensini S (2002) Elementi di semiotica. Carocci editore, Roma
von Glasersfeld E (1981) Einführung in den Radikalen Konstruktivismus. In: Watzlawick P (ed) Die Erfundene Wirklichkeit. Piper, Munich, pp 16–38
von Glasersfeld E (1984) An introduction to radical constructivism. In: Watzlawick P (ed) The invented reality. Norton, New York, pp 17–40
von Glasersfeld E (1995) Radical constructivism. The Palmer Press, London
Gontier N (2006a) Introduction to evolutionary epistemology, language and culture. In: Gontier N et al (eds) Evolutionary epistemology, language and culture. Springer, Dordrecht, pp 1–29
Gontier N (2006b) Evolutionary epistemology. Internet encyclopaedia of philosophy. http://www.iep.utm.edu/evo-epis/#H6. Accessed Jan 2020
Gontier N (2012) Applied evolutionary epistemology: a new methodology to enhance interdisciplinary research between the life and human sciences. Kairos 4:7–49
Gontier N (2017) What are the levels and mechanisms/processes of language evolution? Lang Sci 63:12–43
Gontier N (2018) On how epistemology and ontology converge through evolution: the applied evolutionary epistemological approach. In: Wuppuluri S, Doria FA (eds) The map and the territory. Springer, Dordrecht, pp 533–569
Gontier N, Bradie M (2017) Acquiring knowledge on species-specific biorealities: the applied evolutionary epistemological approach. In: Joyce R (ed) The routledge handbook of evolution and philosophy. Routledge, London, pp 136–152
Gould SJ (1977) Ontogeny and phylogeny. Harvard University Press, Cambridge, MA
Gould SJ, Lewontin R (1979) The spandrels of San Marco and the panglossian paradigm: a critique of the adaptationist programme. Proc R Soc London, Ser B 205:581–598
Hahlweg K, Hooker C (eds) (1989) Issues in evolutionary epistemology. State University of New York Press, Abany, NY
Hoffmeyer J (2010) God and the world of signs: semiotics and the emergence of life. Zygon 45 (2):367–390
Hoffmeyer J (2011) Biology is immature biosemiotics. In: Emmeche C, Kull K (eds) Towards a semiotic biology: life is the action of signs. Imperial College Press, London, pp 43–65
Hoffmeyer J, Emmeche C (1991) Code-duality and the semiotics of nature. In: Anderson M, Merrell F (eds) On semiotic modelling. Mouton de Gruyter, Berlin, New York, pp 117–166
Hoffmeyer J, Kull K (2011) Theories of signs and meaning: views from copenhagen and tartu. In: Emmeche C, Kull K (eds) Towards a semiotic biology: life is the action of signs. Imperial College Press, London, pp 263–286
Hull DL (1988) Science as a process: an evolutionary account of the social and conceptual development of science. The University of Chicago Press, Chicago
James W (1907) Pragmatism: a new name for some old ways of thinking. Project Gutenberg, http://www.gutenberg.org/files/5116/5116-h/5116-h.htm
Kalevi K (1998) Semiotic ecology: different natures in the semiosphere. Sign Syst Stud 26:344–371
Kull K (1999) Biosemiotics in the twentieth century: a view from biology. Semiotica 127 (1/4):385–414
Kull K (2009a) Biosemiotics: to know, what life knows. Cybern Hum Knowing 16(1–2):81–88
Kull K (2009b) The importance of semiotics to university: semiosis makes the world locally plural. In: Deely J, Sbrocchi LG (eds) Semiotics 2008: specialization, semiosis, semiotics. Legas, Ottawa, pp 494–514
Kull K (2009c) Vegetative, animal, and cultural semiosis: the semiotic threshold zones. Cognit Semiotics 4:8–27
Kull K (2014) Zoosemiotics is the study of animal forms of knowing. Semiotica 198:47–60
Kull K (2015) Introduction to biosemiotics. In: Trifonas PP (ed) International handbook of semiotics. Springer, Dordrecht, pp 521–533
Kull K (2018) On the logic of animal umwelten: the animal subjective present and zoosemiotics of choice and learning. In: Marrone G, Mangano D (eds) Semiotics of animals in culture: Zoosemiotics 2.0. Springer, Cham, pp 135–148

Kull K, Emmeche C, Favareau D (2011a) Biosemiotic research questions. In: Emmeche C, Kull K (eds) Towards a semiotic biology: life is the action of signs. Imperial College Press, London, pp 67–90
Kull K, Emmeche C, Hoffmeyer J (2011b) Why biosemiotics? An introduction to our view on the biology of life itself. In: Emmeche C, Kull K (eds) Towards a semiotic biology: life is the action of signs. Imperial College Press, London, pp 1–21
Kull K et al (2011c) Theses on biosemiotics: prolegomena to a theoretical biology. In: Emmeche C, Kull K (eds) Towards a semiotic biology: life is the action of signs. Imperial College Press, London, pp 25–41
Lewontin RC (1982) Organism and environment. In: Plotkin HC (ed) Learning, development and culture. Wiley, New York
Lewontin RC (1983) Gene, organism, and environment. In: Bendall DS (ed) Evolution from molecules to men. Cambridge University Press, Cambridge
Lewontin RC (2000) The triple helix. Harvard University Press, Cambridge, MA
Liu CH, Matthews R (2005) Vygotsky's philosophy: constructivism and its criticisms examined. Int Educ J 6(3):386–399
Lorenz K (1941) Kant's Lehre vom Apriorischen im Lichte gegenwärtiger Biologie. Blätter Dtsch Philos 15:94–125
Lorenz K (1977) Behind the mirror: a search for a natural history of human knowledge. Methuen, London
Löwenhard P (1989) The mind-body problem: some neurobiological reflections. In: Hoyningen-Huene P, Wuketits FM (eds) Reductionism and systems theory in the life sciences. Kluwer Academic Publishers, Dordrecht, pp 85–135
Magnani L (2017) The abductive structure of scientific creativity: an essay on the ecology of cognition. Springer, Heidelberg
Maturana HR (1970) Biology of cognition. In: BCL report 9.0. University of Illinois, Champaign, IL
Maturana HR, Varela FJ (1980) Autopoiesis and cognition: the realization of the living. D. Reidel, Dordrecht
Munz P (1993) Philosophical Darwinism: on the origin of knowledge by means of natural selection. Routledge, London
Nöth W (1998) Ecosemiotics. Sign Syst Stud 26:332–343
Odling-Smee FJ, Laland KN, Feldman MW (1996) Niche construction. Am Nat 147:641–648
Pattee HH (ed) (1973) Hierarchy theory: the challenge of complex systems. Braziller, New York
Pattee HH, Kull K (2011) Between physics and semiotics. In: Emmeche C, Kull K (eds) Towards a semiotic biology: life is the action of signs. Imperial College Press, London, pp 213–233
Peirce CS (1931–1935) Collected papers of Charles Sanders Peirce, vol 1–8. Harvard University Press, Cambridge
Peirce CS (1992, 1999) The essential Peirce, 2 volumes. Indiana University Press, Bloomington
Piaget J (1970) Genetic epistemology (trans: Duckworth E). Columbia University Press, New York
Plotkin H (1994) Darwin machines and the nature of knowledge: concerning adaptations, instinct and the evolution of intelligence. Cambridge University Press
Popper KR (1963) Conjectures and refutations. Routledge & Kegan Paul, London
Popper KR (1972) Objective knowledge: an evolutionary approach. Clarendon Press, Oxford
Quine WV (1969) Naturalized epistemology. In: Quine WV (ed) Ontological relativity and other essays. Colombia University Press, New York, pp 69–90
Riedl R (1984) Biology of knowledge: the evolutionary basis of reason. John Wiley & Sons, Chichester
Riegler A (2001) Towards a radical constructivist understanding of science. Found Sci 6:1–30
Riegler A (2006) Like cats and dogs: radical constructivism and evolutionary epistemology. In: Gontier N et al (eds) Evolutionary epistemology, language and culture. Springer, Dordrecht, pp 47–65
Rorty R (1980) Philosophy and the mirror of nature. Blackwell, Oxford

Rothschild FS (1962) Laws of symbolic mediation in the dynamics of self and personality. Ann N Y Acad Sci 96:774–784

Salthe SN (1985) Evolving hierarchical systems: their structure and representation. Columbia University Press, New York

Sebeok TA (1994) Signs: an introduction to semiotics. University of Toronto Press, Toronto

Sebeok TA (1996) Signs, bridges, origins. In: Trabant J (ed) Origins of language. Collegium Budapest, Budapest, pp 89–115

Shannon CE (1948) A mathematical theory of communication. Bell Syst Tech J 27:379–423 & 623–656

Sharov AA (1999) Pragmatism and umwelt-theory. http://alexei.nfshost.com/biosem/txt/umwelt.html. Accessed 13 Dec 2018

Sharov AA (2016) Evolutionary biosemiotics and multilevel construction networks. Biosemiotics 9 (3):399–416

Sharov AA (2018) Mind, agency, and biosemiotics. J Cognit Sci 192:195–228

Sharov AA (n.d.) What is biosemiotics? http://alexei.nfshost.com/biosem/geninfo.html. Accessed 13 Dec 2018

Sharov AA, Maran T, Tønnessen M (2015) Towards synthesis of biology and semiotics. Biosemiotics 8:1–7

Sharov AA, Maran T, Tønnessen M (2016) Comprehending the semiosis of evolution. Biosemiotics 9(1):1–6

Sharov AA, Vehkavaara T (2015) Protosemiosis: agency with reduced representation capacity. Biosemiotics 8(1):103–123. [Available at (without page numbers): https://www.ncbi.nlm.nih.gov/pmc/articles/PMC4414345/. Accessed 23 Sept 2019]

Simon HA (1962) The architecture of complexity: hierarchic systems. Proc Am Philos Soc 106:467–482

Tinbergen N (1963) On aims and methods in ethology. Z Tierpsychol 20:410–443

Tooby J, DeVore I (1987) The reconstruction of hominid evolution through strategic modeling. In: Kinzey WG (ed) The evolution of human behavior: primate models. SUNY Press, Albany, NY, pp 183–237

Toulmin S (1972) Human understanding: the collective use and evolution of concepts. Princeton University Press, Princeton

von Uexküll J (1921) Umwelt und Innenwelt der Tiere. 2. verm. u. verb. Aufl. J. Springer, Berlin

von Uexküll J (1928) Theoretische Biologie 2, 2nd edn. J. Springer, Berlin

von Uexküll J (1934) A stroll through the worlds of animals and men: A picture book of invisible worlds. Semiotica 89(4):319–391. [Originally published in Schiller CH (ed) (1957) Instinctive behavior. International Universities Press, Madison, CT, 5–80]

von Uexküll J (1937) The new concept of umwelt: a link between science and the humanities. Semiotica 134(1/4):111–123

von Uexküll J (1940) Bedeutungslehre. J.A. Barth, Leipzig

Vehkavaara T (2002) Why and how to naturalize semiotic concepts for biosemiotics. Sign Syst Stud 30(1):293–313

Vehkavaara T (2003) Natural self-interest, interactive representation, and the emergence of objects and Umwelt: an outline of basic semiotic concepts for biosemiotics. Sign Syst Stud 31 (2):547–587

Vollmer G (1984) Mesocosm and objective knowledge: on problems solved by evolutionary epistemology. In: Wuketits FM (ed) Concepts and approaches in evolutionary epistemology: towards an evolutionary theory of knowledge. D. Reidel, Dordrecht, pp 69–121

Vollmer G (1987) [1975] Evolutionäre Erkenntnistheorie, 4th edn. Hirzel, Stuttgart

Whittaker RH, Margulis L (1978) Protist classification and the kingdoms of organisms. Biosystems 10:3–18

Wiener N (1948) Cybernetics: or control and communication in the animal and the machine. MIT Press, Cambridge, MA

Woese CR, Kandler O, Wheelis ML (1990) Towards a natural system of organisms: proposal for the domains Archaea, Bacteria, and Eucarya. PNAS 87(12):4576–4579
Wuketits FM (1989) Organisms, vital forces, and machines: classical controversies and the contemporary discussion 'reductionism vs. holism'. In: Hoyningen-Huene P, Wuketits FM (eds) Reductionism and systems theory in the life sciences. Kluwer Academic Publishers, Dordrecht, pp 85–135
Wuketits FM (1992) Adaptation, representation, construction: an issue in evolutionary epistemology. Evol Cognit 2:151–162
Wuketits FM (1998) Functional realism. La Nuova Crit 31–32:5–21
Wuketits FM (2006) Evolutionary epistemology: the non-adaptationist approach. In: Gontier N et al (eds) Evolutionary epistemology, language and culture. Springer, Dordrecht, pp 33–46

Extended Synthesis and Jablonka and Lamb's Four-Dimensional View of Evolution

Jonathan Luís H. Ferreira

Abstract Within the neo-Darwinian paradigm, as seen in sociobiology (Wilson, Sociobiology: the new synthesis. Harvard University Press, 1975) and evolutionary psychology (Jeffery, Two behavioral hypotheses for the evolution of male homosexuality in humans. In: Shackelford TK, Hansen RD (eds) Evolutionary psychology. The evolution of sexuality. Springer International Publishing, pp 207–219, 2015), evolution and heredity are understood in a gene-centered view. Combined with adaptationism, this view results in what I call reproductionism, the tendency to reduce sexuality to reproduction. This view does not suffice to explain several biological phenomena, especially the human mind and culture. Extended Synthesis and Jablonka and Lamb's idea of evolution in four dimensions (Jablonka and Lamb, Evolution in four dimensions: genetic, epigenetic, behavioral, and symbolic variation in the history of life. Massachusetts Institute of Technology Press, 2005; Evolution in four dimensions, revised edition: genetic, epigenetic, behavioral, and symbolic variation in the history of life. Massachusetts Institute of Technology Press, 2014) emerged as a broader view of evolution, encompassing behavioral and cultural diversity, including sexuality. Along with a phenomenological attitude and Biosemiotics, this view can lead to better evolutionary hypotheses. This paper aims to analyze some impacts of this view on evolutionary biology, especially on understanding social and sexual behavior.

Keywords Evolutionary biosemiotics · Extended synthesis · Four-dimensional evolution · Sexuality · Sexual behavior · Reproduction

---

J. L. H. Ferreira (✉)
Department of Biological Sciences, Federal University of Juiz de Fora, Juiz de Fora, Brazil

© Springer Nature Switzerland AG 2021

E. Pagni, R. Theisen Simanke (eds.), *Biosemiotics and Evolution*, Interdisciplinary Evolution Research 6, https://doi.org/10.1007/978-3-030-85265-8_10

1 Introduction

In *Phenomenology of Perception* (2002), the philosopher Merleau-Ponty argues that, as a work of art, the body is a signification whose expression is not distinct from what is expressed (Merleau-Ponty 2002, p. 175) and that sexuality, always present in life as an atmosphere, assumes several meanings (ibid. pp. 195–196). An evolutionary biosemiotician who accepts these definitions will proceed by asking how such a sexed and meaningful body may have evolved by the Darwinian principle of natural selection. However, most current evolutionary hypotheses, based on the neo-Darwinian version of Darwinism, reduce sexuality to reproduction: sex, at least in evolution, is nothing more than a means of propagating our genes. Before continuing the investigation, the evolutionary biosemiotician must seek another model that can handle the body as an expression in all its forms and meanings.

The problem is that, within the neo-Darwinian paradigm, heredity and evolution are usually understood according to a gene-centered view, which is the idea that genes are the main units of natural selection and that genetics is the only system of inheritance relevant to evolution. However, in the last decades, the Extended Synthesis and new areas of biological research, such as Evolutionary Developmental Biology, have challenged the traditional gene-centered view. Within this framework, Jablonka and Lamb (2005, 2014) propose a broader view of evolution: in addition to the genetic dimension, they present an epigenetic, a behavioral and, only among humans, a symbolic dimension, each one with particular systems of inheritance and different types of heritable biological information. Acknowledging these other dimensions of evolution and their particular inheritance systems reframes the epistemological foundations of evolutionary biology, shedding new light on the evolution of several psychosocial phenomena, such as sexuality.

The gene-centered view is often associated with the adaptationist approach of evolution criticized by Gould and Lewontin (1979), with a particular focus on the reproductive success that a specific biological trait provides to the organism that exhibits it. This association reflects in what I call a reproductionist account of behavioral and social evolution. Reproductionism can be characterized as a tendency to analyze behavior and sociality only in terms of fitness and reproductive success. The conceptual distinction between reproductionism and adaptationism was drawn because one can have one without the other. As we will see, a broader view of evolution allows a broader view of adaptation.

Regarding the evolution of anatomical, physiological, and behavioral traits, the reproductionist approach can be useful and, often, represents successful heuristics (although it is, by no means, a complete explanation). However, the same cannot be easily said regarding complex behaviors and other biosocial and psychological phenomena. A gene-centered view, often followed by a mechanical idea of animals and behavior, is at the center of what Dominique Lestel calls realist-Cartesian ethology, the dominant approach in current ethology (Lestel 2011). By treating animals as ahistorical and generalizable mechanical beings behaving to spread their genes, this approach limits ethology as a science and its use as a comparative method for evolutionary inquiry.

Since reproduction is considered the only relevant way of transmitting biological information useful to the selection, most evolutionary hypotheses to social and sexual behavior focus on reproductive sex. As a consequence, sex and reproduction in biology are frequently mistaken as synonyms. Although it might be true for most living beings that reproduce sexually, this is certainly not the case for the hominid lineage (humans and other great apes). Here sexual behavior appears in a broader social context beyond the original reproductive function, with multiple meanings and motivations, such as tension-reducing, reconciliation, hedonism, dominance display, social learning and playing (de Waal 1995; Dixson 2012; Furuichi 2019), especially throughout human history (Foucault 1984, 1988; Dixson 2012; Kirkpatrick et al. 2000; Gray 2013). In this sense, non-reproductive sex can be described as a form of social communication extremely relevant for Biosemiotics.

Due to its epistemological limitations and the focus on reproductive success, evolutionary thought is full of supposed "paradoxes," especially regarding human sexuality and other psychosocial aspects (e.g., female sexuality, homosexual and other non-conceptive sexual behaviors, non-heterosexual identities, altruistic behaviors, childcare and adoption by non-relatives, etc.). The inheritance of a Christian morality toward sex in Western societies does not help either, and, ironically, evolutionary discourses often end up reinforcing this morality. Within the reproductionist model, as seen in *The Naked Ape* (Morris 1999) and the fields of sociobiology (Wilson 1975) and evolutionary psychology (Buss 2019; Jeffery 2015), scientists cannot successfully answer these so-called paradoxes and the attempts have been charged with speculation and moral preconceptions. A common explanation for the emergence and maintenance of such behaviors and other biosocial phenomena, as seen, for instance, in Alcock's reference to alloparental adoption (adoption of non-relatives), is that they could be maladaptive byproducts of the complex human evolution (Alcock 2011, p. 515). Throughout the history of biology, similar misconceptions about human evolution have led to unfortunate political ideas, and, even today, it is a theme of much controversy. Consequently, these ideas contribute to a problematic public persona of evolution and biology, which might explain, in part, the resistance of human sciences to accept an evolutionary approach to the human mind, behavior, and culture.

The same co-extensiveness between sex and life observed by Merleau-Ponty is valid for all the animals that reproduce sexually, especially for those whose sexual behavior has acquired other meanings beyond reproduction. Merleau-Ponty's phenomenology gives an account of behavior not as a set of discrete actions but as intentionality deeply embedded in nature and projected toward an already meaningful environment. A phenomenological attitude, especially in the tradition envisioned by him, a biosemiotic account on life (for a comparative analysis between Phenomenology and Biosemiotics, see Tønnessen et al. 2018), and the four-dimensional view of evolution can provide a fruitful method of inquiry into evolution and development of both human and non-human social behavior, including sexuality in all of its meanings and manifestations. Such an approach leads to a less restrictive and more inclusive evolutionary biology than that of Neo-Darwinism, generating hypotheses that better represent the vast biodiversity, including cultural diversity. Thus, the so-called paradoxes of evolution, such as human sexual diversity, may be paradoxical only within a limited idea of biology and evolution.

Since the effort required to put an approach like this into practice would be far beyond the scope of this work, here I argue for its need by exploring: some implications posed by the four-dimensional view on the role of reproduction and other concepts (e.g., adaptation, fitness) in evolution; epistemological limitations of traditional evolutionary approaches to socio-sexual behavior, which makes evolutionary biology within Neo-Darwinian paradigm insufficient to understand human sexuality; and how a phenomenological attitude, in addition to Biosemiotics, Extended Synthesis, and the four-dimensional view of evolution might overcome those limitations, contributing to change the problematic image of evolutionary biology in public perception into a less deterministic science. I hope this perspective can provide further insights into the evolution of other psychosocial phenomena.

2 Four-Dimensional View of Evolution: Jablonka and Lamb (2005, 2014)

The rediscovery of Mendelian genetics in the early twentieth century revolutionized biological sciences. For the first time, the mechanism of heredity was unraveled, and it was possible to visualize how the natural selection theorized by Darwin, almost half a century before, could occur. The Modern Synthesis affected several fields of biology and also opened space for new ones. However, the role of development, behavior, and culture was largely ignored. Since then, every attempt to understand living beings, including humans, must pass like a train on the tracks of genetics and reproductive success. Edward O. Wilson's sociobiology (Wilson 1975) and the field of evolutionary psychology (Buss 2019; Jeffery 2015), for instance, attempt to explain psychosocial behavior within the neo-Darwinian perspective.

Indeed, genes have a fundamental role in our organization as living beings and biological inheritance. However, considering the recent discoveries of horizontal gene transfer, epigenetic inheritance, and the idea of behavioral and cultural evolution, it has become evident that there is more in evolution than just genetic inheritance and reproduction. We can no longer explain animal behavior and its evolution exclusively through genetics. Genes do not work alone; their message must be *interpreted* to produce a protein, which also must be interpreted and so on; there is no single unit of life. Besides, complex developmental pathways do not strictly follow the Mendelian laws, nor is there a clear causal relation between genotype and phenotype.

As the discoveries accumulated, the Extended Synthesis emerged. Focusing on heredity, Jablonka and Lamb (2005, 2014) criticize the neo-Darwinian assumption that evolution occurs exclusively through natural selection of random genetic mutations. Their main contribution is the view of evolution in four dimensions: genetic; epigenetic (these first two are shared by all living beings, including unicellular organisms); behavioral (shared by some animals); and human beings who have an additional symbolic dimension. Each of these dimensions has a particular inheritance system, with distinct kinds of constraints and processes generating different types of heritable information useful for natural selection in each level of biological organization. Thus, even for an adaptationist, it is no longer necessary to think of evolution only in terms of random variations blind to function, enabling the formulation of new kinds of evolutionary hypotheses.

Even in the genetic dimension, genes' effect during development is regulated by epigenetic systems, resulting in complex networks of interacting molecular and environmental signs. Therefore, there is no simple relation between genotype, phenotype, and environment. Plasticity (the way development can react to different conditions by changing the phenotype) and canalization (the resistance of development to genetic and environmental changes) make it difficult to predict the pathway the development can take: identical genes can lead to different phenotypes, and different genes can produce the same phenotype (Jablonka and Lamb 2014, pp. 62–64).

In the behavioral dimension, characters like food preferences can be transmitted through behavior-affecting substances via placenta and milk, and behavioral novelties can be acquired through social learning and imitation, even between non-relatives (ibid., pp. 160–177). Thus, behavior can vary and develop even when there are no genetic variations associated with it. Of course, a biologist, accustomed to a gene-centered view, would argue that the learning capacity depends on genetic constraints involved in producing a brain able to learn and other anatomical features that allow the behavior at hand to be expressed. Although this is true, learning makes behavior much more flexible and somewhat independent from its genetic constraints. In humans, the symbolic system of inheritance, within which variations are transmitted through cultural interactions, symbols, and traditions, is indistinguishable from the behavioral one. Another essential feature of the behavioral and symbolic systems is that they can be rich sources of end-directed variations, thus reducing mind and culture to blind genetic mutations, biochemical products, and reproduction is a poor characterization of how they evolve and develop.

The four-dimensional view, then, provides a more sophisticated, holistic, and flexible concept of evolution and heredity, being able to better explain the enormous biodiversity, including cultural diversity, as opposed to the restrictive view of Neo-Darwinism. Curiously, neo-Darwinian evolution is a static process: organisms change, functions change, but the evolutionary process remains the same for all biological phenomena, always acting on the same level and from the same source of variation; in the four-dimensional view, evolution itself evolves, originating different levels of biological organization and different systems of inheritance. Selection can then act both at the level of genes and the individual and the level of cells, organisms' conglomerates, symbiotically and ecologically associated, and groups—each of these levels with specific constraints and possibilities.

From the perspective of Biosemiotics, heredity, then, can be defined as a semiotic process involving a complex network of signs being transmitted at different dimensions over time, vertically (between parents and offspring via reproduction), horizontally (between relatives as well as non-relatives, via social learning), and even between distant generations (through literature, art, philosophy, etc.). Biosemiotic inheritance is not limited to the genetic dimension and can cross even the reproductive barrier between species. Distinct kinds of semiosis can happen throughout evolution, with different types of signs influencing one another's transmission. For instance, a symbol culturally constructed and transmitted, such as the idea of a punitive God commanding "Be fruitful and multiply," may affect an entire population's behavior for several generations, increasing its reproductive success without any genetic adaptation taking place. The increase in the reproductive success of this species and industrial and technological development influences others' survival and may lead them to extinction. Indeed, human population growth and its consequences, such as agriculture and urban expansion, usually appear in red lists of threatened species as the leading causes of extinction (ICMBio 2018; IUCN 2020).

The term *biolons*, which is used by Longo and Montévil in *Perspectives on Organisms: Biological Time, Symmetries and Singularities* (2014), describes biological entities at different levels of organization such as a single cell, a multicellular organism, or a species, maybe encompassing even higher levels of organization: population, community, ecosystem. These entities result from the complex interactions between different kinds of signs at each evolutionary dimension. Hence, phenotypic and biolonic development can follow through a range of possible ontogenetic and phylogenetic pathways depending on different ecological contexts. This multiplicity of pathways results in the structural, behavioral, and cultural diversity of organisms and groups. Each dimension, its respective functioning and the particular types of semiosis involved must be considered when developing an evolutionary hypothesis. Obviously, greater biological complexity requires more factors to be considered, making the development of evolutionary hypotheses about a species as diverse as *Homo sapiens* extremely difficult.

This is an overwhelming realization, and the most straightforward path offered by Neo-Darwinism may look like a tempting alternative. However, although practical, generalizable evolutionary models, for all types of living beings and ecological contexts, turn out to be extremely limited. Also, the emphasis on the genetic aspects hinders this kind of understanding, also contributing to a problematic public image of biology and evolution which, in turn, occludes any possible dialogue between the biological and human sciences due to the mistaken opinion that the naturalistic approach necessarily implies a deterministic model applied to a restrictive and limited idea of biological evolution. By expanding these concepts, like the four-dimensional view and Biosemiotics allow us to, subjectivity and historicity appear as fundamental aspects of the evolutionary process. This seems to be the key to a less restrictive and more inclusive evolutionary biology, which can better dialogue with other sciences.

3 On Reproduction

Reproduction is not the only way of transmitting heritable variations. Even genetic variations can be transmitted horizontally. Horizontal gene transfer is a prominent mechanism of transmission of genetic material among organisms not necessarily kin to each other and a significant one for prokaryote evolution, the basis of the evolutionary tree. Some studies also show that horizontal gene transfer occurs even between different species of eukaryotes, an essential key to endosymbiosis: the sea slug *Elysia chlorotica* can synthesize chlorophyll and realize photosynthesis for at least six months when deprived of the algae *Vaucheria litorea*, the food source from which they acquire chloroplasts via kleptoplasty (a symbiotic relationship where the host organism sequesters chloroplasts from algae) and horizontal gene transfer (Pierce et al. 2009). In biosemiotic language, that means genetic and physiological semiosis between different species throughout their evolution.

Moreover, in addition to genetics and epigenetics, there are also behavioral and symbolic systems of inheritance, within which animal traditions and culture are immeasurable ways of transmitting the information. Learning became a fundamental system in social evolution, especially for primates. Merleau-Ponty wrote that to learn is to acquire a new use for the body (Merleau-Ponty 2002, p. 177). By allowing the flexibility of the body, the behavioral dimension plays an important role in animal evolution. The flux of information involved in social learning, between parents and offspring, and between non-relatives and different cultures improves individual development and reinforces sociability. This positive feedback loop between learning and sociability was essential for hominid lineage and, consequently, for human history. Behavioral and symbolic systems can even cross the reproductive boundaries between species. Human culture can interact with the behavioral systems of other animals. For instance, English tits developed the habit of opening milk bottles delivered by the milkmen, a habit that rapidly spread across England, even between different bird species. Such a habit cannot be due to genetic mutation, but rather to a new invention and its social dissemination when naive birds learned by watching experienced ones (Jablonka and Lamb 2014, pp. 167–168).

The examples above show the interconnectedness of the four dimensions of evolution and the different inheritance systems within and between species. Thus, reproduction, through which information is transmitted in the form of DNA, is one process of transmission among others. Moreover, it is a fundamental one since there would be no evolution without genetic material inheritance. However, as biological complexity increases and new inheritance systems evolve, reproduction shares its fundamental status in evolution with other processes, such as social learning.

Besides, at the ecosystemic level, one species cannot reproduce indefinitely. In some cases, like in overpopulations, too much reproduction even becomes a problem. Nature itself controls the species birth rates. Like adaptations, such control is a consequence of evolution and selection, not a goal *per se*. Selective pressures generated by intraspecific and interspecific interactions, such as competition and predation, prevent a particular group from growing beyond the limit, which would compromise the group itself and the whole ecosystem due to the limitation of resources. For instance, top predators are fundamental to the integrity of the ecosystem since they control the population growth of their prey (and, in turn, predator populations are also controlled by the same fluctuation in prey populations, alongside other factors). Nature's logic seems to be "each one gets a little" rather than "one should take as much as one can," even if, on the individual level, it looks only like a competition sometimes.

Therefore, even though reproduction is essential to the species' survival, it is limited by ecological constraints and is often accompanied by other systems of inheritance. In the behavioral and symbolic dimensions, social learning and cultural transmission are as crucial as reproduction. Perhaps even more important, since a human individual can live perfectly well without reproducing, but would not properly develop social and cognitive skills without social learning. At the species level, a whole social species is sustained by intricate networks of systems of inheritance within its populations, which means that the existence of non-reproductive individuals (whether by choice, infertility, or any other cause) is not detrimental to the species as a whole group, as long as some individuals are reproducing. These intricate networks that can compensate for some non-reproductive individuals' presence remind us of the process of canalization at the epigenetic level, through which development remains unaffected by genetic and environmental changes. Even when knockout experiments are used to silence a particular gene, the epigenetic landscape compensates for its absence (Jablonka and Lamb 2014, p. 64). Analogously, at the species level, reproductive individuals within a population compensate for non-reproductive ones, similarly to how the epigenetic landscape compensates for a gene's inactivation. Thus, at the supra-genetic (e.g., at the group and ecosystem) levels, this compensation provided by other systems of inheritance could impact the evolution of sexual behavior, making it more flexible to produce different meanings and functions other than reproduction. If this is the case, traditional evolutionary assumptions must be reviewed.

4 Reproductionism

Science is a cultural product. It deals with culturally constructed representations about the natural world and pre-established notions of normal/abnormal, natural/unnatural, and nature/culture; it means that science is highly influenced by the cultural and political minds operating within. This statement sounds so obvious that many scientists take it for granted and overlook the socially constructed nature of scientific conceptions.

Discourses about sex are produced in all academic fields, especially in Biology, Medicine, Psychology, Sociology, and Philosophy. Each area of knowledge presents a distinct aspect of sex, producing discourses that can sometimes be complementary, sometimes different and even opposing ones, all due to the interdisciplinary nature of sex. Sex inspires every form of art; it is a theme of significant political, moral, and religious concerns and motivates heated debates. It also attracts much curiosity, making evolutionary accounts on sex a bestselling theme that promises a lot, although it accomplishes very little. Consequently, everything we know about sex and sexuality is the result of these distinct discourses and is permeated with scientific, moral, political, and religious elements, complexly combined, making it difficult to theorize about it without being influenced by pre-established notions.

Although scientists like to think that they are unbiased, it is difficult to be so, considering how the notions about sex run subtly, nonetheless profoundly, through the cultural background. In evolutionary biology, the ultimate meaning of sex is reproduction, the inheritance of Christian morality that scientists do not even recognize. According to the gene-centered view of evolution, random genetic mutations are the primary source of variations, and reproduction is the only way of transmitting biological information relevant to evolution. Associated with adaptationism, which aims to explain the origin of phenotypic variations as discrete adaptive traits, based merely on plausibility and speculation, without scrutiny and without considering possible non-adaptive alternatives (Gould and Lewontin 1979), this restricted concept of heredity results in what I call a reproductionist approach to behavioral and social evolution.

Here, reproductionism refers to the tendency to analyze behavioral and social evolution only regarding reproductive success. Its logic is, thus, the following: since the gene is the unit of selection and the only source of adaptive variations, and reproduction is the way through which these variations are transmitted across generations, evolution can be explained in terms of the adaptive value of a particular behavior, and the reproductive success it provides to the animal that exhibits it. Any physiological, behavioral, or social variation that increases reproductive success tends to be selected and transmitted through differential reproduction; if it reduces reproductive success, it tends to be eliminated by selection; if it is not eliminated, it will require adaptive explanations that do not harm the initial assumptions.

This logic is likely to be the case for most anatomical and physiological traits since they depend on transmitting latent information in the DNA. Still, even in such cases, this is not the complete answer: Jablonka and Lamb argue that even genetic mutations can result from instructive processes in response to the environment (Jablonka and Lamb 2014, pp. 92–102); besides, the epigenome, which regulates development, is extremely sensitive to environmental signals. Concerning induced genetic and epigenetic variations, environmental signals play a role as fundamental as reproduction in the generation, transmission, and selection of these variations. It is also worth considering that the signs involved in recognizing potential reproductive partners are essential for reproduction to occur.

When extended to behavioral and social evolution, the limitations of reproductionism become more evident, especially concerning socio-sexual phenomena that may not be inextricably associated with reproduction. Some examples are the homosexual greeting behaviors involving the genitals, such as the genito-genital rubbing between female bonobos (*Pan paniscus*), the penile fencing and rump-to-rump contact between males of the same species (Furuichi 2019), all of which have social meanings rather than reproduction; affection, pleasure, and other aspects of human sexuality, among others. Due to the high status attributed to the role of reproduction in the inheritance of morphological phenotypes, most evolutionary hypotheses about human affection and sexuality, even those designed to explain the non-reproductive aspects of sex, tend to focus on reproductive success, as in *The Naked Ape* (Morris 1999), in sociobiology (Wilson 1975) and evolutionary psychology (Jeffery 2015). Although I will briefly mention female sexuality, the main focus will be on homosexual behavior since it is the most evident non-reproductive meaning of sex. The terminological problem of using "homosexuality" (i.e., a current identitary category that has been around since the nineteenth century) in evolution will not be addressed here. I will be referring to sexual behavior between same-sex individuals and those who practice it, regardless of identity or sexual orientation, although these categories indeed emerged later in our evolutionary history to designate previous behavioral expressions.

For Desmond Morris (1999), human cultural diversity is a problem; his solution is to elect a society that he considers most representative of human evolution: North America. For him, it is enough to take the results found in what he considers the most successful societies, while those he considers "backward" and "bizarre" may be ignored (Morris 1999, p. 51). This attitude only reveals the author's ideological biases rather than any evolutionary reality. What he perceives as a trend in human evolution seems to be nothing more than the effects of agriculture, colonization, and industrialization.

Morris sees the differences between human sexual activity and physiology and other primates as adaptations for monogamy due to the need for prolonged parental care (ibid., p. 83). Examples of these differences are that most primates do not form prolonged pair-bonds, the continuous sexual receptivity of human females, and the reduced amount of hair that favors the erogenous zones development. Despite this, Morris is a supporter of non-reproductive sex since, as he observes, most human sexual activity is not directed to produce offspring but to cement the pair-bond. For him, however, this characteristic seems to make sense only for heterosexual and monogamous couples (ibid., pp. 65–66).

According to Morris's view of sexual evolution, the continuous female receptivity and orgasm have to do more with the man and the reproduction, rather than with the woman herself; she remains receptive even during pregnancy, so the man is not frustrated, and the pair-bond is not at risk (ibid., p. 66). It is not because she is herself a desiring, sexual being. According to him, the female orgasm may have evolved because of the exhaustion it causes, keeping the woman in the horizontal position, preventing the seminal fluid from flowing back down the vaginal tract (ibid., p. 79). He also speculates that voyeurism, pornography, and prostitution help satisfy the demand for sexual curiosity without damaging the pair-bond (ibid., pp. 92–93).

Nothing in his approach is more charged with moral elements than his view on homosexuality, which he considers "aberrant." According to him, the only morality that one can apply to the question would be the "biological morality," which means reproductive success. He applies this logic not only to homosexuals, but also to "monks, nuns, long-term spinsters and bachelors" (ibid., p. 98). However, it is worth noting that there is no biological morality; seeing "morality" in nature reveals more about the scientist's moral codes than a real biological phenomenon.

The Naked Ape was widely criticized since its publication (originally in 1967), and I doubt that any serious biologist today would publicly advocate for his arguments. However, I chose to explore them because they still echo today in more cautious forms and represent the reproductionism in its most radical form. Based on the traditional neo-Darwinian assumptions (to be clear, I do not believe that all neo-Darwinians today have the same thought) and on a reckless reflection on human sexuality, it is easy to reach this kind of conclusion.

Today, most hypotheses about the evolution of sexuality take a somewhat less problematic form and are less charged with moral prejudices. However, most of them still fall under the reproductionist model. For instance, the kin selection hypothesis of homosexuality, initially proposed by Wilson (1975), is one of the most popular until today. He proposes that, by forgoing their own reproductive success, ancestral homosexuals could improve their siblings' reproductive success through inclusive fitness, assisting the nurture of their nieces and nephews, with whom they share some of their genes.

Jeffery (2015) offers two evolutionary hypotheses for the continuity of human homosexuality: paternal investment and sneak copulation hypothesis. The paternal investment hypothesis speculates that males with "reduced gynephilia" (i.e., reduced interest in women) signal to women heterosexual fidelity, long-term mate quality, and increased paternal investment, even if it also causes some extent of androphilia (Jeffery 2015, pp. 209–210). In the sneak copulation hypothesis, males with increased androphilia seduce and stimulate a target male, making him tolerant and less likely to inseminate his female partner, "being sexually satisfied by the male." As he does that, the hypothesis moves forward, the sneaky male reduces the degree of sperm competition and improves his chances of fertilizing the female. (ibid., pp. 211–212). In both hypotheses, homosexual behavior is really about reproduction and heterosexual (or at least bisexual) behavior.

Two paradoxical problems characterize all of these hypotheses: on the one hand, a lack of imagination prevents the proponents from thinking outside reproductionism; on the other hand, too much imagination leads to arguable attempts to use reproductionist explanations for all sorts of behaviors and subjectivities. The assumption of a general model of evolution that might explain the psychological and cultural diversity creates several epistemological gaps. Hypotheses are formulated on speculative grounds, filled with moral biases and relying on data that could have several interpretations; simultaneously, historical and scientific facts are neglected or ignored to protect the initial assumptions.

Primatological and anthropological data (de Waal 1995; Furuichi 2019; Gray 2013; Kirkpatrick et al. 2000) suggest that non-reproductive aspects of sexuality play an essential role in the evolution and social organization of the entire hominid lineage, a role that has been maintained throughout human history. Perhaps this statement sounds strange, or even like a mistaken heresy, for many evolutionary biologists accustomed to the idea that increased reproductive success is the ultimate meaning of sexual behavior and that cultural and sexual diversity between human groups and societies gets in the way of developing evolutionary hypotheses more than it helps them. However, for social scientists and philosophers, who are familiar with works such as Plato's *Symposium* and Michel Foucault's *History of Sexuality* (1984, 1988), that is not a shocking statement. Suppose only the data we have now were available back then. In that case, Foucault could have gone even further beyond Ancient Greece with his archeology and realized that the dynamics between power, sex, and society are not, exclusively, human artifacts, but a major trend in hominid evolution that had acquired names and institutional organization later during human history.

5 Some Remarks on Ethology

The behavioral system of inheritance and the different levels of organization can improve the way we approach evolutionary explanations for behavioral strategies. For instance, alternative strategies into the same population and the maintenance of some behavioral variants are essential elements for survival and stability within this population, given the inconstancy of the environmental and ecological conditions. Even if an individual is genetically or ecologically disadvantaged, he may resort to a conditional mating strategy in order to take some benefit in a non-ideal situation; he will not necessarily die without the chance to mate and reproduce. For example, among anurans, satellite males wait for dominant males to attract females and then ambush them and try to mate with them, often successfully. Such behavioral plasticity may generate good behavioral variations upon which selection can act. When everyone else is behaving exactly in the same way, a change in the behavior may end up being enormously advantageous, even if it does not provide optimal reproductive success in all cases.

These variations are context-dependent and manifest as a response to specific environmental signs and conditions. For instance, ecological factors have a significant influence on territory size and the distribution of males and females, which, in turn, influence the mating system and the set of strategies within it. Male mammals often tend to behave monogamously when females are dispersed from each other, each occupying a small territory, and the Galapagos hawk's mating system (*Buteo galapagoensis*) may vary from monogamy to extreme polyandry depending on territory conditions (Alcock 2011).

Among dunnocks (*Prunella modularis*), a varying mating system (i.e., monogamy, polygyny, polyandry, or polygynandry) depends on the ecological conditions. If the female territory fully overlaps the male territory and has a particular extension that he can defend alone, the mating system tends to be monogamy. However, if the female territory is small and abundant in food resources, allowing the male to divide its parental care between more than one female and offspring, polygyny tends to occur. In another case, if the female territory is too large to be defended by the male alone, polyandry may occur: the female mates with two males and can rely on both to defend the territory and feed the shared offspring, being the amount of parental care spent by each male, proportional to its expected genetic share. Lastly, two males may cooperate to guard the overlapping territories of two females, leading to polygynandry: both males and females have several mates, and the males cooperate to feed the shared offspring (Avital and Jablonka 2000). Thus, four different mating systems may coexist within the same population among dunnocks depending on environmental and social signs. Which is the most adaptive strategy? All of them are. Each strategy is the most adaptive one, given the ecological conditions within which it occurs. Who benefits the most from each strategy, males or females? Even though sexual conflict is essential for a focal explanation, in the grand scheme of things, it does not seem to matter which specific sex benefits the most since the whole population adaptively benefits from the behavioral plasticity.

At one level, it seems that the individual is the target of natural selection, but on a broader level, the whole population benefits from the resulting network of strategies. The whole ecosystem will benefit from the stability within the populations. A multilevel selection seems to be in action, selecting not one strategy to the detriment of others but rather the whole set of strategies to maintain the ecosystem's stability. Whether evolution is some tug of war in which someone will necessarily lose or, as Sapolsky (2017) metaphorically suggests, a complex kind of rock-paper-scissors game, it does not matter who is winning (if there is an actual winner), as long as the game keeps going on. Understanding the environmental and semiotic processes involved in all organization levels provides better evolutionary explanations for behavioral variation and stability.

Another implication of the four-dimensional view is that a variation originated in one dimension, whether it provides an apparent advantage or a disadvantage in this origin dimension, could, in a similar way, be adaptive or not in other dimensions. Thus, "adaptation" and "fitness" may gain a broader meaning. Perhaps we do not need to redefine those concepts to look differently at them or at least include some other concepts alongside them. "Adaptation" could easily assume different shapes in different dimensions, namely, a behavioral or a cultural adaptation. As described above, in its usual definition, fitness could be a final product of the relations between the variation in question and each dimension of evolution. Alternatively, fitness could still be used in the traditional sense but understood as one evolutionary factor working alongside others. For instance, a cultural or behavioral variation could affect the fitness or not; if it does, it can be adaptive, neutral, or maladaptive, but this outcome cannot be easily predicted. The final balance of any variation that emerged in any dimension must be, at that moment, meticulously measured as the product of these interdimensional interactions. This view could be a useful tool in the ethological approach, providing an insightful look at behavior as a biological phenomenon with a multilevel organization and multilevel causes and consequences. It is an active element in the evolutionary process rather than a mere way of spreading selfish genes. A hen is, after all, more than just an egg's way of making another egg.

Lestel (2011) proposes an ethology that is both biosemiotic and phenomenological. With his bi-constructivist ethology, Lestel stresses the role of intention and interpretation, both by the animal about its environment and by the researcher, who is never an external observer, disconnected from the experience as in usual ethological approaches. Besides, acknowledging the behavioral system of inheritance and its interactions with the environment, and the other systems at different levels of organization, expand the range of approaches in behavioral sciences and the questions ethologists may ask. With this phenomenological attitude, comparative ethology and Biosemiotics can provide better evolutionary hypotheses than the traditional ones.

6 Toward the Method

An entire book would be required to have a phenomenological approach to evolution in four dimensions, as a semiotic process that evolved and was modified in distinct ways for different groups of animals. With that in mind, the only thing I intend to do here is to outline some points necessary for this task to be efficiently accomplished. Since it is essential to consider how the different dimensions of evolution and development, with their particular systems of inheritance and different types of information, interact with each other, creating the extraordinary biodiversity found in nature, a metaphor will be valuable.

An evolutionary biosemiotician must consider evolution almost as a work of art, extended over time, where each element, with its specific meaning, combines with the others, creating more complex meanings. Like a work of art, evolution is not reducible to its parts. Art as a metaphor to understand natural phenomena is not an original idea, but other theorists inspire it. Jablonka and Lamb (2005, 2014) use music to illustrate the genetic and non-genetic dimensions: the genetic dimension is like the score, which is copied as faithfully as possible over time so that the music can be performed and can accumulate some transmissible errors in the process; non-genetic dimensions are like new technologies that allow the transmission, not of the score, but of the performance, the phenotype (Jablonka and Lamb 2014, pp. 107–108). Merleau-Ponty uses different forms of art and literature to describe the body as a dynamic phenomenon in which expression is not distinct from what is expressed (Merleau-Ponty 2002, pp. 171–177). The combination of these metaphors makes it possible to describe how a phenomenological attitude toward evolution must take place, treating the development, the evolutionary process, and its four dimensions as a work of art, i.e., an ongoing film, written, produced, and directed, not by an active conscience, but by nature itself.

The genetic dimension would be like a molecular script, transmitting the guidelines and information on how the story and the other parts should be produced and organized. However, the script is still a long way from faithfully representing the final product. Several other production steps are still required; reading a script (the genome) and watching the film (biolon, i.e. the organism, the species, etc.) are two very different experiences. Like the film director, epigenetics interprets and develops the molecular script, regulating its development through its processes, adding or removing some elements, and giving images and shape to it. Thus, the filming process (the development) hardly deviates too much from the original planning on the script, but it is also not entirely determined by it.

The cast's performance would count as the behavioral dimension. Film enthusiasts know that sometimes the actors are free to improvise some things—facial expressions, behaviors, gestures, and lines that were not in the script—based on their subjective experiences, just as animals improvise some aspects of their behavior. As in the behavioral dimension, interaction with other actors involves perception, interpretation, communication, and learning. In the symbolic dimension, no one would deny that improvisations are everywhere. Among human beings, symbolic culture is a complementary dimension indistinguishable from the behavioral one; everything we think, say, and do is biological and symbolic, always starting from an ontogenetic, phylogenetic, and cultural context larger than ourselves and projecting into the future. Evolution and development are some kinds of push-and-pull processes between constraints and open possibilities.

Evolution is not just the story told, but the entire production process behind the film, always involving countless agents and interactions. As a result of these different stages, cinephiles will have different experiences when focusing on just one of them. Even when watching the finished film, they will not decipher all the detailed semiosis involved in the process. Besides, two different viewers will always have different experiences. Like a critic who specializes in some aspect of production, a typical biologist can focus on some biological organization level to interpret it: a geneticist's focus will be different from that of an ethologist, for example. However, evolutionary biologists, especially if they are biosemioticians, must analyze the complete phenomenon, the film as a whole, even considering the processes they cannot experience with their own eyes, and how these processes work together to produce the final result. More than that, evolutionary biosemioticians must consider themselves a fundamental part of the analysis, being both a product of the evolutionary process and the analysts. However, being attached to their cultural background, with their prior knowledge, biases, and prejudices, good evolutionary biosemioticians must always scrutinize and question their premises and assumptions, as well as their entire role during research.

This film metaphor is intended to illustrate how the evolutionary process should be viewed by evolutionary biosemioticians rather than to explain precisely how evolution occurs. Although it is possible to use it for educational purposes, especially for non-biologists, biological reality will always be more complicated than any explanation we can give about it.

7 Conclusion

Sexuality is not merely a reproductive strategy; instead, it is a complex biosemiotic phenomenon with multiple social meanings, being always context-dependent. The rigidity of the traditional Neo-Darwinian approach to evolution and its need for a general evolutionary model that would explain all living forms make it insufficient to accurately describe several biological phenomena, let alone those in the realm of the mind and culture. Reproductionism, when addressing sexuality and other psychosocial aspects, is incompatible with human reality. A less restrictive and more inclusive view of biology is needed for a better dialogue between biological and human sciences. With Biosemiotics as a starting point, a phenomenological attitude combined with the four-dimensional view of evolution proposed by Jablonka and Lamb (2005) and with comparative ethology, psychology and ecology as a methodology may prove to be a fruitful path toward this kind of biology.

Also, historicity and subjectivity are central elements of the evolutionary process. As we reflect on the human mind and culture, this addition becomes a *sine qua non* condition for a complete evolutionary approach. It provides more representative descriptions of human reality than the unfounded belief that cultural diversity is contrary to biological evolution. By enabling new types of hypotheses and questions and also promoting self-criticism (which is often neglected in evolutionary biology), as a fundamental part of its methodology, this phenomenological attitude seems to be an excellent alternative to the traditional Neo-Darwinian model, marked by biased assumptions and generalizations incompatible with the very notion of biodiversity. Therefore, evolutionary biosemioticians recognize that their existence does not switch between nature and culture: it is both all at once.

References

Alcock J (2011) Comportamento animal: uma abordagem evolutiva. Artmed editora

Avital E, Jablonka E (2000) Animal traditions: behavioural inheritance in evolution. Cambridge University Press

Buss D (2019) Evolutionary psychology: the new science of the mind. Routledge, London

de Waal FB (1995) Bonobo sex and society. Sci Am 272(3):82–88

Dixson AF (2012) Primate sexuality: comparative studies of the prosimians, monkeys, apes and humans, 2nd edn. Oxford University Press, Oxford, UK

Foucault M (1984) História da Sexualidade 2: O uso dos prazeres. Graal, Rio de Janeiro

Foucault M (1988) História da sexualidade. Vol. 1: A vontade de saber. Graal, Rio de Janeiro

Furuichi T (2019) Bonobo and chimpanzee: the lessons of social coexistence. Springer Nature

Gould SJ, Lewontin RC (1979) The spandrels of San Marco and the Panglossian paradigm: a critique of the adaptationist programme. Proc R Soc London Ser B Biol Sci 205(1161):581–598

Gray PB (2013) Evolution and human sexual behavior. Harvard University Press

ICMBio/MMA (2018) Livro Vermelho da Fauna Brasileira Ameaçada de Extinção: volume 1, 1st edn. ICMBio/MMA, Brasília

IUCN (2020) The IUCN red list of threatened species. Version 2020-2. https://www.iucnredlist.org. Accessed 12 Sept 2020

Jablonka E, Lamb MJ (2005) Evolution in four dimensions: genetic, epigenetic, behavioral, and symbolic variation in the history of life. Massachusetts Institute of Technology Press

Jablonka E, Lamb MJ (2014) Evolution in four dimensions, revised edition: genetic, epigenetic, behavioral, and symbolic variation in the history of life. Massachusetts Institute of Technology Press

Jeffery AJ (2015) Two behavioral hypotheses for the evolution of male homosexuality in humans. In: Shackelford TK, Hansen RD (eds) Evolutionary psychology. The evolution of sexuality. Springer International Publishing, pp 207–219

Kirkpatrick RC, Blackwood E, Dickemann JM, Jones D, Muscarella F, Vasey PL (2000) The evolution of human homosexual behavior. Curr Anthropol 41(3):385–413
Lestel D (2011) What capabilities for the animal? Biosemiotics 4(1):83–102
Longo G, Montévil M (2014) Perspectives on organisms: biological time, symmetries and singularities. Springer Science & Business Media
Merleau-Ponty M (2002) Phenomenology of perception. Routledge, London
Morris D (1999) The naked ape: a zoologist's study of the human animal. Random House
Pierce SK, Curtis NE, Schwartz JA (2009) Chlorophyll a synthesis by an animal using transferred algal nuclear genes. Department of Integrative Biology, University of South Florida
Sapolsky R (2017) Behave: the biology of humans at our best and worst. Penguin, New York
Tønnessen M, Maran T, Sharov A (2018) Phenomenology and biosemiotics. Biosemiotics 11 (3):323–330
Wilson E (1975) Sociobiology: the new synthesis. Harvard University Press

Part III
Physics, Medicine, and Bioenergetics

Physical Intentionality: The Phenomenological Roots of Biosemiotics

Roberta Lanfredini

Abstract The concept of intentionality is traditionally ascribed to a state of consciousness or a mental state. The thesis that intention is the mark of the mind implies a *representational* model, fundamentally connected to the crucial notion of determination or characterization. This essay aims to investigate the seemingly oxymoronic notion of physical intentionality, which implies an *expressive* model, and the centrality of the notion of forceful (or powerful) property. The reason for this profound paradigm shift is the replacement of the notion of invariance with the notion of tension. This change marks the transition from a phenomenology of the inert to a phenomenology of the living.

Keywords Intentionality · Phenomenology · Forceful property · Embodiment

1 Mental Intentionality

Having Brentano's formulation as its starting point, the concept of intentionality stands as the mark of the mental. We will call this thesis IMM (Intentionality Mark of the Mental) (Crane 1998), in the sense that intentionality is the necessary and sufficient condition for a state to be called mental and differentiated from a physical state. With this understanding, intentionality functions as a demarcation criterion of the physical and mental realms. In this sense, Brentano's theory of intentionality differs from the Scholastic tradition. Indeed, while for Scholasticism, intentionality fits into the frame of the philosophy of knowledge, for Brentano, it belongs to an introductory chapter about psychology, being a criterion for *defining* psychic phenomena.

> Every mental phenomenon is characterized by what the Scholastics of the Middle Ages called the intentional (or mental) inexistence of an object, and what we might call, though

R. Lanfredini (✉)
Department of Humanities (DILEF), University of Studies of Florence, Florence, Italy
e-mail: roberta.lanfredini@unifi.it

© Springer Nature Switzerland AG 2021

E. Pagni, R. Theisen Simanke (eds.), *Biosemiotics and Evolution*, Interdisciplinary Evolution Research 6, https://doi.org/10.1007/978-3-030-85265-8_11

> not wholly unambiguously, reference to a content, direction toward an object.... This intentional in-existence is characteristic exclusively of mental phenomena. No physical phenomenon exhibits anything like it. (Brentano 1995: 92–93)

In Husserl's reformulation of Brentano's theory, intentionality is the mind's distinguishing feature in the weaker sense that it is a necessary (though not sufficient) condition for a state to be called mental. Indeed, for Husserl (as for Kant), there exists a passive or pathic dimension (we might specifically think of sensory states such as pain, or mood-states such as anxiety, depression, or happiness, which infuse the whole of experience with a certain tone) which does not, at least in the first instance, show an intentional *direction* toward any object. In this regard, Husserl anticipates the distinction between mind and phenomenal consciousness in the philosophy of mind (Chalmers 1995, 1996, 2018), laying the ground for what will become phenomenal intentionality (Kriegel 2009).

In the contemporary debate, intentionality is often referred back to as the problem of its possible naturalization. The dispute between so-called *tracking theories* (TT) (Dretske 1981, 1995; Millikan 1989, 2009, 2017) and the so-called *Phenomenal intentionality research program* (PIT) (Kriegel 2013; Loar 1987) takes place within this general picture. TT attempts to neutralize qualitative states' content, reducing them to a *tracking relation* that holds sway between the phenomenality of consciousness and physical properties, framing intentionality in terms of causal relations between mental states and the environment. The underpinning assumption is that intentionality should by no means be conceived as a unique condition proper to elusive mental states, but as a natural interaction with the environment. For TT, the content of our conscious mental states strongly depends on objective external properties. Indeed, the *tracking approach* maintains that the content of qualitative states can be reduced to a monitoring relationship conceived as a function of the correspondence with certain external reality data that are capable of revealing information relating to the environment. In contrast to TT, phenomenal intentionality proposes an anti-naturalization of intentionality. According to this approach, experience manifests phenomenal properties that intrinsically characterize experience itself and, as such, are not reducible to naturalistic data.

A clear example of antinaturalism can easily be discerned in the phenomenological approach. Despite the attempts, some of them significant (Petitot et al. 1999), to naturalize phenomenology and to maintain full compatibility between the phenomenological method and the cognitive sciences, and more generally between the fundamental concepts of phenomenology (lived experience [*Erlebnis*], consciousness, subject) and the fundamental concepts of the natural sciences, it cannot be denied that the central approach of phenomenology is of a specifically and irreducibly antinaturalistic type. This general stance has repercussions for how intentionality is conceived.

For Brentano, the theses which lead to the formulation of the concept of intentionality are as follows:

(a) Every psychic phenomenon is directed toward an object.
(b) The object of a psychic phenomenon is a physical phenomenon.
(c) The physical phenomenon is not a psychic phenomenon, nor part of it.

From these three theses derives the fact that, as we have seen, the characteristic sign of the mental is intentionality or, which is the same thing, that intentionality is the necessary and sufficient condition for something to be considered a mental state.

Physical phenomena, which constitute the terminal point of the "intentional arrow," are not, for their part, directed to something, at least, not in the sense of intentional direction. Indeed, we can say that if a stone is dropped, it is directed toward the earth, but this "directionality" has nothing to do with intentional direction. Therefore, what characterizes the specificity of the intentional relation? In other words, what makes it "mental" compared to physical relations?

Two answers can be given to this question. According to the first, the distinctiveness of intentional mental states (perceiving, imagining, desiring, being afraid), as opposed to non-intentional states (walking, dropping books, sitting on a chair, lifting the arm), can be attributed to the distinctiveness of the *intended object*. We will call this thesis the *ontic theory of intentionality*. According to this thesis, intentionality is a *sui generis* relationship, not so much because of the link it sets up between a mental state and the object toward which it is directed, as because of the nature of the intended object. This object turns out to be ontologically distinct from the entities which are involved in non-intentional relations. The objects of intentional relations are independent of existence and, even in the case where the intended object can be accepted as actually existing (as when I perceive the pen with which I am writing, and there is a strong probability that it is not the product of a hallucination), the ontic theory of intentionality requires the intentional object to be distinct from the actually existing object. We can therefore conclude that the intentionality of mental states consists, in any case, in their being related to a sort of ontology that is parallel to the ordinary or real one.

The second answer goes in the direction of an *epistemic theory of intentionality*. According to this theory, intentionality is not a property of the object but a property of the conscious state (*Erlebnis*). More precisely, intentionality stands for the internal structure described in detail by Husserl in the *Fifth Logical Investigation* (Husserl 2001), which makes the intentional act capable of addressing the objects of experience. Based on this new and more complicated idea of intentionality, which is founded on the (originally Kantian) notions of the constitution and the transcendental, the object is such (*objectum*: i.e., "thrown before," "placed in front") in that it contains the reference to a content of consciousness (*Erlebnis*). The object contains *not* the content but only a reference to it. Otherwise, we would fall into that indistinction typical of classical empiricism (of Locke, Hume, and Berkeley) and the neutral monism of Russell, Mach, and James, in which experience presents the collapse of that bipolarity between the subjective (noetic) component and the objectual (noematic) component that is crucial for Husserl, following Kant. However, the object does contain the necessary reference to content to satisfy that *principle of manifestation*, which characterizes Husserl's phenomenology.

Therefore, an object understood as a manifestation or phenomenon must contain not the lived experience but the reference to a lived experience. This reference translates into a founding or dependency relation that, if read in the opposite direction, exactly expresses the notion of intentionality. The mirror image of the thesis that every object necessarily contains the reference to a lived experience is in fact that, if the experience is intentional, it is directed toward an object through its content: therefore, according to specific qualifying modalities (e.g., perceptual or imaginative), but also attributive or determinative ones (such as having a particular color, or a particular sonic pitch).

The experience is, therefore, the condition by which the object is given. This topic of the *condition* is clearly inspired by Kant but becomes re-proposed by Husserl in descriptive and not prescriptive terms: the condition does not lie in a categorical system *a priori*, but in the internal structure of the experience. The conditions should therefore be read not as *conditions of possibility* but as *conditions of correlativity*. Phenomenology is a science of the necessary correlation between phenomenon and (underlying) experience or between the world and (underlying) consciousness. Every object is an intentional object and must only be investigated within the limits (and strictly within those limits) in which it is given in the intentional act.

2 Intentionality and Determination

The epistemic theory of intentionality relates to three central theses.

We will call the first one the *thesis of the perspectival character of intentionality*. According to this, every phenomenon that is not, in turn, an experience is always and necessarily given based on determined points of view. For Husserl, the object's perspectival character is a fundamental assumption: every object manifests itself only through perspectives and never in its entirety: that is, completely, from all points of view. The perspectival character is thus incorporated into the notion of the object as a defining condition. The intentional relation set up between a state of subjectivity and an object is always linked to (the act of) conceiving the object in specific ways (and not in others). Husserl denominated the intentional object as an object intended *in the how* of its determinacies and indeterminacies. For example, a thing's perception is always partial (from below, from above, from the side). It is never a perception of the thing in its entirety, whereas throwing something to the ground does not involve any intentional modality: that is, it has no perspectival character.

That intentionality is not confined to indicating *the what* (the intended object) but also *the how* (the way in which we intend it) is well expressed by the story of Oedipus. Only if we accept the thesis of intentionality's perspectival character, Oedipus can despise the man he kills on the road to Delphi without despising his father, or desire Queen Jocasta without desiring his mother; or again, hate the murderer of Laius without hating himself. In this case, moreover, the difference between intentional and non-intentional states is shown in all its tragic clarity:

Oedipus kills the man he meets on the road to Delphi and, in doing so, also kills his father; he marries Queen Jocasta and, in doing so, lies with his mother. Intentionality is, therefore, always partial or aspectual.[1]

Focusing on an intended object, the statement "in the how of its determinacies (and indeterminacies)" also indicates the existence of a halo or indistinct background which constitutes the horizon (not attentional, but intentional) concerning which intentionality delimits and brings focus. For Husserl, appearance is always surrounded by a halo or horizon functioning as a non-attentional background. This background merely allows the appearance itself to be given by contrast. This horizon presupposes a permutation through space (the background, the tacit horizon) and a permutation through time (the retentional past). In *Ideas II*, Husserl refers to a "dark background" very remote from the attentional present, which nevertheless constitutes the foundation of experience.

> What is specific therein is motivated in the obscure background and has its "psychic grounds" about which it can be asked: how did I get there, what brought me to it? That questions like these can be raised characterizes all motivation in general. The "motives" are often deeply buried but can be brought to light by "psychoanalysis." A thought "reminds" me of other thoughts and calls back into memory a past lived experience, etc. In some cases, it can be perceived. In most cases, however, the motivation is indeed actually present in consciousness, but it does not stand out; it is unnoticed or unnoticeable ("unconscious"). (Husserl 1989: 235)

The thesis of tacit or background intentionality confirms the idea that the primary function of intentionality is to detach an object from a background, determining its characteristics and properties. Thanks to the very notion of intentionality, phenomenology is a discipline that describes the *determinations* of the phenomena constituted by states of consciousness.

We will call the second thesis *the thesis of the synthetic character of the intended object*. The perspectives by which the object is manifested are predominantly given as unitary. In the object, the manifestations are coherent and connected around a unitary, though indefinitely open, pole. This identificatory synthesis of appearances is an integrative part of the experience, except when the cohesion breaks or fragments. This fragmentation happens in hallucinations or, generally, in the interruption of the cohesion of experience in illusory states.

We will call the third thesis, which we have already encountered, the *thesis of the intended object's independence from existence*. Whereas a non-intentional relation (like riding a horse, being shorter than Tom, dropping a book) subsists between two entities only if both exist, the relationship between intention and intended object holds independently of the latter's existence. We can *see* an oasis in the desert, *fear* encountering the abominable snowman, *hope* that the fountain of youth will be discovered independently of these entities' existence.

---

[1] The example of Oedipus also shows the possibility of re-signifying both the object and the mental state through time, so introducing into the notion of intentionality the dynamic character we will find in physical intentionality.

All the theses summed up so far imply this primary activity of "decoupage" from lived experience. Indeed, it is impossible to speak about perspective or viewpoint except based on many objectual determinations, just as it is not possible to refer to a synthesis of perspectives except based on coherence or coexistence of two or more "agglomerates" of determinations. The same suspension of existence strengthens the notion of determination by freeing it from the existential bond. Lastly, the implicit character of intentionality and the notion of horizon or background offer the intentional act the (indistinct, confused) terrain based on which it can distinguish an object of experience.

This centrality of determination, expressed by the notion of intentionality, is founded on the concept—a crucial one for Husserl—of invariance. Invariance results from a unifying function performed by consciousness on the variations, the incessant fluctuation, which experience presents by its very nature.

One of Husserl's most frequently used formulae is, not by chance, *the object in the how of its determinations.* In this formula, the close nexus between the expressions *in the how* and *determinations* implies the modality's indispensability and the need for this modality to be crystallized into properties or characteristics.

Without determinations, it would be impossible to speak about perspectives on the thing (the perspective being nothing but a "bunch" of determinations of the thing); or of synthesis and unity of perspectives (understood as cohesion and regularity between bunches of determinations); or about the implicit or background character of intentionality (understood as that which is not yet determined). In brief, intentionality converges on the notion of determination (or attribution), which in turn relates to the concept of representation.

My thesis is that this mental model is complementary to the conception of matter as essentially characterized by extension. The idea that the mind, as a cognitive modality, is a representational activity is the reverberation of conceiving matter as essentially extended (hence fragmentable, divisible, countable) and vice versa. Conceiving the mind as essentially intentional (that is, representational) and conceiving matter as essentially extended (object of representation) are two sides of the same coin.

In this model of the mind, in which representation becomes the mold of material extension (and vice versa), representational and extensive structure become primary (essential), at the expense of the qualitative dimension both of the physical body (secondary properties) and the mental (qualia).

This model's source lies misconception of the notion of organism (and of life) in favor of "functional" notions such as mind, representation, cognition, and intentionality. This misconception extends to the notion of the body understood predominantly as a kinesthetic body, the primary function of which is to unify the object's perspectives. Indeed, for Husserl, the living body (*Leib*) fulfills predominantly (though not exclusively) constituent purposes: through its kinesthesis, the body can accomplish that synthesis of perspectives that is an integrative part of the intentional function.

The result is an adaptation of the living to the inert (Bailly and Longo 2011; Longo and Montévil 2011, 2014) in a double sense: on the one hand, what emerges from this perspective is an artificial conception of the mind, understood as emptily representational and a merely kinesthetic conception of the living body (the *Leib* as a kinesthetic automaton); on the other, a reductive conception of matter, understood as a mere extension.

An alternative path can be traced in Husserl's acknowledgment that "the fully comprehensive essential attribute of material being is not mere extension but is, instead, materiality, so far as this latter in itself requires temporal as well as spatial extension" (Husserl 1989: 31, 1997).

It is permissible at this point to pose the question: what does it mean, and what does it entail to see the essence of materiality in temporality?

3 The Forceful Qualities

The Husserlian notion of invariance in variation implies, as its theoretical nuclear vehicle, a particular way of understanding essence. For Husserl, essence (*Eidos, Wesen*) is the product of that fundamental phenomenological mechanism that is eidetic reduction. The theory of ideation proposes, first of all, an explanation of the notion of datum and of what is effectively, analytically contained in such a notion. No datum could be understood/intended if it had not been ideated: indeed, the individual turns out to be such (that is, individuated) only by being an element in a field of possible variations limited by eidetic boundaries. If this were not the case, we would be dealing with a "formless substrate" lacking intuitive determination. Husserl's idea is that if deprived of its essentiality (phenomenologically understood as the unification of the possibility of its variations), the datum could not be discriminated.

Conceiving the qualitative in terms of categorical qualities has a long tradition. Onto it, Descartes and Husserl graft their idea of quality as invariance in variation (of states for Descartes, of adumbrations for Husserl). This conception is based on the fundamental, though not exhaustive, the character of extension. In fact, to be able to discriminate, we need to unite, and to unite, we must be able, even if only virtually, to "fragment" experience into unity. However, this fragmentation presupposes something anterior to extension, which is the terrain on which invariance can come into effect.

Conceiving extension as the essence of the matter is a mirror image of conceiving representation as the mental essence. In this sense, the idea that intentionality (and therefore representation) is a necessary (though not sufficient) condition of the mental fuels the idea that extension is the essence of matter, and vice versa.

However, maintaining that an essential condition for being material is *also* temporality entails escaping the circle of representation/extension by introducing a new actor: time understood as duration or history.

In this new scenario, the static concept of determination (*qualitative/categorical dimension*) is replaced by the dynamic and dispositional concept of *forceful qualities*.

The description of qualities in terms of *force* or *power* also has a tradition, albeit a lesser one, which counts Herbart and Mach among its most significant exponents (Blackmore 1972). Herbart originally presented his doctrine of quality in *Psychologie als Wissenschaft.* He proposed an *Elemententenlehere* in which elements are conceived as similar to the Newtonian forces and qualities as the direction of the force itself (Banks 2005). One of the main consequences of this approach is a profoundly changed theory of essence (*Wesen*). For Husserl, as we have seen, the essence is attributable to the notion of invariance, and hence of identity. For Herbart, on the contrary, the essence is attributable to the notion of an equilibrium of forces. As he claims, the elements:

> (...) press one another. For in the world of the senses we find resistance in pressure, where nothing gives way although each is supposed to move. Pressure is rest, through reciprocal endurance against another. (Herbart 1828–1829: 103)

And Banks comments, in response:

> When Herbart's qualities exactly cancelled one another in magnitude and direction they were in spaceless equilibrium and could group together into stable point-like objects, which he called *Wesen* ("beings," "objects"). (...) *Wesen* were really point-like nodes held in position by the mutually constraining forces acting there and not by their mere existence. (...) Herbart claimed that each *Wesen* conserved itself against the others by "pressing back" through its qualities. In fact, he described two sorts of pressures. (Banks 2005: 210)

Resistance originates from the forces ("first forces must act, then there is resistance to them, thanks to other forces" (Herbart: 103) and on the other hand, force originates from resistance "force is that which overcomes resistance" (*ibidem*), since force is that which tends to change the state of a thing.

In the Herbartian scenario, essences are not the fruit of invariances but of *forces* or *equilibria.* In turn, qualities, far from being a "qualitative patina" that spreads over an extension—or invariance in variation, as Husserl maintains in his explication of eidetic variation—can be likened to *powers*. In this sense, properties traditionally considered primary, such as size, position, duration, divisibility, and solidity, are attributable to qualitative properties. We find something very similar in Mach in that he claims "the thing is (...) a mental symbol for a relatively stable complex of sensational elements" (Mach 1976: 322). Banks has shown how Mach's approach derives from the legacy of Herbart's thought, as well as from Bernard Riemann and Hermann Grossman, further connecting it to the neutral monism of James and Russell. The Machian elements can, in fact, be conceived as *forceful qualities*—"Elements are manifested forces in events" (Banks 2003, 2014: 49)—, an idea which Herbart himself traced in Leibniz's outlook.

The qualities are therefore not static but dynamic. Like Mach, Herbart proposes a distinctive theory of the datum in which it is constituted by a multiplicity of simple substances existing in a dynamic relationship with one another.

This theory of the datum can easily be read in terms of dispositions. These dispositions, in turn, are understood as "power or capacity" (Heil 2005: 343). The dispositions satisfy the following theses (Heil 2005, 2010; Martin 1993):

1. They are real characteristics of objects. What is merely potential is the manifestation of the disposition (e.g., the breaking of the glass) and not the disposition itself.
2. They are intrinsic properties of the objects which possess them. Most dispositions could never be manifested.
3. Their nature is not entirely reducible to conditional analysis. The glass would be fragile even if the conditional "the glass is fragile if it breaks when struck by something solid" were false.
4. They are not contingent characteristics of the world.

In conclusion, every property is dispositional and qualitative at the same time. Therefore, the primary properties are also qualities, and the qualities, in turn, are *powers*.

Hence, the reference to the distinction between primary and secondary properties is in a certain sense turned on its head. While tradition (which finds its most significant expression in the Galilean perspective) portrays the material world as consisting solely of primary properties and relegates the qualities to conscious observers' minds, in the view just considered the qualities, understood as *forceful qualities* are in fact primary. In contrast, the properties initially considered primary (extension, size, form) are limited cases of those "disturbances and self-preservations" that constitute the ontological basis from which all the rest, including the concept of representation on the one hand and extension on the other, have their origin.

4 Physical Intentionality

To give priority to the qualitative dimension of experience over its character as extension, and to define this dimension in agentive rather than categorical terms, implies a transformation of the underlying ontological vision: the qualitative dimension is no longer understood in a functional sense (as invariance in variation) but in a dispositional sense (as powers, or forces). Ontological priority, according to this theory, belongs not to determination but to power, not to invariance but to force.

The concept of power is, according to Molnar 2003 and Munford (1998), closely connected to the concept of intentionality. Its fundamental characteristics are, in fact, five: *independence* with respect to its effective manifestation (power is ontologically independent with respect to the eventual phenomena in which it is realized), *actuality* (powers are "fully actual properties of their bearers"), *intrinsicality* (powers are intrinsic and not extrinsic properties of their bearers), *objectivity* (powers are endowed with an objective ontological existence and are not secondary, for example, with respect to microphysical structure) and, finally, *intentionality*. The last concept is the most important: the essential characteristic of power—what distinguishes it prima facie from non-powers—is its directness. Power is essentially connoted by directionality: that is, by being directed toward something.

But what exactly is meant here by intentionality? We have seen how there are three theses in this tradition that characterize the concept of intentionality: (1) the distinctive direction toward something: that is, the reference to an intended object, structurally different from a physical relationship); (2) the intended object's independence from existence (unlike a physical relationship, in which this independence does not exist); (3) the partiality and perspectivability of the intentional relationship (unlike the physical relationship, which is neither partial nor perspectival).

The first thesis is the one that allows Brentano to conceive of intentionality as a criterion of demarcation between psychical and physical phenomena (Brentano 1995) and that allows Husserl to conceive of intentionality as an essential feature of consciousness (Husserl 2001). For Husserl, in fact, not all states of consciousness (Erlebnisse) are endowed with an intentional structure: neither are sensations such as pain or pleasure, for example, nor moods, nor temperamental traits; nor are some emotional states such as panic or anxiety.

However, despite the recognition of a non-intentional dimension, Husserl conceives intentionality as characterizing consciousness. In fact, a consciousness that is not endowed with intentionality cannot be properly defined as such. The distinctive relation that consciousness has with its objects is confirmed by two further assumptions: the intended object is in fact *independent of existence* (I can imagine, fantasize, even perceive objects that do not exist, as in hallucinations or illusions), but necessarily *dependent on points of view* or perspectives.

It is not in principle possible to perceive, imagine, judge, feel something without incorporating a point of view. An absolutely independent object is nonsense, just like a round square. Furthermore, intended objects can be fuzzy objects. For example, I may perceive something in a vague way, as when I hear an indistinct noise or see a figure without identifying its outlines. But vagueness in phenomenology is not the same as indeterminacy. Being intentionally addressed toward an object means not only determining the intended object, but also the way in which it is intended. Intentionality, to use Nagel's words, is not a "view from nowhere" (Nagel 1974), a "naked" perception of the object, but a perspectival slant on things. And the perspectival slant is closely linked to the set of determinations in which it is realized.

This specific sense of *directness* interprets intentionality as closely connected to representational activity. The principle according to which "Every act is representation or founded on a representation" is considered by Husserl, and before him by Brentano, as a founding thesis of phenomenological investigation.

In physical intentionality what is lacking is the representational structure of mental intentionality. It is interesting to note, however, that despite this profound change in conception, the formal structure of the definition of intentionality remains unchanged. The three theses that define the concept of mental intentionality are in fact also maintained in the case of physical intentionality.

Physical intentionality is indeed directed toward something. The intentional object of a physical power (such as solubility or electromagnetic charge) is its manifestation.[2] The link between power and its manifestation, just like the link between the intentional act and intended object, is not contingent but necessary and essential. Physical intentionality, like mental intentionality, is also independent of existence. Something may be soluble without ever being dissolved, or fragile without ever breaking. The manifestation of physical power may exist or not without detriment to the existence of the physical power itself. Finally, physical intentionality, just like mental intentionality, can be vague (Martin and Pfeifer 1986). Physical powers can also have fuzzy objects; such as the propensity of unstable elements to decay, which is indeterminate as to timing. What is missing is, despite the structural identity between the two theories of intentionality, the link (necessary for mental intentionality) with representation. The nexus between power and its manifestation is not representational but expressive and agentive.

The identification between dispositional properties and powers radically transforms the theory implied by the distinction between primary and secondary properties (Lanfredini 2018). Dispositional qualities or powers are not identified with primary properties simply because they are not properties. Powers are autonomous with respect to what is traditionally considered primary (that is, extension, figure, motion—in a word, what is measurable) simply because they are not on the same plane: if the distinction between primary and secondary properties has to do with the notion of determination, the distinction between power and non-power has to do with the notion of action. In this sense, if it is plausible to adopt a reductionist perspective and identify the dispositions with underlying physical structures, we may also quite plausibly adopt an anti-reductionist perspective and consider the powers as totally *groundless*: that is, ontologically independent of non-powers.

This difference implies a radical paradigm shift: time (understood not as succession—spatialized time—but as duration, or as history) is now primary to space (understood as extension). From this perspective, time is not a lack as far as stability and fixity (invariance in variation) are concerned, but it is efficacious or creative (stability or continuity in variation); it has, in fact, power. If dispositional qualities cannot be identified even partially with secondary properties, how can they be defined?

In this case, there are two strategies we could adopt. According to the first, powers and the qualitative/categorical dimension are two sides of the same coin (Martin 1994): that is, "a power is only a face/facet/side of a property that also has a qualitative face/facet/side" (Molnar: 159). According to this hypothesis all properties have something about them that is irreducibly and ineliminably powerful in the qualitative/dispositional sense, and something else about them that is irreducibly and ineliminably non-dispositional in the qualitative/categorical sense. From this perspective, the qualitative/categorical and the qualitative/dispositional are "parts" or "aspects" of the single underlying property.

[2] "Of the many ways of characterizing a power, the only one that reveals the nature (identity) of the power is the characterization in terms of its manifestation" (Molnar: 63).

The second hypothesis is that secondary properties in the qualitative/categorical sense and powers in the qualitative/dispositional sense are not being thought of a part or an aspect of the property, but are thought of the whole property *in a certain way*. Equally, to think of the property is not to think only of an aspect of the property but again to consider the whole property in a certain different way (Taylor 2013, 2018). A good example of this idea is the case of the Gestalt shift, as in the example of the famous ambiguous figure which can be seen as a duck or a rabbit. When we consider the figure as a rabbit or as a duck, we are not considering only a part or an aspect of the figure. Rather, we are considering the whole figure in a certain way. However, there is a further possibility: that powers are considered as neither different aspects nor different modes of the same thing. From this perspective, which we can call neutral and monistic, disposition or power is the only reality, the only true, vital dimension (Bergson 1998).

Seen in this light the qualitative/categorical dimension or secondary property is neither an aspect nor a certain mode of an underlying, whole property, but a limiting case of the original expression of qualitative dimension: the case, that is, in which this property is spread over an extension. According to this third hypothesis, the link between power or qualitative/dispositional dimension and secondary property or qualitative/categorical dimension is not mereological (i.e., between part and whole) but derivative: the latter would in fact be founded on the former, the former being more original.

5 Embodied Cognition

Interpreting the notion of intentionality not as "the mark of the mental" but as "the mark of the physical" entails a radical shift of viewpoint, resulting in a total paradigmatic reversal.

In the current debate about the philosophy of mind, it is customary to speak of an incarnate conception of the mind, guided by what is called in the literature *the four Es*, by which the mind would be *Embodied*, *Embedded*, *Enactive*, and *Extended* (Varela et al. 1991; Lakoff and Johnson 1999; Shapiro 2014; Clark and Chalmers 2003; Thompson 2009).

These words converge on a general and partially revolutionary thesis: that the mind is not a mere process elaborating information proceeding from the outside world (according to the well-known metaphor of the mind as the software of a piece of hardware) but, on the contrary, an open system endowed with self-organization and constant interaction and integration with the world; elaborating propositions directed toward the environment and dealing with challenges posed by the environment, in a constant reorganization and reconversion of the system: a system that is customarily called homeostatic.

This system corresponds to the paradigm of *Embodied cognition* in the philosophy of mind, by which the motor capacities that can be ascribed to a body should be considered indispensable factors for the development and functioning of a cognitive system. This attribution leads the supporters of this view to maintain that, from both epistemic and ontological viewpoints, the definition of processes like perception, reasoning, and language depends on the bodily properties that can be situated beyond the central nervous system's established boundaries.

In particular, the paradigm of embodied cognition emphasizes the possibility of interaction with the environment over the processes that transform the information.

The best explanation of the systematic character of cognition must be sought not so much in the possession of computational and representational abilities or algorithmic processes within the system, as in morphological properties of the body and its interactions with the environment.

So, according to this theory, cognition is not a computational process. The idea that mental processes have a syntactic nature—that is, the thesis that posits a cognitive system and, on the same level, an elaborator capable of working on symbols and transforming input into output by using logical operators and rules for their application—becomes superseded.

In other words, cognition can no longer be likened to a calculus in which strings of symbols are elaborated according to definite formal rules within the cognitive system itself. In this highly delicate complex of equilibria, the world is by no means a reservoir of information from which to draw the material we need for "giving birth" to univocal representations of an objective reality which is independent of our way of observing it, but an environment which is the product of a co-creation: that is, of a reciprocal creation between mind, body, and environment, a sort of *structural coupling* which simultaneously modifies both the organization and the environment, and is compared by some phenomenologists (Merleau-Ponty 1968, 2002; Merleau-Ponty and Séglard 2003) to the coupling which occurs in dancing.

Therefore, the mind is *Incarnate* in the sense that it is necessarily inscribed in a body. It is, for that reason, always situated. *Integrated* in the sense that it is not a closed system, but an open one, and, as Sartre saw clearly, not *inside*, solipsistically closed (as in the Cartesian paradigm) but *outside* from the start. *Agentive*, in the sense that the organism's essential property is not representation, which is entirely secondary, but action. *Extended*, in the sense that the organism's fundamental property is that of integrating and incorporating objectual and external elements. Just as a blind person incorporates her cane into her perception and *feels* the tip of the cane resting on the ground, in the same way, we incorporate external elements and make them ours, interpreting them as our effective prolongations.

The concept of openness, together with the concepts of retroaction, generativity, and aleatory character, allows the formulation of the thesis according to which no objects exist separately from their observers and the surrounding environment. There are only aleatory events emerging from complex systemic interactions. This idea performs a thoroughly Copernican reversal from the paradigm of control to recognizing the environment as a source of disturbances rather than information. This shift constitutes the basis of what Maturana and Varela called autopoiesis: that is, the definition of the living being as a system that continually redefines itself, maintaining and reproducing itself from within.

This thesis obliges us also to consider the world as a universe of participation, a world which, as Edgar Morin puts it, "is in us, just as we are in it," and also, on substantially renewed bases, to rethink the concept of the body. The body (as biosemiotics has clearly shown) is, first of all, a means of expression. Alternatively, if we will, expression is achieved through the body before we acquire speech. Expression and perception, activity and passivity come about only through movement; more precisely, they *are* movement. Movement and expression can be considered (as Merleau-Ponty clearly revealed) synonymous: in moving, we explore the environment around us, and exploring the world, we simultaneously explore ourselves, our interior. Moving means bringing the inner toward the external, as Merleau-Ponty meant when he spoke of the body as flesh and flesh as dehiscence: the internal emerging into the external and, vice versa, the incorporation of the external into the internal.

Now, this circular movement is achieved in the flesh as the result of the theoretical passage from intentionality understood as the mark of the mental to intentionality as the mark of the material. More specifically, it is a passage from the static concept of determination to the dynamic concept of disposition (understood as a forceful or powerful property) or, if we will, from a conception of the body as a device for the constitution and objectual representation to a conception of the body as an organism whose essential task is the expression and the formation of sense.

References

Bailly F, Longo G (2011) Mathematics and the natural sciences: the physical singularity of life. Imperial College Press, London

Banks EC (2003) Ernst Mach's world elements: a study in natural philosophy. Kluwer, Dordrecht

Banks EC (2005) Kant, Herbart and Riemann. Kant Studien 96:208–234

Banks EC (2014) The realistic empiricism of Mach, James, and Russel: neutral Monism reconceived. Cambridge University Press, Cambridge

Bergson H (1998) Creative evolution. Dover, New York

Blackmore JT (1972) Ernst Mach: his work, life, and influence. University of California Press, Berkeley

Brentano F (1995) Psychology from an empirical standpoint. Routledge, London

Chalmers D (1995) Facing up to the problem of consciousness. J Conscious Stud 2(3):200–219

Chalmers D (1996) The conscious mind: in search of a fundamental theory. Oxford University Press, Oxford

Chalmers D (2018) The meta-problem of consciousness. J Conscious Stud 25:6–61

Clark A, Chalmers D (2003) The extended mind. Analysis 58(1):17–19

Crane T (1998) Intentionality as the mark of the mental. In: O'Hear A (ed) Contemporary issues in the philosophy of mind. Cambridge University Press, Cambridge, pp 229–252

Dretske F (1981) Knowledge and the flow of information. MIT Press, Massachusetts

Dretske F (1995) Naturalizing the mind. MIT Press, Massachusetts

Heil J (2005) Dispositions. Synthese 144(3):343–356

Heil J (2010) Powerful qualities. In: Marmodoro A (ed) The metaphysics of powers: their grounding and manifestations. Routledge, New York

Herbart JF (1828–1829) Allgemeine Metaphysik nebst den Anfängen der philosophischen Naturlehre. Erster historisch-kritscher Teil, SW VII, Königsberg

Husserl E (1989) Ideas pertaining to a pure phenomenology and to a phenomenological philosophy, second book, studies in the phenomenology of constitution. Boston, London, Kluwer Academic Publishers, Dordrecht
Husserl E (1997) Thing and space. Lectures of 1907, Collected works, Vol VI, Springer, Dordrecht
Husserl E (2001) Logical investigations, vol II. Routledge, London and New York
Kriegel U (2009) Subjective consciousness: a self-representational theory, Oxford. Oxford University Press
Kriegel U (2013) The phenomenal intentionality research program. In: Kriegel U (ed) Phenomenal intentionality. Oxford University Press, Oxford
Lakoff G, Johnson M (1999) Philosophy in the flesh: the embodied mind and its challenge to western thought. Basic Books, New York
Lanfredini R (2018) Categories and dispositions. A New Look at the Distinction between Primary and Secondary Properties, in Special Issue "Contemporary Natural Philosophy and Philosophies", "Philosophies", G. Dodig-Crnkovic, Marcin J. Schroeder (Eds), Philosophies, pp.1–12
Loar B (1987) Subjective intentionality. Philos Top 15(1):89–124
Longo G, Montévil M (2011) Protention and retention in biological systems. Theory Biosci 130 (2):107–117
Longo G, Montévil M (2014) Perspectives on organisms: biological time, symmetries and singularities. Springer, Berlin
Mach E (1976) Knowledge and error: sketches on the psychology of enquiry. Reidel, Dordrecht
Martin CB (1993) Power for realists. In: Bacon J, Campbell K, Reinhardt L (eds) Ontology, causality and mind. Cambridge University Press, Cambridge, pp 175–194
Martin CB (1994) Dispositions and conditionals. Philos Q 44:1–8
Martin CB, Pfeifer K (1986) Intentionality and the non-psychological. Philos Phenomenol Res 46:531–554
Merleau-Ponty M (1968) The visible and the invisible. Northwestern University Press, Evanston
Merleau-Ponty M (2002) Phenomenology of perception. Routledge, London
Merleau-Ponty M, Séglard M (2003) Nature: course notes from the Collège de France. Northwestern University Press, Evanston
Millikan R (1989) Biosemantics. In: McLaughlin BP, Beckerman A (eds) J Philos. Oxford University Press, pp 281–297
Millikan R (2009) Biosemantics. J Philos 86(6):281–297
Millikan R (2017) Beyond concepts: unicepts, language, and natural information. Oxford University Press, Oxford
Molnar G (2003) Powers. Oxford University Press, Oxford
Munford S (1998) Dispositions. Oxford University Press, Oxford
Nagel T (1974) What is it like to be a bat? Philos Rev 83(4):435–450
Petitot J, Varela J, Pachoud J, Roy JM (eds) (1999) Naturalizing phenomenology: issues in contemporary phenomenology and cognitive science. Stanford University, Stanford
Shapiro L (ed) (2014) The Routledge handbook of embodied cognition. London, Routledge
Taylor JH (2013) In defence of powerful qualities. Metaphysica 14(1):93–107
Taylor JH (2018) Powerful qualities and pure powers. Philos Stud 6:1423–1440
Thompson E (2009) Making Sense of sense-making: reflections on enactive and extended mind theories. Topoi 29(1):23–30
Varela F, Thompson E, Rosch E (1991) The embodied mind: cognitive science and human experience. MIT Press, Cambridge

Cancer and Cell Death: A Biosemiotic Perspective

Rogério Estevam Farias

Abstract The cellular systems related to programmed cell death are complex and essential for the maintenance of tissue homeostasis. The integration of signaling molecules, intracytoplasmic enzyme apparatus, and repairing genes is related to eliminating damaged cells. Any change in this dynamic results in many pathologies such as cancer, neurodegenerative, autoimmune, and developmental diseases. The cell has a system of codes capable of assembling and preserve its codes through various levels of ontogenetic processes. The cell is an autopoietic system, establishing its structures from genetic signs interpreted and translated by tRNA. In this environment, cancer can be seen as an enhancement from biosemiotic entropy with corrosion of the genomic/epigenetic machinery and the molecular repair. Assuming the Somatic Mutation Theory, many genes such as the BCL2, TNF, and p53 are related and contribute to cancer's pathogenesis, resulting in informational errors and consequent aberrant phenotypes. Tumor progression and oncogenesis would then depend on the cell's ability to turn off apoptosis, and we can observe an alteration of p53 gene function on the majority of tumorous cells. We can discuss, however, other theories that explain the pathogenesis of cancer in addition to this theory focused on informational errors and aberrant phenotypes; the Cancerous Tissue Organization Field Theory and the Cell Reversal Theory appear as an alternative theoretical field for explaining oncogenic processes, escaping the genetic determinism and the monism of the individual cell. In the Tissue Organization Field Theory, the parenchymal–stroma interaction in a given morphogenetic field is mandatory in carcinogenesis. The Cell Reversal Theory indicates an epigenetic or microenvironmental modified state that can result in the inversion of differentiation, inducing cellular dedifferentiation, and modifying the cell's signaling.

Keywords Cancer · Apoptosis · Biosemiotics · Programmed cell death · Cancer immunology · Genome · Caspase · Extrinsic apoptotic pathway · Cytokines · Organic code

---

R. E. Farias (✉)
Morphology Department of the Federal University of Juiz de Fora, Juiz de Fora, Brazil

© Springer Nature Switzerland AG 2021

E. Pagni, R. Theisen Simanke (eds.), *Biosemiotics and Evolution*, Interdisciplinary Evolution Research 6, https://doi.org/10.1007/978-3-030-85265-8_12

1 Introduction

Biology experienced a great revolution when it assimilated the genetic code to its lexicon concepts. That is explained by the fact that this code's operational method allows a semiotic analysis, with which the process can be understood as informational language and signs (Sebeok 2001; Peirce and Barbieri 2007; Emeche et al. 2011).

A sign-based establishment verging on a genetic logic can be better applied when analyzing an informational effort through DNA. This operation occurs when specific genetic information is translated through adaptors into structures called amino acids, protein-forming elements. This dynamic affirms: the discovery of the genetic code and DNA allowed an approximation of biological knowledge and semiotic debate. However, it is necessary to assert that biosemiotics is not restricted to genetic code or inheritance information analysis. On the contrary, semiotic terminology can be applied to life as a whole. (Sebeok 2001; Peirce 1931–1958; Barbieri 2007; Emeche et al. 2011).

In biology, the existence of a factual genetic code is the proof of semiotics on a cellular level and confirms an organic semiosis defined by a code. An organic code is a combination of rules that establishes the correspondence between two organic molecular systems, independently mediated by structures called adaptors, which comprehend both organic universes. The adaptors are molecules that convey the specificity of coherence between the elements, signaling a code's existence (Barbieri 2007).

Code biology is the study of all life-related codes, displaying new investigation fields, conceiving new concepts, and opening new research areas. It can observe, for instance, the possibility of the employment of this semiotic reasoning, concerning the interaction of genetic information, with extra- and intracellular signaling. It is possible to assert that the cell has its own sign and directional selectiveness framework through membrane receptors and ligands. The process in which cells transform a great scope of foreign signals (first messenger) into intracellular communication (second messenger) is called signal transduction, a process through which a signal is converted into another associated sign to adjoined sequences and a molecular cascade of events culminating in enzymatic actions. This process, which can be called signal transduction code (Barbieri 1998, 2003, 2007; Sutherland 1972; Trifonov 1996, 1999), requires adaptors that communicate at least two different, independent molecular systems.

Programmed cell death is an excellent example of semiotic actions involving extracellular signaling (first messengers), characterized by chemical signs that bind to membrane receptors and intracellular signs triggered by intern alteration, developing a cascade of events composed of various mechanisms and substances. Hence, this mechanism requires intra- and intercellular communication in multicellular organisms (Linda 2007).

2 Programmed Cell Death Mechanisms

Programmed cell death is the elimination mechanism of damaged or undamaged cells, and all organismic cells seem to be able to undergo programmed cell death. Many cells enter programmed cell death in an organism when complex, diverse cell mechanisms activate their death course. The incidence of this death conserves other cell elements, indicating its controlled, intended effect (Elmore 2007).

Kerr et al. (1994) described apoptosis (a type of cell death) as a rigid mechanism controlled by genetic expression triggered by intrinsic factors or modulated by cell and extracellular matrix interaction, which activates many molecules with specific activity, resulting in morphological modifications associated with the process. The defect in apoptosis might result in abnormal cell proliferation, development, and cancer progression, being itself pivotal in the oncogenic problem. Cancer treatment depends on strategies involving tumorous cell sensibility to therapeutic agents by inducing the death of cells that are resistant to the treatment.

It must be noted that the first definition of apoptosis was based on a specific morphological pattern characterized by cellular contraction, cytoskeleton disorganization, condensation (pyknosis), nuclear fragmentation (karyorrhexis) preceded by chromatin margination, and formation of cytoplasmic bodies (Kerr et al. 1972). These bodies contain nuclear fragments and dismantled components of the cytoplasm through an enzymatic system called caspase, hence assembling the apoptotic bodies—intact phospholipid structures containing the result of the process mentioned above. After its formation, the apoptotic bodies are phagocytized and dismantled by macrophages, despite their structural integrity and intern osmotic gradient maintenance. Because of this structural integrity, there are no immunological or inflammatory responses, differing from the other type of cell death, i.e., necrosis—a nonphysiological process initiated by lesion and exposure to cellular material resulting in inflammation.

A family of proteases called caspases initially performs apoptosis. Apoptosis caspases are divided into two groups: initiator caspases (2, 8, 9, and 10), and effector ones (3, 6, and 7) (Cohen 1997; Rai et al. 2005), the latter being responsible for autoproteolysis, i.e., cellular material cleavage, resulting in singular morphology of condensed chromatin and cytoplasmic bodies. It is worth highlighting that this sequence of events depends on specific adaptive molecules, whose interaction enables its manifestation. The proteolytic activity, when initiated, is reversible and might occur through the extrinsic or intrinsic pathways. The first case is represented by cytoplasmic Tumor Necrosis Factor Receptors (TNFR) and its respective ligand. In the second case, it is mediated by intracellular signals converging at the mitochondrial level.

Apoptosis is, therefore, the death of an individual cell that is physiological or involved in the pathogenesis of different morbidities. It develops when signaling molecules induce the activation of specific genes, turning them on or off. The recognition of signaling and genic activation are two different procedures. Being connected by an apoptotic code system that determines which signaling molecules activate which apoptotic gene in its respective tissue, one can evaluate that there are two different, independent systems related by an association of rules of apoptotic apparatus. Moreover, it is possible to assess the intersection of communication systems of the signaling apparatus with molecular interactions and genetic activation, which is, conclusively, biosemiotics.

The extracellular signaling apparatus receives chemical signals, such as hormones, neurotransmitters, paracrine, and autocrine stimulation, from the matrix, activating proteins called receptors on the cell membrane. These proteins must be conjoined with other molecules, establishing intracellular communication, which is also possible through gap junctions that allow the passage of chemical and electrical signals. Thus, the intracellular signaling is responsible for unleashed apoptosis by external elements—known as the extrinsic pathway—or by internal ones—known as the intrinsic pathway. The cell, reacting to various physical and chemical signals, turns on or off the expression of specific genes. The interaction of the extracellular space with genes does not occur directly, but rather indirectly after being preceded by messengers, i.e., the external signals are transformed into internal ones, known as secondary messengers, and, thus, allowing the interaction to occur. The transmission of foreign information up to the secondary messengers is called transduction, and the transmission of information by the secondary messengers to genes is called integration. Hence, the first and the second messengers are from different universes, conversely integrated by an organic code. Therefore, different first messengers might activate the same secondary messengers, or identical first messengers might activate different secondary messengers. The signals of transduction, which are based upon organic codes, require adaptors or analog molecules. This system can be observed to be composed of three molecular components: receptors for the first messenger, amplifiers for the secondary messenger, and a mediator between them. Conclusively, for the transduction signal's occurrence, the code must present correspondence between two independent worlds by an adaptor that provides meaning to both worlds by an association of rules, which specify its biological action (Barbieri 2007).

Thus, apoptosis depends on biochemically regulated processes contained by tissue environmental events, which associate transmembrane receptors, adaptors molecules, and intracellular communication through secondary messengers and signal integration with the nucleus. In light of biosemiotics, this apparatus can be understood as life-inherited systems of integrated communication and signification. It is possible to associate tissue and cellular biological events to semiotic language, as Peirce, Sebeok, and Barbieri described, since this field of knowledge can connect a myriad of fundamental biological phenomena by applying semiotic concepts. Apoptosis, which sustains multicellular organisms' lives, employs the biochemical cell apparatus to ease organismic development and sustain life.

The genes that encode proteins are more straightly related to biosemiotics. What should be thence highlighted is not that entropy is its genomic totality but rather the specific protein codification location. This process is the opposite of biological fitness and the supercritical distortion, like the ones that occur in ribosomes and spark cell death by maximum biosemiotic entropy. In this context, it is essential to point out the importance of epigenetics, given that it is the machinery responsive to environmental damage presented to the cell—drugs, hormones, and radiation, for instance, acting rapidly to matrix pressure and generating a health prospect to the cell. Epigenetic modifications that occur in a fetus might impact its susceptibility to disease and cancer in adulthood so that they have a significant impact on genetic expression, being impossible to detach the two elements. Their nonseparability is due to the fact that genes might be silenced, under expressed, or super expressed, ergo altering cellular phenotype.

3 Cancer, Apoptosis, and Entropy

An essential characteristic of the apoptosis process is that it concerns cells involved in mammalian ontogenetic phenomena, in normal physiology and morbid states, including cancer. Cells in precancerous lesions might progress into malignant forms when followed by recurrent mutations and are not eliminated.

Cancer might also be evaluated by the augmentation of biosemiotic entropy when errors occur in genetic information—e.g., randomly accumulated mutations or epigenetic modifications—resulting in malignant phenotypes. Cancer is not directly related to toxicity but to genetic informational alterations, enhancing system entropy with lethal consequences. Therefore, entropy opposes the idea of order and requires informational fidelity addressed to various cellular functioning levels. Its impediment results in degradation to molecular language necessary for adequate performance. Genome entropy contributes to many human diseases by enhancing oncogenes' expression or diminishing the expression of tumor suppressor genes, evolving to genetic code degradation.

Among the genetic language aberrations, one can include:

(i). Single Nucleotide Polymorphisms (SNP)
(ii). Insertions and deletions
(iii). Copy Number Variation (CNV)
(iv). Aneuploidy
(v). Chromosomal rearrangement

If found at the germline level, these alterations may affect an entire population and even evolutionary processes resultant from deletional mutations inheritance. Another example is HIV, which increases lymphocyte turnover and mutational potential due to the increase in cell cycles. It is known that malignant tumors appear more often in lineages with more cell cycles throughout life. Mutations on the codifying region of RNA or the translational stage are potentially lethal to the cell. Biosemiotic entropy in the formation of proteins is frequently the result of amino acids degradation and recycling. However, the cell might be induced to death if the entropy, by proteotoxic stress, induces too many abnormal proteins (Gryder et al. 2013).

Proliferative signals generated by oncogenes, on the other hand, induce an intense apoptotic system response. Hence there is an intimate correlation between these two elements. These signals might activate tumor suppressor genes inducing cell death and hindering tumoral growth. The concurrent action of other oncogenes and apoptosis inhibitive proteins might alter the suppression and induce tumoral progression. The proliferative factors require concomitant signals to deflect the death system, thus, augmenting its role as promotor. When conjointly evaluated, the results suggest that every tumor has its own characteristic concerning susceptibility to apoptosis, which is also, generally, involved in response to chemotherapy (Short and Johnstone 2012).

For the most part, carcinogenesis was explained from the Somatic Mutation Theory, centered on cells and their molecular changes (Boveri 2008), privileging the control of cell proliferation and inhibition of apoptosis. In 1999, however, a new theory pointed for alterations in the regular interactions of an organ's various components, such as epithelium and stroma, being, therefore, a relational problem linked to cellular interaction, involving physical forces of the most varied types. This view changes the cellular perspective, making tissues and development central problems; we can now envision cancer as a matter of development and organogenesis. In the postulate of the Somatic Mutation Theory, a neoplasm develops at the expense of accumulations of multiple stable mutations, whereas in the Field Theory, the microenvironment (niche) is fundamental for cell differentiation and carcinogenesis. This theory makes possible the idea of tumor regression, as seen in childhood tumors responsible for apoptotic signs (Soto and Sonnenschein 2011).

Another theory that opposes the Somatic Mutation Theory is Cell Reversal Theory, based on the dedifferentiation capacity of tumor cells both by the intrinsic epigenetic state and by the microenvironment disturbance that generates gene instability. This idea could also work as possible treatment and tumor reversibility strategies induced by the modification of the epigenome and the tumor microenvironment. Therefore, a viable cell is a stability of its epigenomic state in a given phenotype that is established at a given time and place. The lack of a viable program would indicate cell death. Stem cells are present in environments that we can call specific niches, interacting with differentiated cells, balancing proliferation, and differentiation. This microenvironment is complex, always interacting with the epigenome and the cellular genome. The apoptotic apparatus must always be ready to eliminate unwanted cells, which is difficult in states of epigenomic instability. This theory places cellular and tumoral ontogeny in the concept of individuation and differentiation apart from substantialist models (Carvalho 2020).

4 Conclusion

Thus, the organic code is a series of rules which establish a correspondence between two independent worlds composed of organic molecules. Studies also reveal that genes and proteins are not spontaneously formed, but their manufacture depends on molecular machinery based mainly on RNA. The more significant part of cell structures and their biological function are managed by proteins, being their inherited instruction the translation of a myriad of rules that constitute the genetic code. Therefore, genes and proteins are the fundamental molecules of life, whose existence depends on a constructive apparatus (molecular machinery), differing it from the inorganic world, which is determined by internal and spontaneous factors, whereas organic molecules are manufactured, hence "artificial."

RNA constitutes the leading contender for the first replication machinery of the first autopoietic life system: the supermolecule of primitive life toward autopoietic beings. Therefore, it is worth remembering that the necessity of an individualized system is separated from the surroundings by a membrane, creating an isolated thermodynamic subsystem from the environment. Consequently, greater integration levels appear from prokaryotes to multicellular organisms. Arguably, life itself consists of chemical preservation in a universe tending toward heat loss and disintegration (Margulis and Sagan 2000).

Survival has a resistance-to-death character (entropy), and its maintenance paradoxically involves the loss of cells, which is determined by intrinsic and extrinsic mechanisms that involve intra- and extracellular integrative communication. Apoptosis is the cell death pathway unleashed by a biochemical system, in which genes related to cellular suicide are activated, being linked to cancer, embryologic development, and the formation of all organs and organic systems. Signaling molecules, which trigger programmed cell death, depend on adaptor molecules responsible for the activation of intracellular communication and genic expression, considering that these molecules do not have a direct relationship and that these genes require the apoptotic code to establish this connection.

Perhaps, associated with cellular ontogeny, we can speak of individuation as dependent on its provisionally stable molecular machinery. We observed that the phenotype, the epigenome, and even the gene expression carry a time-and-place subjected instability. For its survival, the cell depends on viable structures both in temporal succession and in the individual-environment pair.

References

Barbieri M (1998) The organic codes: the basic mechanism of macroevolution. Rivista di Biologia–Biology Forum 91:481–514

Barbieri M (2003) The organic codes: an introduction to semantic biology. Cambridge University Press, Cambridge, UK

Barbieri M (2007) Is the cell a semiotic system? In: Barbieri M (ed) Introduction to biosemiotics: the new biological synthesis. Springer, Dordrecht

Boveri T (2008) Concerning the Origin of Malignant Tumours by Theodor Boveri. Translated and annotated by Henry Harris. J Cell Sci 121(Suppl 1):1–84. https://doi.org/10.1242/jcs.025742

Carvalho J (2020) Cell reversal from a differentiated to a stem-like state at cancer initiation. Front Oncol 10:541

Cohen GM (1997) Caspases: the executioners of apoptosis. Biochem J 326(Pt 1):1–16

Elmore S (2007) Apoptosis: a review of programmed cell death. Toxicol Pathol 35(4):495–516

Emeche C, Kull K, Hoffmeyer J (eds) (2011) Towards a semiotic biology life is the action of signs. Imperial College Press, London
Gryder E, Nelson CW, Shepard SS (2013) Concept paper biosemiotic entropy of the genome: mutations and epigenetic imbalances resulting in cancer. Entropy 15:234–261
Kerr JF, Wyllie AH, Currie AR (1972) Apoptose: um fenômeno biológico básico com amplas implicações na cinética do tecido. Br J Cancer 26(4):239–257. https://doi.org/10.1038/bjc.1972.33
Kerr JF, Winterford CM, Harmon BV (1994) Apoptosis. Its significance in cancer and cancer therapy. Cancer 73(8):2013–2026
Linda EF (2007) Biosemiotics and survival: life and Death, genes and signals. In 32nd Annual Meeting of the Semiotic Society of America New Orleans, Louisiana, October 4–7
Margulis L, Sagan D (eds) (2000) What is life, 1st edn. University of California Press, Los Angeles
Peirce CS (1931–1958) Collected papers of Charles Sanders Peirce. Harvard University Press, Cambridge, USA
Rai NK, Tripathi K, Sharma D, Shukla VK (2005) Apoptosis: a basic physiologic process in wound healing. Int J Low Extreme Wounds 4:138–144
Sebeok TA (2001) Biosemiotics: its roots, proliferation, and prospects. Semiotica 134:68
Short J, Johnstone RW (2012) Oncogenes in cell survival and cell death. Cold Spring Harb Perspect Bio 4:12
Soto AM, Sonnenschein C (2011) The tissue organization field theory of cancer: a testable replacement for the somatic mutation theory. BioEssays 33:332–340
Sutherland EW (1972) Studies on the mechanism of hormone action. Science 177:401–408
Trifonov EN (1996) Interfering contexts of regulatory sequence elements. CABIOS 12(5):423–429
Trifonov EN (1999) Elucidating sequence codes: three codes for evolution. Ann N Y Acad Sci 870:330–338

Biosemiotics and Bioenergetics: Two Perspectives Compared

Giulia Degl'Innocenti

Abstract The purpose of this work is to show how the body, understood as a biological body or as the expression of psychic phenomena, represents a dimension that is open to the world, expressive and inhabited by meaning. To this end, two disciplines will be juxtaposed: biosemiotics and bioenergetics. They seem alien to each other and come from two different disciplinary fields; nevertheless, they present some interesting similarities. According to biosemiotics, the living world is generated, structured, and evolved through semiosis, and the biological body is, therefore, the place of signification. At the same time, bioenergetic analysis shows how the body is inscribed in particular psychic meanings related to the individual's relational and affective experience. In both, therefore, the body is delineated as a reality characterized by meaning.

Keywords Biosemiotics · Bioenergetic analysis · Semiotic relations · Umwelt · Living world · Constraints · Character analysis

1 Introduction

Although they stand as two disciplines with separate origins and development, the biosemiotic paradigm and bioenergetic analysis present many similarities and offer an original contribution to an interpretation of the body as a dimension inhabited by meaning. Therefore, this study aims to do a comparative survey that will bring to light the common features of the two approaches.

In the first place, biosemiotics and bioenergetics share the idea that the body, which may be understood as a biological body or as the expression of psychic phenomena, constitutes a creative dimension open to the world and not reducible to a simple pre-established mechanism. In this context, relationship, the rapport with

G. Degl'Innocenti (✉)
Department of Humanities, University of Studies of Florence, Florence, Italy

© Springer Nature Switzerland AG 2021

E. Pagni, R. Theisen Simanke (eds.), *Biosemiotics and Evolution*, Interdisciplinary Evolution Research 6, https://doi.org/10.1007/978-3-030-85265-8_13

what is external, plays a primary role and is the mode in which the body lives and defines its structures.

In the second place, the possibility that the body represents a dynamic, not predetermined, dimension is given in both approaches by the specific link that they maintain with meaning. Indeed, for biosemiotics, the living world generates itself, evolves, and creates its own forms through semiotic processes. In this sense, every structure represents the result of a relationship that is semiotic in nature.

In parallel to this, bioenergetics interprets an individual's physical attitude as the outcome of psychic processes and affective and relational experiences. Thus, the body is capable of producing a system of symbols and transforming into a totality of meanings formed over time in a dimension that bears on itself the inscribed signs of a narrative, of its relation to the world, like a map that helps guide therapists.

The present study is divided into two parts: the first concentrates on analyzing the biosemiotic paradigm, the second on bioenergetics. Both disciplines are situated in a historical, ontological, and epistemological framework.

In the first part, an initial section is devoted to locating biosemiotics in a historical context, illustrated with precursors and pioneers (1.1). In a second section, the concepts of "*Umwelt*" and "*Innenwelt*" are analyzed and reread in the light of studies of biosemiotics (1.2) in an attempt to understand the ontological considerations that these concepts may suggest. By contrast, the third section will try to outline, from a biosemiotic viewpoint, the modes by which life structures, experiences, and knows itself (1.3). In this context, the concepts of relationship and limit will emerge and occupy a central place, being essential for delineating the organism as an open and relational dimension, but at the same time defined by its own boundaries.

The second part will deal with bioenergetics and the original interpretation of the body that it offers. As before, the first section provides some historical notions aiming at contextualizing the discipline (2.1). The second section investigates the concept of bioenergy and analyzes the connected functions of charge and discharge (2.2). By contrast, the third and final section studies the expressive modes in which the body gives voice to its own psychological experience, with particular attention to the analysis of characterological structures (2.3).

2 Biosemiotics: Body and Meaning

2.1 Historical Hints: Precursors and Pioneers

The paradigm of biosemiotics offers an original contribution to the idea that the body weaves a privileged relationship with meaning. According to this perspective, life emerges, generates, and evolves through semiotic relations: semiotic processes characterize the living world, and the biological body (in the broad sense of the term, as living matter) constitutes the epicenter of meaning. At the same time, meaning is not the prerogative of a disembodied mind, as a human privilege, but is rooted in the body, in its organic processes: life has always been meaning and meaning exists only as a living product.

The main idea behind biosemiotics is the belief that biological processes are semiotic processes. Therefore, meaning is how life expresses itself: "biosemiotics," writes Maurita Harney (Harney 2007, p. 134) "is premised on the idea that all living organisms engage in processes of signification and communication by means of signs."

If signs are the foundation of organic life, communication, and the relationship with the other represent the conditions of life for any organism. Therefore, biosemiotics is concerned with analyzing how this happens, showing how, if we deal with life, both at a macroscopic and microscopic level, we necessarily talk about meaning.

Admitting the existence of an indissoluble link between life and meaning involves establishing a union between two subjects such as semiotics and biology: biosemiotics represents the possibility of endorsing the union since it constitutes the point of arrival of two disciplines which, structured as separate, realize that they cannot live without each other, having always been intimately interconnected.

Although biosemiotics appeared relatively recently in the scientific panorama as a defined and specific field, the idea that life is meaning right from the outset starts to take shape between 1961 and 1966 (Barbieri 2008a) with the discovery of the genetic code. The theorization of the genetic code—and hence of a linguistic principle, of a "communicative" order in the organism's depths—opens up new research possibilities. However, the idea was not enthusiastically accepted within the scientific community. In fact, communication has always been the concern of a different field from biology, a field with different methods and objectives. Thus, after such a discovery, how can a linguistic principle be combined with the life sciences without abandoning the quantitative methods and categories of scientific investigation? The answer comes quite easily. The term "genetic code," it is said, must be used in a metaphorical sense, as "a linguistic expression that biologists have adopted just because it was intuitively appealing. Deep down, according to this view, the genetic code is but a metaphor because all its features must be completely accounted for by physical quantities" (Barbieri 2008a, p. 578). Therefore, "code" simply means a sequence of "information" reducible to a set of biochemical processes. With this interpretation, the biological body continues to be conceived as a quantitative dimension that responds to mechanical principles.

In parallel to the discovery of the genetic code, something starts to move in the field of semiology. The Hungarian semiologist Thomas Sebeok realizes the limitations of a purely semiological approach compromised by an excessive anthropomorphism and proposes a widening of the subject's boundaries.

In 1963, inspired by the contents of Jakob von Uexküll's work, *Theoretische Biologie*,[1] Sebeok suggests the hypothesis that even communication between

[1] Here begins a long collaboration between Sebeok and von Uexküll's son, the physician Thure von Uexküll. Cf Barbieri (2008a); cf. Petrilli and Ponzio (2013). Von Uexküll is considered, together

animals is founded on the sign and introduces the term "zoosemiotics" (Barbieri 2008a). Subsequently, detecting the further limitedness of the term, which confines meaning to the animal world, he proposes using "biosemiotics" to indicate how semiotic processes concern life in general and are indeed its foundation.[2]

In the 1980s, the idea that semiotics might apply to more varied research fields begins to make powerful advances, but it will only be in the 1990s that biosemiotics will find a way to assert itself as a specific and defined field. In 1992 Jasper Hoffmeyer met Kalevi Kull and Thomas Sebeok in Glottertal (Petrilli and Ponzio 2013), and the first avowedly biosemiotic research group is formed. As the years' pass, it is configured more and more as a discipline defined by its own principles, though these are polyvocal and varied.

Sebeok is today considered the father of biosemiotics. He suggested the idea of a "global semiotics,"[3] which promotes an interactive, interdisciplinary, and cooperative approach to the subject.

Although Sebeok is the initiator of biosemiotics, he acknowledged Charles Sanders Peirce, Charles Morris, and Jacob von Uexküll as worthy precursors (Petrilli and Ponzio 2013). As an example, biosemiotics inherits the Peircean concept of interpretation and readjusts it to the world of life. Organisms inhabiting a semiosphere move between signs and are involved in processes of interpretation.

> Peirce claimed that "the universe is perfused by signs, if not entirely composed of them", thereby indicating that the locus of meaning in the case of the sign is not the human mind, but rather processes in nature. (Harney 2007, p. 133)

However, what precisely interpretation means?

According to Peirce, the sign has a triadic structure that consists of a "sign vehicle" (sometimes called "representamen"), an "interpretant," and an "object." The "sign vehicle" represents the sign, something that stands for something else and refers to an "object." The interpretant is the most interesting figure of the three in that it constitutes the action that connects sign, vehicle, and object, which implements signification. Let us take an example provided by Maurita Harney:

> Suppose, looking at the horizon, I see smoke and I take this as a sign that there is a fire in the vicinity. Here, the relation between the *sign vehicle* (smoke) and its *object* (fire) is mediated by a third term, the *interpretant* (my thought of the fire). The sign relation here cannot be

---

with Charles S. Peirce, Juri M. Lotman and Charles Morris, the biosemiotics' precursor. Cf. Ibid., p. 382.

[2] On the question of the year when the term was coined, cf. Barbieri (2008a): "I am not quite sure that I can pinpoint Tom's original coinage of the term "biosemiotics"—I think it is 1975, but Kalevi Kull has suggested 1972′ (Ibid). Other authors differ on the date. Cf. Petrilli and Ponzio (2013, p. 380). In fact, it was a 'term that had already been proposed by Juri Stepanov in 1971, but which had appeared for the first time in 1962, when Friederich Rothschild used it to illustrate a new approach to psychology' (Ibid.).

[3] "After a paper of 1994 entitled 'Global semiotics,' Sebeok consolidated this expression, proposing it as the title of his book of 2001 (the last published before his death that same year)" (Petrilli and Ponzio 2013, p. 375).

reduced to two dyadic, causal relations, one between *object* and *sign*, and the other between *sign* and *interpretant*. (Harney 2007, p. 138)

Therefore, we encounter meaning where there is a process of interpretation that associates a particular sign to a specific object. From a biosemiotic perspective, the hypothesis of an "interpretant" guarantees the irreducibility of biological systems to pure mechanisms; it makes them semiotic.

Admitting the presence of interpretative procedures in the living world implies breaking free from the idea that the exchange of information that characterizes biological processes is regulated by a cause-effect relationship, where the relationship between the sender and the receiver is previously determined.

On the other hand, the receiver is always an interpreter. According to the biosemiotic hypothesis, the processes that involve the organism do not follow a predetermined path but are defined by an act of signification: the interpretant is what establishes the meaning of the sign and therefore what provides an "interpretative decision" on every occasion, a "response" that is never the only one possible.

In addition to Peirce, another author recognized as a worthy precursor of the biosemiotic paradigm is the biologist Jacob von Uexküll, to whom we owe the concept of "*Umwelt*" elaborated in 1909 (the concept was later developed in 1920 in *Theoretische Biologie*). *Umwelt* means the subjective and qualitative environment in which each animal lives, by which it is "wrapped" and maintains a relationship that, in the light of biosemiotic studies, can be defined as semiotic. It is defined as the "perceptual world," the world perceived by the animal, and as "an effector world," a world in which the animal does not assimilate external stimuli passively and mechanically but "acts" on its environment and contributes to its definition.

This world environment is formed by a series of components that von Uexküll defines as "marks" (*Merkmalsträger*) and which are the only elements that make sense to the animal. Therefore, the concept subverts the anthropocentric belief that every living being has access to the same world, understood as an objective totality regulated by quantitative principles.

2.2 Biosemiotics and the Concept of Umwelt: Some Ontological Considerations

The biosemiotic paradigm inherits and adapts the concept of *Umwelt,* conferring a specifically semiotic connotation upon it. According to this perspective, the *Umwelt* is defined as "a qualitative and meaningful model of a species' significant surrounding" (Sebeok 1986, p. 23) and also "Umwelt is not a set of objects in the environment but rather a system of signs interpreted by an organism" (Sharov 2001, p. 211). Therefore, the organism and its environment are not linked by predetermined relationships, by physical laws or quantitative principles, but move in a semiotic and qualitative relationship of reciprocity. The specific and experiential relationship that the organism establishes with its surroundings and the system of symbols created determine the Umwelt.

"The experience," Hoffmeyer writes, "is at each moment the superior, immediate, and unconditional interpretant in the ongoing biosemiosis of the organism [...]. In other words, we suggest that the phenomenon of experience has primitive parallels all over the lifeworld" (Hoffmeyer 2008, p. 188).

Furthermore, the *Umwelt* does not represent a neutral dimension, a "blank sheet of paper" on which the living being inscribes its meanings, just as the organism is not parachuted into its environment as into a structure already pre-ordered or involved in an exclusively perceptive dimension. There is no priority of one of the two terms over the other, but they are in an experiential relationship of co-determination and co-specification. In this sense, *Umwelt* is not defined as an accidental feature of a living being but represents its foundation, how an organism is structured as such (Tønnessen et al. 2018).

The animal body, its sensorial apparatus, which takes the name of *Innenwelt*, is structured by the meanings of its world environment (i.e., the marks). The *Umwelt*, in turn, is delineated thanks to the action of the animal body. *Umwelt* and *Innenwelt*, therefore, represent two different notes within the same musical score. The organism confers the former's meanings, and the latter takes its form thanks to its relationship with the environment.

Varela, Thompson, and Rosch provide an example of the co-determinant relationship between living beings and their environment in analyzing the relationship between bees and flowers.

> It is well known that honey bees are trichromats whose spectral sensitivity is shifted toward the ultraviolet. It is also well known that flowers have contrasting reflectance patterns in ultraviolet light. [...]. It is therefore interesting to observe that the colors of flowers appear to have coevolved with the ultraviolet sensitive, trichromatic vision of bees. (Varela et al. 1993, p. 201)

The bees' sensory-motor apparatus and the flowers' chromatic structure evolved and were structured together through what Varela, Thompson, and Rosch define as a history of "structural coupling." This coupling is responsible for forming the various structures, both of the spectral sensitivity of the bees and the ultraviolet reflection patterns of the flowers. The example sheds light on the fact that *Innenwelt* and *Umwelt* are two dimensions enveloped in each other and developed from each other, established together in a relationship of co-determination.

A central notion in biosemiotics and linked to the concept of *Umwelt* is that of "dialogue." As it is understood here, the concept dates back to the Russian theorist Mikhail M. Bakhtin but is adapted by biosemiotic studies and provided with biological bases. The main idea is the belief that the semiosis of biological systems is articulated in a dialogical form.

Meanings are continually inscribed in the living world and arise from a dialogic relationship, a continuous dialectical comparison between what is "proper" and what is "other," between "inside" and "outside." Therefore, dialog does not merely represent a means of interaction, but rather the a priori that makes life possible, allowing forms, stabilizations and therefore organisms to be created. It is the basis of every individual formation.

The organism always falls into a dialogical context. This condition means that the living world takes the form of intercorporeity, of a dimension with evanescent borders and connoted by otherness.

> Dialogism and intercorporeity denote interconnectivity among bodies in the great sign network that is life. The mere fact of being alive already places living beings, including human beings, in a sign network, or to evoke an organic metaphor dear to Sebeok, in a semiotic web, that is, in a situation of interconnectivity with the other, including other forms of life not human (nonhuman animals, plants) as well as the inorganic. To set aside the other, to ignore the other is impossible, just as to ignore dialogue is impossible. (Petrilli and Ponzio 2013, p. 386)

The concept of Umwelt, as well as the hypothesis of an intercorporeity, suggests some ontological considerations. The various *Umwelten*, the many semiotic worlds in which organisms are involved, and their relative interaction constitute what some authors define as "semiosphere." This concept was first developed by Juri Lotman in 1982, as distinct from that of "biosphere" formulated by Vladimir Vernadsky (Kotov and Kull 2011). Thomas Sebeok then introduced the concept into biosemiotics and extended its meaning (Petrilli 2003). From a biosemiotic perspective, the biosphere and the semiosphere coincide.

Considering the living world as a semiosphere, as a semiotic dimension, means abandoning a conception of reality understood in an objective sense, as a set of predetermined and universal rules. It means understanding the living world as a result of semiotic exchanges and interactions. It means giving relationships a primary role.

However, does breaking free from the realm of objectivity entail looking out to that of subjectivity? Does *Umwelt* represent the subjective ontological space that every living being occupies? What is the position of biosemiotic thought about the possibility of attributing a form of subjectivity to the living being? In this regard, it is necessary to suggest some points for reflection. As Tønnessen emphasizes, "biosemiotics depicts a world of the living in which all living beings are to be regarded as true subjects actively engaged and involved in (and through) their life-worlds" (Tønnessen 2009, p. 58). Therefore, according to this perspective, it is appropriate to extend a principle of subjectivity to the whole living world. However, Tønnessen also reminds us to be careful not to confuse a biosemiotic interpretation with an anthropocentric conception, which provides for the attribution of mental states to every living being. In this regard, he proposes a distinction between "mental states" and "semiotic states:"

> In analogy with notions such as "mental state" and "cognitive state", I suggest to use the term semiotic state. While in philosophy of mind a 'mental state' refers to a certain performance which is unique to rational or sentient beings, a 'semiotic state' is to be understood as the state of a sign process, an inner state (in its most general form) which might be said to be distinctive not only for all living beings, but for all living systems, as far as they possess what we could call semiotic integrity, defined as an autonomous (self-organized) coordination of semiosis. (Tønnessen 2009, p. 63)

In conclusion, biosemiotic thought and its concept of *Umwelt* allow a rethinking of the living world, freeing it from its objective and deterministic interpretation, which understands it as a uniform structure regulated by mechanical principles instead of understanding it as a dynamic, creative, and relational dimension. Furthermore, according to some authors, every living being can be conceived as an acting subject. However, this consideration does not imply the "humanization" and attribution of a form of consciousness to the entire biological world.

2.3 The Living World Between Relationships and Constraints

The ontological considerations raised above and the biosemiotic readjustment of the Umwelt concept enable some reflections. There is a relationship of co-specification and co-determination between the living being and its environment, and the exchange of information takes place through semiotic processes. As a result, the definition of a permanent structure comes after the establishment of a relationship. Therefore, a biosemiotic perspective suggests an interpretation of the living being according to which the relational element represents the primary element and structures life.

Since symbolic processes have a bodily matrix and the biological world is connoted by meaning, adopting a biosemiotic perspective means definitively moving beyond the Cartesian dualism of mind/body, *res cogitans* and *res extensa*. The consequences of this dualism, on an epistemological level, are the reduction of semiosis to an incorporeal mental act and of the body to a quantitatively determinable physical matter regulated by mechanical and quantitatively determinable laws. According to biosemiotics, "life and semiosis are coextensive" (Kull et al. 2008, p. 43).

> [...] dualism, the idea that *soma* and *sema* represent equally inescapable but incompatible dimensions (substances, properties or whatever) of our world. [...] The main problem is that it is not obvious what the matter-spirit distinction is all about. The idea of passive matter as ruled by natural laws (or by the heavenly ruler) has long ago lost its credibility. Instead, modern conceptions of physical nature make ample space for the vision of the world as an emergent process in which those peculiar things we call living systems and their bodies might well have evolved as genuinely semiotic creatures. (Hoffmeyer 2008, p. 170)

Overcoming a dualistic logic has two implications. First of all, the abandonment of reductionist biology; secondly, the redefinition of semiosis as a non-purely mental prerogative of human beings.

Reductionist biology reads the processes that underlie life in quantitative terms, as physical-chemical processes regulated by physical laws. Thus, living beings are reduced to objective entities, to surfaces that can be analyzed *partes extra partes* and function mechanically. Furthermore, the organism is conceived as "the sum of its parts," so that the whole's properties represent the direct expression of its constituent elements, and each component corresponds to a particular function. Giorgi (Giorgi 2017) says in this regard that the strictly reductionist analysis represents the functions as attributes of the parts. For example, the heart is used to pump blood, the lung to breathe oxygen, and the brain to think. According to a biosemiotic perspective, however, a reductionist biology addresses the phenomenon of the living being only a posteriori, positing it in terms of a re-composition of its functions, as if they constituted predetermined entities. Therefore, its *modus operandi* consists in isolating the single parts and subtracting them from the set to which they belong.

According to Giorgi, while this strategy can be valid on an epistemological level and adopted as a cognitive procedure, it is not ontological. The error is to consider the parts' properties as predefined and previous to the relation that binds them, conceiving the mechanism as what determines the interaction, not as its result. Thus, the subjective nature of the living and the relative behavioral autonomy it enjoys compared to the physical-chemical mechanisms are disregarded. In fact, when one observes an organism or its parts without considering the whole they belong to, they are subtracted from all the potential interactions that define the context within which their functions manifest themselves (Giorgi 2017).

Therefore, analyses of this kind apply to the world of objects, whose behavior is determined by the law that describes it and is therefore predictable. However, they do not apply to the living world, where the interactive element is essential: every property emerges from relationships and is configured as the realization of one alternative among the many available.

The second implication resulting from overcoming a dualistic logic consists in the need of redefining semiosis. Biosemiotics, connecting meaning to the sphere of life, offers an appropriate response to the problem: meaning is not reduced to a verbal reality but becomes the prerogative of the living being understood as an organism. Consequently, the processes of signification do not concern only the human being but every living being. After dismantling anthropocentrism, "human language" is therefore configured only as a specific modality, "a species-specific modelling device" (Petrilli and Ponzio 2008, p. 34), of a much wider semiotic process that involves the whole biological world.

From an epistemological perspective, it is interesting how the living being known itself, creating and defining its own structures. As previously mentioned, organisms are structured and take their shape through a semiotic exchange and thanks to a relationship. Therefore, the consolidation of a structure can be understood as the establishment of semiotic boundaries, as a negotiation between inside and outside.

The genotype/phenotype relationship can provide a classic example. According to a gene-centered view, the phenotype is expressed deterministically from the genotype and "the organism is left with the purely passive role of container of genes" (Giorgi 2017). However, from a biosemiotic perspective, the genotype does not contain its phenotypical realization within itself, and any coding of "information" takes the form of an interpretation. In turn, the interpretation occurs based on a "context," which provides the key to reading the message.

Therefore, in the passage from genotype to phenotype, the transmission of characters does not take place directly, but thanks to the intercellular universe's interpretative context. In this regard, Pagni writes:

> (...) [the] genotype does not contain a complete description of the phenotype that genetic constraints do not exert unidirectional (linear) causal influence from DNA to RNA [...]. There is therefore no simple relation between genome and the construction of the organism: biological information is inseparable from its context, meaning that it needs to be interpreted [...] the genetic code cannot operate out of its coextensive array of cellular and molecular mechanisms, which, in turn, are determined by specific historical and functional contexts. (Pagni 2016, p. 63)

In this sense, according to Giorgi (Giorgi 2017), the information transmitted has a "semantic" rather than "syntactic" character: it is not merely reconstructed from the phenotype in its structure but instead interpreted in its content. In fact, the phenotypic realization does not represent the direct execution of a pre-established message in the genetic structure. The information "is formed" during its transmission.

Therefore, the reception of the message takes place actively, based on a shared repertoire of which the source is already in possession and which the receiver must only interpret. Cellular differentiation and embryonic development do not occur because of pre-established information, but, on the contrary, because there is information that is gradually being constructed according to stage-by-stage choices (Giorgi 2017). Furthermore, the gene-centric vision leads to a deterministic conception of the living being, for which the genetic structure directly causes the organism. On the contrary, the biosemiotic believes that even equal-level relations intervene even from the hierarchically inferior levels.

The interpretative capacity of the living being, expressed at each level (Barbieri 2008b), makes the organism a semiotic dimension formed step by step through "choices" and whose path cannot be entirely predictable. In this perspective, the cells "respond" to signals with interpretations, and the body acts by taking account of its personal history, its registered meanings, its structures. Therefore, every living being, at whatever level of complexity—from cells to human beings—is endowed with the capacity to interpret its own environment and to choose among the alternatives that the environment makes available to it (Giorgi 2017).

A further contribution to the idea that the living being is structured by semiotic "choices" comes from Sebeok and his theory of the immune system as a defining system of the self. According to this perspective, the organism's ontogenesis begins with distinguishing between the self (*ego*) and the other (*alter*). This distinction is obtained through the recognition by the immune system of the antigen as something foreign, a non-self (Petrilli and Ponzio 2013).

The fact that the distinction "ego"—"alter" is not an a priori, something defined once and for all even within the organism, supports the hypothesis that the living being does not represent a stable dimension but is the temporary result of a continuous and inexhaustible tracing of borders. Therefore, as something distinct from otherness, the organism as unity is never a predetermined and immutable structure. Instead, its evanescent boundaries result from a series of interactions, negotiations, and exchanges between an "inside" and an "outside."

The "skin" metaphor, proposed by Jesper Hoffmeyer, can be useful to show the semiotic essence of the living world. "On one hand," Hoffmeyer writes, "the skin thus serves us as a kind of topological boundary; while, on the other hand, its semiotic capability opens up the world to us" (Hoffmeyer 2008, p. 174). The skin is configured as a border, a liminal layer, an interface between an outside and an inside. It represents the semiotic space par excellence, where the dialog with the outside takes on a form in which the first meanings are assigned.

> It reminds us how indispensable the skin is in semiotic terms as well. The skin keeps the world away in a physical sense but present in a psychological sense. It is the skin that gives us the experience of belonging—it allows us to feel the world. But the very fact that the world can be felt is already a complex phenomenon that doesn't just presuppose that there are receptors (sensory cells) in the skin that register touch, pressure, pain, cold, warmth, pH, and various chemical influences, but also that biological meanings are assigned to these sensations. (Hoffmeyer 2008, p. 172)

"Skin" generally means the outermost layer of the body. However, a particular type is also present in the body's depths, both as a cellular membrane and as a coating inside the cell (think of the structure of the "organelles"). The epidermis, understood as a surface, a membrane and a place of exchange between outside and inside, affects the organism at every level. It best defines this organism as an interface, a decentralized semiotic unit where there is a constant reference to an outside, to its surroundings.

Thus, the skin is configured as the place of passage between an outside and an inside. It follows that the living being is also a closed unity, identifiable from otherness, despite being "open" to a continuous exchange with the exterior. "While living entities," writes Sebeok, "are, in a commonly recognized sense, open systems [...], they are at the same time closed systems, in the sense that they make choices and evaluate inputs, that is to say, in their semantic aspect" (Sebeok 1989, p. VII).

The concept of limit, therefore, plays a fundamental role in biosemiotics. The limit represents the constitutive principle of life, of how the biological world is organized. There are no unlimited organisms.

However, the hypothesis of a binding principle is not contrary to the relationality that characterizes the living, but merely the "form," the "limit" as per our previous definition. The "constraints" are therefore configured as necessary elements to ensure certain stability and permanence to the organisms, to guarantee their equilibrium and keep them cohesive under an identifying sign (Pagni 2016).

Nevertheless, the limits are also the premise for living beings to experience certain freedom. In fact, since the boundary between self and other is not clearly marked in the biological world, the organism must use a mobile and flexible principle that allows it to measure itself against what is happening around it. In this sense, in biosemiotics, the concept of "limit" replaces that of "law." The law appears as a definitive enunciation, established once and for all. On the contrary, the "limit" represents an approximate principle that does not mark any definitive attitude.

> Contrary to physical objects, a biolon (a cell, an individual, a species) is far from the state of equilibrium and nested within an unpredictable and contingent phenomenon [...]. Moreover, whereas the physical objects are characterized by the preservation of laws during

> transformations, the global structural stability of biological systems is associated to variability and permanent changes of symmetries. (Pagni 2016, p. 67)

The limit, thus understood as a regulating principle of the exchange between organism and environment, has a semiotic nature. Therefore, the living being is defined as a structure "bound" by semiosis, by interpretative processes that establish its form. In this sense, the organism can be understood as "semiotic closure." Since an organism is situated, limited, and the result of a singular history, not everything is possible. Only in this context can meaning emerge. In artifacts, by contrast, there are no mechanisms of self-regulation that would limit their functioning (Benasayag 2016).

In this perspective, the concept of "interpretation" undergoes a restriction. The interpretative processes in which the living being is continuously involved are only possible within a context, a range of limited possibilities. Giorgi (Giorgi 2017) claims that "interpretation" means the living being's capability to explore the surrounding environment actively and, therefore, be constantly confronted with the interactive possibilities offered by it.

Therefore, the interpretative margin available to the organism is restricted by the presence of constraints of meanings formed over time that have established its operating rules. Thus, the previous choices constrain future ones by restricting the field to grant the body a limited degree of freedom of action. Once a semantic *habitus* has been established, this becomes binding for all the interpreters who share the same semiotic system as a map to orient the exploration of the territory (Giorgi 2017).

In this regard, a relevant notion in the biosemiotic paradigm is the "scaffolding," a term coined by Hoffmeyer in 2000. The scaffolding is a kind of "structure," which ensures the organism's equilibrium. This structure is not predetermined and given once and for all but has crystallized over time due to its interactions and semiotic processes. The temporal element is fundamental in this perspective.

> This network of semiotic controls establishes enormously complex semiotic scaffolding for living systems. Semiotic scaffolding safeguards the optimal performance of organisms through semiotic interaction with cue elements which are characteristically present in dynamic situations. History thus not only matters to the cell, but literally operates inside the cell through the structural couplings—or semiotic scaffolds—that it has served to build into the system. And this is exactly what distinguishes living systems from non-living systems: the presence in the former of historically created semiotic interaction mechanisms which have no counterpart in the latter. (Hoffmeyer 2015, p. 149)

Hoffmeyer (2015) reports the example of the *Escherichia coli* bacterium. This species can detect the concentration of nutrients in the surrounding environment and record any changes. Any change in the concentration of an edible amino acid causes the bacterium to swim upstream towards the amino acid source. This behavior depends on a sophisticated interaction of different proteins that cooperate to compare two measurements taken at different times. The result of the comparison is transmitted to the surface of the cell where the flagella, responsible for the movement, are located. Interpreting the message, they move the body of the bacterium in the indicated direction. Therefore, the *E. coli* has a structure, a "scaffolding," in which the sensor apparatus and the motor system are connected. This structure also refers to an "external" element present in the bacterium's *Umwelt* (the concentration gradient of nutrients in the liquid).

The behavioral pattern of *E. coli* has evolved over time, establishing itself as a typical structure of the species: history thus not only matters to the cell but literally operates inside the cell through the structural couplings—or semiotic scaffolds—that it has served to build into the system (Hoffmeyer 2015, p. 151). The relational processes that involve the organism, therefore, sculpt its biology in various scaffoldings.

In this context, another central notion is that of "memory." Memory understood as biological or corporeal memory, both ontogenetic and phylogenetic, represents the context in the light of which every interpretation, which makes it possible, takes place.

Therefore, each structure derives from an experiential, qualitative relationship that does not respond to any law, to any principle that can be quantitatively expressed. In fact, scaffolding has a semiotic and relational nature: it is a reality which "derives" from dialog with the outside world, not a pre-programmed operation.

The living organism, the biological body, is a qualitative dimension that is formed through relationships, semiotic experiences, which is "equipped" and structured in patterns, in models of orientation that are not definitively established but respond to the environment with their constraints, with their history, with their experience. In short, to put it in the words of Varela, "we are always constrained by the path we have laid down, but there is no ultimate ground to prescribe the steps that we take" (Varela et al. 1993, p. 214).

3 Bioenergetics: Body and Meaning

3.1 Historical Notes: From Reich to Lowen

As a generating principle of meaning and dimension characterized by it, a further look at the body is offered by bioenergetic analysis. The bioenergetic analysis is a form of psychotherapy conceived by Alexander Lowen, a physician and psychotherapist, who puts the body at the center as an energy dimension. In its founder's words, it is "the study of human personality in terms of the energetic processes of the body" (Lowen 1994, p. 45).

Lowen elaborated and developed the bioenergetic method after meeting Wilhelm Reich, his teacher from 1940 to 1952 and his analyst from 1942 to 1945. Lowen met Reich in New York, at the New School for Social Research, on the occasion of a course entitled "Biological Aspects of Personality Formation." The encounter with the master immediately proved fruitful. "I sensed," Lowen reports, "Reich was introducing me to a new way of thinking about human problems, and I was immediately excited" (Lowen 1994, p. 15). The theories of Wilhelm Reich (physician, psychotherapist, and student of Freud) focus on the hypothesis that there is a functional unity between a person's character and their bodily attitude, so that body and psyche are intimately connected.

His investigation starts from the Freudian principle of "repression," developed starting from the studies on hysteria. According to Freud, the hysterical pathology is generated following a sexual trauma repressed and converted into a physical symptom. Reich's criticism of the master concerns the fact that the concept of repression only explains the *cause* of the illness, without providing any information on its formation process, on *how* a repressed psychic phenomenon translates into a bodily manifestation.

Therefore, Reich's work is an attempt to fill this gap. To do so, he introduces an energy principle. The body, says the Austrian psychiatrist, is an "energetic" dimension, "maneuvered" and structured by the individual's personal experience. Consequently, neurosis can only be understood by analyzing the distribution of energy within the body, thus making use of an "economic" principle: "the energy or sexual economy of an individual concerns the balance that he maintains between energy charge and discharge or between excitement and sexual discharge. The symptom of hysterical conversion develops only when this economy or this balance is upset" (Lowen 1994, p. 15).

Moreover, Reich noticed the correlation that exists between muscular tension and inhibition of emotional expression. He hypothesized that a precise psychic attitude corresponded to a specific bodily attitude and determined muscular tensions.

The bioenergetic analysis is, therefore, grafted onto this background. However, although Lowen maintains the Reichian analysi's central concepts, he develops, deepens, and partly departs from them until he develops his independent thought.[4]

At a therapeutic level, the bioenergetic approach attempts to combine an analytical methodology, dear to traditional psychoanalysis, with work on the body, interpreting the individual as a unitary complex. Bioenergetic work acts simultaneously on two levels: the somatic one, through the elimination of chronic muscular tensions, which leads to a particular mobilization of energy, and the psychic one, through the awareness of one's own body history. Lowen believes that no lasting change is possible if the therapist does not work on both levels.

The bioenergetic analysis is defined as a bodily mediation therapy, in which the body, as an energy complex, becomes the bearer, with its expressive baggage, of the personal psychic past. "The life of an individual," declares Lowen, "is the life of his body" (Lowen 1994, p. 54).

[4]Lowen's theories differ from the Reichian analysis in several respects. First, Lowen distances himself from the orgonic drifts of his master's thought, according to which the energy of the body would be part of a cosmological energy dispersed in the universe. For Lowen, the energetic principle of the body acquires meaning and must be understood only within the clinical practice; secondly, according to Lowen, Reich's analysis does not consider the ego and the reality principle, focusing only on the body as a vegetative-energetic complex. Instead, the purpose of Lowen's therapy is an integration of the ego with the body (Helfaer 2013).

3.2 An Energetic Principle: The Functions of Charge and Discharge

Lowen claims that the body is suffused with a vital energy called "bioenergy." Although rooted in Freudian *libido*, the concept combines the idea of transferable psychic energy with that of physical body energy. Indeed, bioenergy is responsible for both psychological and somatic phenomena.

> We work with the hypothesis that there is one fundamental energy in the human body whether it manifests itself in psychic phenomena or in somatic motion. This energy we call simply "bioenergy". Psychic processes as well as somatic processes are determined by the operation of this bioenergy. All living processes can be reduced to manifestations of this bioenergy. (Lowen 1971, p. 18)[5]

Agreeing with Reich, Lowen argues that an individual's psychic attitude is determined by his energy economy, i.e., by the amount of bioenergy employed in the body and the way it is distributed. An individual in good psycho-physical condition lives a body with a high energy level and whose bioenergy is adequately distributed; consequently, one can understand mental disorders in the light of energy disturbances as dysfunctions in the body's bioenergetic organization.

> The relationship of energy to personality is most clearly manifested in a depressed individual. Although the depressive reaction and the depressive tendency result from the interplay of complicated psychological and physical factors, one thing is abundantly clear. The depressed individual is energetically depressed. Cinematic studies show he makes only about one-half the spontaneous movements usual in the nondepressed individual. [...] He generally feels that he lacks the energy to get moving. [...] The depression of his level of energy is seen in the decrease of all energy functions. His breathing is depressed, his appetite is depressed, and his sexual drive is depressed. In this state he should not possibly respond to our exhortations that he interests himself in some pursuit: he literally *does not have the energy* to develop an interest. (Lowen 1994, p. 47)

More specifically, defining a psychic attitude from a bioenergetic point of view means putting it in terms of a relationship between the charge function and the discharge function. Therefore, a body with a high energy level and psychic harmony is one in which the activities of charge and discharge are balanced. Conversely, a psychic problem must be read in the form of imbalance and dysfunction of charge and discharge.

Therefore, clarification is needed regarding the body dynamics that affect the charge and discharge functions, showing in what terms a balance is possible. Lowen believes that we can distinguish an upper and a lower part of the human body, ideally divided by the diaphragm. From a bioenergetic point of view, the first is in charge of

---

[5]However, as Elizabeth Michel points out, one must understand the concept of energy within the therapeutic practice and following the meaning attributed to it by common sense, i.e., as "vigour," "vitality," and "strength," rather than with a scientific meaning, which would require further clarifications. Bioenergetic analysis talks about "core energy" and says that some structures are "high energy charge," and others, instead, are "low energy charge." Using "energy" in this way, Michel declares, is right when you look at the patients' bodies (Michel 2014).

the charge functions, the second of discharge functions. The charge function supplies the body with energy through food, oxygen, or sensory excitations. Dynamically, it is represented by an upward flow of energy. The discharge function affects the lower part of the body: it is described as a downward movement aimed at discharging the excess energy through two exit channels, the legs and the genital apparatus.

Regarding the discharge function, Lowen analyzes the importance of an anatomical element, such as the pelvis. It fulfills a storage function, an energy accumulator preceding the discharge. According to Lowen, the presence in the human body of this element constitutes the bodily foundation of the psychic principle of reality (which for Lowen corresponds to the "grounding,"[6] the ability to remain "well-rooted to earth").

In fact, the pelvis acts as a "temporizer," an element capable of postponing the energy discharge and hence the pleasure connected to it. The reality function:

> [...] demands the acceptance of a state of tension and the postponement of pleasure in accordance with the demands of an external situation [...]. The essence of this function is the interposition of a time interval between the impulse and its expression in overt action. (Lowen 1971, p. 56)

Therefore, pleasure can be postponed since the human body is provided with "containers." Within these containers, the charge accumulates before it is discharged to the outside in a discharge.

The possibility of a deferral does not depend on the voluntary exercise of an ego that, by contracting the musculature, represses the impulse, but by the capacity of the reserves, by their amplitude and flexibility. In an individual in physical and psychic equilibrium, the two reserves function adequately, making it possible to defer the impulses without inhibiting them.

Assigning a somatic base to the reality function, Lowen maintains the close connection between reality and pleasure. From a bioenergetic perspective, a strong sense of reality and a good grounding ensure a fuller experience of pleasure.

Therefore, an individual with a good sense of reality does not have an inhibition on the discharge (perceived as pleasure); he can simply postpone it. Moreover, since the charge is directly proportional to the discharge, the more a body can accumulate energy, the greater the pleasure gained. The reality principle develops from the pleasure principle. It is its modification, its "temporalization." The reality function, in fact, "[...] cut off from its motivation and energy source in the pleasure principle becomes sterile" (Lowen 1971, p. 67), dependent on a disembodied ego.

Following Reich, Lowen pays particular attention to the study of muscular tensions in which he identifies repressed impulses at an unconscious level. Every chronic muscle contraction, in fact, constitutes an impediment to the free flow of energy directed upwards and downwards. More precisely, "muscle block" means a set of chronic muscle contractions that prevent a natural and integrated body movement (Michel 2014).

---

[6]For a discussion on the reality principle in bioenergetics, cf. Lowen (1971); Lowen (1993, 1994).

For example, anger suffocated at an early age can cause chronic contractions in the shoulders' muscles or tension in the arms and hands. The repression of the anger impulse, which at the conscious level is operated by the ego by blocking the outgoing energy, can establish itself unconsciously in the musculature. The adult in question will have difficulty expressing feelings of anger since the muscles in charge of the choleric outlet will find themselves in a state of contraction.

> Anger as expressed in striking out can be inhibited by contracting the muscles of the shoulder girdle, thereby pulling the shoulders back. At first the inhibition is conscious and aims to spare the person further conflict and pain. However, the conscious and voluntary contraction of muscles requires an investment of energy and cannot therefore be maintained indefinitely. When an inhibition against some feeling must be maintained indefinitely because its expression is not accepted in the child's world, the ego surrenders its control over the forbidden action and withdraws its energy from the impulse. The holding against the impulse then becomes unconscious, and the muscle or muscles remain contracted because they lack the energy for expansion and relaxation. (Lowen 1994, p. 144)

Michel (2014) explains that the muscles present a rhythmic pattern of low-frequency contraction in a state of relaxation. On the contrary, in a stressful situation, the body produces a fight-or-flight response that activates many muscles. If the tension is not resolved, this widespread activation will remain and acquire chronic features.

Therefore, every muscle contraction represents an energetic block that delimits areas in which the body's motility is reduced, identifiable by their "inanimate" aspect or by palpation. Furthermore, it is an indication of an emotional conflict present at an unconscious level. In this sense, the muscular tension model is a mirror of the individual's psychic attitude, and bioenergetic therapy aims at retracing the etiology of the block and dealing with the related emotional disorder.

The body, therefore, registers its own relationship with the world, its interpretation. In this perspective, one can represent the body as a tangle of meanings of which physical evidence and psychological attitudes are the signs.

3.3 *The Language of the Body: The Formation of Characters*

Therefore, the correlation between bodily attitude, muscular tensions, and psychic attitude emerges more clearly in Lowen's analysis of the various "characters."

Lowen inherits from Reich and his 1933 work *Analysis of character*, the notion of "character." By "character," Reich means a defensive psycho-physical structure developed in childhood after traumatic relational experiences.[7] The term "structure" indicates a certain fixity in the form: it occurs following freezing of the natural energy movement and has a pathological significance.

[7] The organization of Reich's therapy derives from the analysis of resistance to therapy. In this way, he realizes that these resistances represent personality structure, a structure developed during childhood and defensive by nature (Reich 1980).

Following the master, Lowen defines the "character" as "the unitary expression of the individual's function on both the psychic and in the somatic realm" (Lowen 1971, p. 119) and distinguishes five fundamental "types:" the schizoid character, the oral character, the masochistic character, the rigid character and the psychopathic character. According to Lowen, character constitutes a sort of behavioral model, a habitual orientation in which the individual finds himself trapped.

Without entering the thorny and delicate question of the relationship between pathology and state of health, the study of character is useful here to bring out the close connection between bodily behavior and psychological behavior and show the body's tendency to express itself in meanings structured following precise relational experiences.

First of all, Lowen stresses that the elaboration of five "kinds of character" is motivated by functional rather than theoretical reasons. Identifying characterological types is useful for the therapist to carry out the work of analysis, tracing the patient's defensive structures and related bioenergetic dynamics, and therefore has an orientation function rather than a taxonomic value. In other words, the therapist is dealing with people, not structures.

> In bioenergetics, the different character structures are classified into five basic types. Each type has a special pattern of defense on both the psychological and the muscular levels that distinguishes it from the other types. It is important to note that this is a classification not of people but of defensive positions. It is recognized that no individual is a pure type and that every person in our culture combines in different degrees within his personality some or all of these defensive patterns. [...] Nevertheless, it is necessary to speak in terms of types for the sake of clarity in communication and understanding. (Lowen 1994, p. 151)

Each character, therefore, presents a specific energy distribution, specific muscular tensions and distinctive psychic attitudes.

The schizoid character is defined by the tendency to dissociate, to "split in two." The subject does not feel connected to his own body, the perception of which is extremely reduced; the result is a split between the upper and lower part. This splitting is embodied in the body of the schizoid, which has a disproportion between the upper and lower half. Bioenergetically speaking, energy is held in the body's center and does not flow to the extremities. The bodies in charge of contact with the world do not receive energy, which strongly limits the individuals' sensitivity and communicative capacity. The energy charge, blocked by muscular tensions at the head's base, in the shoulders and at the pelvis, is frozen in the nucleus region. The impulses are retained and compressed in the center.[8]

Michel (2014) explains that a pattern of blocks in the schizoid body prevents the energy of the core from reaching the peripheral body areas that allow contact with the world. These blocks serve to separate the individual from the world, which has been the setting of some traumatic experience. The schizoid body often presents

---

[8]However, if the body were subjected to such a quantity of charge that the defenses could not block it, the energy could pour out violently, without the subject being able to exercise conscious control over it.

contractions in the micro-musculature around the joints. These contractions have the symbolic function of "holding together" the body parts to counteract the schizoid tendency to psychic disintegration, generated by a terrible fear experienced during childhood.

Since the extremities are lacking in charge, Lowen compares the face of the schizoid subject to a mask in which the eyes appear devoid of life and communication. The feet and hands are often cold, with little sensitivity. In the etiology of the schizoid character, the mother's rejection is identified, experienced by the child in the first years of childhood. The maternal hostility, manifest, or hidden behind compensatory attitudes, leads the subject to believe that any request or initiative is rejected. To avoid pain and distress, the child, therefore, avoids facing the outside world. The outwardly directed charge suffers inhibition through the intervention of the musculature, which functions as a defense and contracts, thus repressing the impulse. To survive terror, the child makes the body insensitive.

Therefore, the adult's behavior will be characterized by a poverty of feelings and difficulty in establishing intimate relationships. Lowen defines the attitude of the schizoid character as an "as if" attitude: "that is, as if it were based on feelings, but the actions themselves are not expressive of feeling" (Lowen 1994, p. 154).

The second type, the oral character, is defined by the presence of oral elements related to early childhood in the psychological behavior and the physical structure. The body has immaturity and weakness features, such as a depressed chest and a belly with an inanimate and empty appearance. The legs tend to be rigid and not very sensitive, being used to compensate for the back's weakness, which is usually responsible for the support function.

An early deprivation of nourishment, contact, or maternal affection represents the determining historical factor, which leads to the conscious renunciation by the child of explicit requests for satisfaction. However, the repressed desire is imprinted on an unconscious level, creating a state of perpetual dissatisfaction. Due to his history of "malnutrition" (understood in a broad sense), the oral individual is bio energetically characterized by a reduced charge. Energy flows weakly from the center to the extremities, preventing the musculature from forming adequately and giving it a hypotonic appearance. This condition gives the oral personality a low level of aggressiveness. For example, the collapsed chest affects the energy level of the whole body: breathing is superficial and reduced and not able to provide enough energy for substantial efforts.

Thus, the oral character has weak muscles and does not provide the muscular system with sufficient oxygen to produce a vigorous and integrated aggressive movement (Michel 2014). Basal metabolism is slow, blood pressure is low and genital function is reduced: "orality and genitality," in fact, "are antithetical tendencies. One is related to the function of charge, the other to discharge. The sexual drive of the oral character [. . .] represents the need to take in, to feed from the partner" (Lowen 1971, p. 175).

The masochistic character is frequently generated by an asphyxiating mother, who attributes excessive importance to food and hygiene, resulting in attitudes aimed at the compulsion to eat and control over evacuation. At first, the child responds to impositions with open hostility and heated expressions of anger. In a second moment, however, the threat that the aggressive expression causes the loss of maternal love and, therefore, pain leads to suppressing the impulse and directing it towards the interior.

From a bioenergetic perspective, the masochistic structure works "with a high energetic charge." It is characterized by inverting the aggressive drive towards the inside, with consequent compression of the tender feelings.[9] The two instincts, the aggressive and the tender, are antagonistic, and the second is engaged in a perennial struggle to break the compression inflicted by the first. However, every attempt at affirmation fails, causing the individual to sink into a state of profound depression that is called the "masochistic swamp."

The masochistic individuals present a corpulent, massive, and squat physical structure since the pressure exerted at the two ends prevents the body from lengthening. The musculature, contracted for retaining aggressive impulses (for example, the limbs express in their appearance both aggressiveness and inhibition) and regulating the natural ones, is exaggeratedly developed. There is strong muscular tension in the shoulder girdle (whose muscles hold back resentment) and the pelvis.

The mother's control over natural functions also causes the child to have an intense fear of discharge and an early regulation of the evacuation movements. According to Lowen, "premature insistence upon excremental cleanliness forces the child to employ the *levator ani* muscle, the gluteals and the hamstrings to gain anal control since the external sphincters have not yet come under voluntary control" (Lowen 1971, p. 214). The result is the aspect of a "dog with the tail between the legs," due to the contraction of the pelvis that is pushed forwards.

The masochistic individual is an individual who has been denied the right to independence and self-assertion. Therefore, his manifest behavior is compliance and eagerness to please (renunciation of the self). Behind this renunciation, however, there is intense hostility and resentment. He often assumes a provocative attitude with which he induces others to express negative feelings towards him. In this way, the masochistic subject justifies his own aggressive or violent reactions: he seeks, in fact, the possibility of "discharging" himself, but since he cannot authorize himself to do so, he provokes the phenomenon indirectly by pushing responsibility for it onto others.

The rigid individual's history is characterized by "the experience of frustration in the striving for erotic gratification, especially at the genital level" (Lowen 1994, p. 169). In fact, at the genital stage, the child received a refusal of his/her search for erotic and sexual pleasure, to which he reacted by stiffening.

Bioenergetically, the rigid structure is highly charged, and the discharge activity is working. However, the aggressive component's prevalence over the tender

[9]From this idea of an internal upheaval of the aggressive instinct, Freud theorizes the existence of a death drive as characteristic of the human being. On the other hand, to Lowen, what Freud calls the death drive is simply the result of a specific character structure, the masochistic one. Generally, the human body's energy is not "programmed" to turn against itself (Lowen 1971).

component prevents the body from reaching significant energy levels. The body presents rigidity elements in the front part and takes on an "armored" shape.[10] This structure was erected as a defense to eliminate the possibility of a new rejection.

> Dynamically, the tension in the front is produced by pulling back the shoulders and the pelvis, thus putting all the front muscles on the stretch at the same time that they are contracted. When the front and back of the body are thus encased in a rigid sheath of tight muscles, we can say that the organism is armored. (Lowen 1971, p. 258)

The fear of rejection has been denied on the unconscious level, covered with an attitude of pride and aggression.

Finally, the psychopathic character is defined by a denial of feelings, a dissociation of the ego from the body and an attempt to dominate it. The same attitude of domination manifests in psychopathic individuals' behavior towards others, a behavior that can take two forms: that of arrogance or that of seduction.

The psychic structure's energy is concentrated in the upper part of the body, which assumes larger dimensions than the lower one. The diaphragm area is taut, preventing the energy flow from passing freely downwards. The eyes are wary and suspicious, and tensions in the occipital area are frequent. Moreover, the excessive concentration of energy in the upper part of the body causes a disproportionate load on the head, which is strongly contracted.

The etiological reconstruction of the psychopathic character identifies the presence of a seductive parent: "seduction is covert and is done to meet the parent's narcissistic needs. It aims to tie the child to the parent" (Lowen 1994, p. 162). However, in terms of support and affection, the seductive parent rejects the child. Moreover, the seductive element's presence gives rise to a form of competition towards the parent of the same sex, preventing identification. As a result, the child is emotionally isolated. In response to this, in defense, he rises above his feelings, emotional needs, and, therefore, his body.

A look at character analysis is useful for understanding the close connection between psychic attitude and body expression. From the analysis of the character emerged the body's tendency to inscribe on itself in a bioenergetic way, traceable in the bodily and psychological attitude assumed by the subject. Therefore, the character allows us to see the significant relationship between muscle blockage and emotional conflict: a particular psychic dynamic corresponds to every energetic tension within the organism.

In this perspective, the body is configured as a narrative offered to the expert eye of a therapist.

> A person is the sum total of his life experiences, each of which is registered in his personality and structured in his body. Just as a woodsman can read the life history of a tree from a cross section of the trunk showing its annual growth rings, so it is possible for a bioenergetic therapist to read a person's life history from his body. Both studies require knowledge and experience, but they are based on the same principles. (Lowen 1994, p. 57)

[10]The concept of "armor" was developed by Reich to describe a defense mechanism that, on a bodily level, is expressed as a muscular armor.

The history of an individual, therefore, is recorded on his own body. Like the personality, the physical structure contains a meaning, reflects an experience of life, and is the sign of a particular psycho-somatic interpretation of the world. Thus, the therapist, reading the body, interpreting the signs that it bears, can retrace that history and help the patient do the same.

Therefore, the body constitutes a colorful and expressive panorama in which each region corresponds to an emotional meaning. The topography of the body reflects how it has interpreted the surrounding environment. It indicates the attitude with which the organism faces the world, of a specific aspect of life. As a result, a muscle block located in a particular area or the dysfunction of an organ contains emotional meanings (Lowen 1995).

From this viewpoint, the concept of "self-expression" is central. Lowen argues that the body expresses itself mainly through three channels: movement, voice, and eyes. Attention to these aspects provides a useful tool for the therapist to ascertain the patient's body condition.

The voice's quality depends mainly on three elements: the flow of air under pressure, the vocal cords, which allow vibration, and the resonating chambers, which determine the volume. Any interference in their regular functioning influences vocal expression. The contractions of the neck and throat, for example, generate head or chest sounds. Usually, a high-pitched voice, which is associated with a difficulty in producing deep notes, is an indication of an expressive block of feelings of sadness; on the other hand, a low and chest voice is a sign of a denial of the feelings of fear that are expressed in the cry. The schizoid individual, for example, tends to have a flat voice caused as much by tensions in the throat as by a limited breath that prevents the flow of air passing through the vocal cords from being full and spontaneous.

Lowen observes that in the vocal apparatus, there can be three main rings of tension: the first around the mouth, the second in the area where the pharynx is connected to the esophagus and trachea, and the third between the neck and the chest. Among the three, the second ring, set up in the area where we witness the passage from voluntary to involuntary control of swallowing, has a significant psychological significance. It represents an unconscious resistance to the passage of something considered unacceptable, be it an external element that one does not want to swallow (frequent in masochistic individuals) or an internal feeling that one is afraid to express.[11]

Therefore, if the body always contains an affective and emotional meaning, a bioenergetic reading of physical disorders is also possible. A specific case is represented by myopia, to the analysis of which Lowen dedicates copious pages (Lowen 1994). From a bioenergetic perspective, the myopic eye is wide open and fixed, therefore marked by low mobility. The etiological factor of the disorder is found in the presence of a feeling of fear on an unconscious level fixed in the eyeball as a muscle contraction. Observe in detail the genesis of this phenomenon:

[11] Furthermore, the work of the jaw reinforces tension, and the jaw tightens to prevent passage.

> Wide eyes [. . .] enlarge the field of peripheral vision but reduce central vision. To regain its visual acuity, the child will forcibly constrict its eyes, creating a condition of rigidity and strain. There is another element. Frightened eyes tend to roll upwards. This tendency, too, must be overcome by an effort of will if the child is to maintain its ability to focus. Now the strain of these efforts cannot be maintained indefinitely. At some point, the eye muscles tire. and the child gives up the effort to look out. Myopia sets in when this compensation breaks down. (Lowen 1994, p. 290).

Lowen also devotes a few pages to the "headache" disorder. Headache is described as the result of interference of the aggressive upward flow. Similarly, according to Lowen, migraine is caused by a block of desire and the upward flow of excitation. In every migraine, one can notice the development of a strong tension on one side of the neck, under the jaw, functional to the block of desire, the palpation of which causes a sharp pain behind the eye.

Therefore, considering the body from a bioenergetic perspective means avoiding an objectual and mechanistic conception of it and espousing the idea that the body is a dimension that generates meaning and is inhabited by it, an iridescent reality that inscribes its own relational experiences on itself. "Bioenergetics," Lowen writes, "does not see the body as a machine, not even as the most complex and beautiful machine ever created" (Lowen 1994, p. 84).

4 Conclusions

The journey that is now complete has aimed to take the form of exploration across the range of biosemiotics and bioenergetics in the attempt to show how two disciplines that are so different in scope, objectives, and methods present certain similarities.

Both perspectives offer an interpretation of the body as a dimension open to the world that defines its own structures thanks to a relationship, an exchange, with the external. At the same time, however, they recognize the limits and constraints that an organism or an individual establishes based on its experiences and history: life is, therefore, as much a continuous creative possibility as it responds to rules set up over time.

However, the most original and substantial aspect of our research consists in that both biosemiotic and bioenergetic thought acknowledge the indissoluble link that life maintains with meaning, with a system of signs through which and thanks to which it communicates, defines itself, and takes on forms. In this sense, the body defines itself as the semiotic dimension *par excellence*.

To conclude, it is important to recall how the work completed here represents only a limited perspective, the attempt to carve out one route through a much broader and more complex panorama. Nevertheless, taking some first steps in this journey may help suggest reflections on the subject and promote a dialog between two disciplines that, originating and developing differently, offer a similar and original interpretation of the body.

References

Barbieri M (2008a) What is biosemiotics? Biosemiotics 1:1–3
Barbieri M (2008b) Biosemiotics: a new understanding of life. Naturwissenschaften 95:577–579
Benasayag M (2016) Il cervello aumentato, l'uomo diminuito. Erikson, Trento
Giorgi F (2017) Un mondo di relazioni. La logica del vivente: dalla certezza dell'essere alla esplorabilità del divenire, Tangram, Trento
Harney M (2007) Merleau-Ponty, ecology, and biosemiotics. In: Cataldi SL, Hamrick WS (eds) Merleau-Ponty and the environmental philosophy. Dwelling on the landscapes of thought. State University of New York Press, Albany, pp 233–148
Helfaer PM (2013) I fondamenti dell'analisi bioenergetica. In: Heinrich-Clauer V (ed) Manuale di Analisi Bioenergetica. FrancoAngeli, Milano
Hoffmeyer J (2008) The semiotic body. Biosemiotics 1:169–190
Hoffmeyer J (2015) Semiotic scaffolding of living systems. Biosemiotics 8:153–158
Kotov K, Kull K (2011) Semiosphere is the relational biosphere. In: Emmeche C, Kull K (eds) Towards a semiotic biology. Life is the action of signs. Imperial College Press, London
Kull K et al (2008) Biosemiotic questions. Biosemiotics 1:41–55
Lowen A (1971) The language of the body. First Collier Books Edition, New York
Lowen A (1993) Depression and the body. The biological basis of faith and reality. Penguin Books, New York
Lowen A (1994) Bioenergetics. Penguin Books, col. Compass, New York
Lowen A (1995) Joy. The surrender to the body and to life. Penguin Books, New York
Michel E (2014) Bent out of shape. Anatomy and alignment for bioenergetic trainees, trad. It. In: Filoni MR (ed) L'emozione e la forma. Manuale di anatomia e postura per analisti bioenergetici e conduttori di classi. FrancoAngeli, Milano
Pagni E (2016) Why explicit semiotic grounding is essential to biology as a science? The point of view of biosemiotics. HumanaMente Journal of Philosophica Studies 31:57–72
Petrilli S (2003) Sebeok's Semiosic universe and global semiotics. Cybernetics And Human Knowing 10:61–69
Petrilli S, Ponzio A (2008) A tribute to Thomas a. Sebeok Biosemiotics 1:25–39
Petrilli S, Ponzio A (2013) Review article biosemiotic scenarios. Semiotica 195:373–408
Reich W (1980) Character analysis. Farrar, Straus and Giroux, New York
Sebeok TA (1986) I think I am a verb. More contributions to the doctrine of signs. Indiana University Press, Bloomington
Sebeok TA (1989) Preface. In: Lee B, Urban G (eds) Semiotics, self, and society. De Gruyter Mouton, Berlin, New York
Sharov A (2001) Umwelt-theory and pragmatism. Semiotica 134:211–228
Tønnessen M (2009) Umwelt transitions. Uexküll and Environmental Change Biosemiotics 2:47–64
Tønnessen M et al (2018) Phenomenology and biosemiotics. Biosemiotics 11:323–330
Varela et al. (1993) The embodied mind cognitive science and human experience. MIT Press, Cambridge

The manufacturer's authorised representative in the EU is Springer Nature Customer Service Centre GmbH, Europaplatz 3, 69115 Heidelberg, Germany. If you have any concerns regarding our products, please contact ProductSafety@springernature.com

Printed and bound by CPI Group (UK) Ltd, Croydon, CR0 4YY
28/09/2026
02224090-0015